Geometric Dimensioning and Tolerancing for Mechanical Design

Geometric Dimensioning and Tolerancing for Mechanical Design

Gene R. Cogorno

Third Edition

New York Chicago San Francisco
Athens London Madrid
Mexico City Milan New Delhi
Singapore Sydney Toronto

Library of Congress Cataloging-in-Publication Data

Names: Cogorno, Gene R., author.
Title: Geometric dimensioning and tolerancing for mechanical design / Gene R. Cogorno.
Description: Third edition. | New York : McGraw-Hill, [2020] | Includes index.
Identifiers: LCCN 2019041952 | ISBN 9781260453782 | ISBN 1260453782
| ISBN 9781260453799 (ebook)
Subjects: LCSH: Engineering drawings—Dimensioning. | Tolerance (Engineering) | Machine design.
Classification: LCC T357 .C64 2020 | DDC 620/.0042—dc23
LC record available at https://lccn.loc.gov/2019041952

Geometric Dimensioning and Tolerancing for Mechanical Design, Third Edition

1 2 3 4 5 6 7 8 9 LWI 25 24 23 22 21 20

ISBN 978-1-265-82145-6
MHID 1-26-582145-3

Sponsoring Editor
Robert Argentieri

Editing Supervisor
Stephen M. Smith

Production Supervisor
Pamela A. Pelton

Acquisitions Coordinator
Elizabeth M. Houde

Project Manager
Sarika Gupta,
Cenveo® Publisher Services

Copy Editor
Bruce Owens

Proofreader
Upendra Prasad,
Cenveo Publisher Services

Art Director, Cover
Jeff Weeks

Composition
Cenveo Publisher Services

About the Author

Gene R. Cogorno is a professional educator, speaker, and author with more than 30 years of experience in education and training. He earned both his Bachelor's and Master's degrees in Industrial Education from San Jose State University. Mr. Cogorno developed and taught a practical training program in industrial education, and taught machine technology at San Jose State University. In 1984, he joined FMC Corporation as a Senior Technical Trainer. In 1992, Mr. Cogorno founded Technical Training Consultants, where he teaches courses in geometric dimensioning and tolerancing, tolerance analysis, and blueprint reading.

Contents

Preface

This book is written primarily for the learner who is new to the subject of geometric dimensioning and tolerancing (GD&T). The purpose of this book is to teach this graphic language in a way that the learner can easily understand and use in practical applications. This work is intended as a textbook to be used in colleges, universities, technical schools, and corporate training programs. It is intended for use in engineering, design, manufacturing, inspection, and drafting curriculums. This book is also appropriate for a self-study program.

The material in this book is written in accordance with the latest revision of the geometric dimensioning and tolerancing standard, ASME Y14.5-2018. GD&T is a graphic language; in order to facilitate the understanding of this subject, there is at least one drawing to illustrate each concept discussed. Drawings in this text are for illustration purposes only. In order to avoid confusion, only the concepts being discussed are completely toleranced. All of the drawings in this book are dimensioned and toleranced with the inch system of measurement because most drawings produced in the United States are dimensioned and toleranced with this system. The reader is expected to know how to read engineering drawings.

Organization

The discussion of each control starts with a definition and continues with how the control is specified, interpreted, and inspected. There are a sequential review, a series of study questions, and problems at the end of each chapter to emphasize key concepts and to serve as a self-test. This book is logically ordered so that it can be easily used as a reference text.

A Note to the Learner

To optimize the learning process, it is important for the learner to do the following:

1. Preview the chapter objectives, the subtitles, the drawing captions, and the summary.
2. Preview the chapter once again, focusing attention on the drawings and, at the same time, formulating questions about the material.
3. Read the chapter completely, searching for answers to your questions.
4. Underline or highlight important concepts.
5. Answer the questions and solve the problems at the end of the chapter.

Comprehending new information from the printed page is only part of the learning process. Retaining the new information in long-term memory is even more important. In order to optimize the learning process and to drive new information into long-term memory with the least amount of effort, it is suggested that the learner follow these steps:

1. Review all new information at the end of the day.
2. Review it again the next day.
3. Review it again the next week.
4. And, finally, review the new information again the following month.

Review is more than just looking at the information. Review includes rereading main ideas, speaking them out loud, and/or writing them. Some learners learn best by reading, others by hearing, and still others by writing or doing. Everyone learns differently, and some students may learn best by employing a combination of these activities or all three. Learners are encouraged to experiment to determine their own best method of learning. The answers to the questions and problems at the end of the chapters are available on the publisher's and author's websites shown below.

A Note to the Instructor

An Instructor's Guide is available. It includes the following:

1. A course calendar
2. Suggested lecture topics
3. Answers to questions and problems at the end of each chapter
4. Midterm examinations
5. A final examination
6. The answers for the midterm examinations and the final examination

Also, this book is organized in such a way that the instructor can select appropriate material for a more abbreviated course. This text can also be used as supplementary material for other courses, such as mechanical engineering, tool design, drafting, machining practices, and inspection. Using the Instructor's Guide with this text will greatly facilitate the administration of a course in GD&T.

Access to the Instructor's Guide

- Publisher's website: https://www.mhprofessional.com/GDTMD3
- Author's website: http://www.ttc-cogorno.com

Gene R. Cogorno

Acknowledgments

I would like to express particular gratitude to my wife, Marianne, for her support of this project and for the many hours she spent reading and editing the manuscript. Also, thanks go to my son, Steven, for his efforts toward shaping the style of this book. I would like to acknowledge Robert Argentieri, my editor, and the McGraw-Hill Professional staff for their technical contributions and editorial comments. A special thanks goes to James D. Meadows, my first GD&T instructor, for his guidance and support throughout the years. Finally, thank you to the American Society of Mechanical Engineers for permission to reprint excerpts from Dimensioning and Tolerancing, ASME Y14.5-2018. All rights are reserved. No further copies may be made without written permission.

Geometric Dimensioning and Tolerancing for Mechanical Design

CHAPTER 1

Introduction to Geometric Dimensioning and Tolerancing

For many in the manufacturing sector, geometric dimensioning and tolerancing (GD&T) is a new subject. During World War II, the United States manufactured and shipped spare parts overseas for the war effort. Many of these parts, even though they were made to specifications, would not assemble. The military recognized that defective parts caused serious problems for military personnel. After the war, a committee representing government, industry, and education spent considerable time and effort investigating this defective parts problem; this group needed to find a way to ensure that parts would fit and function properly every time. The result was the development of GD&T.

Ultimately, the USASI Y14.5–1966 (United States of America Standards Institute, predecessor to the American National Standards Institute) document was produced based on earlier standards and industry practices. The following are revisions to that standard:

- ANSI Y14.5–1973 (American National Standards Institute)
- ANSI Y14.5M–1982
- ASME Y14.5M–1994 (American Society of Mechanical Engineers)
- ASME Y14.5–2009
- ASME Y14.5–2018

The 2018 revision is the current, authoritative reference document that specifies the proper application of GD&T.

Most government contractors are now required to generate drawings that are toleranced with GD&T. Because of tighter tolerancing requirements, shorter time to production, and the need to communicate design intent more accurately, many companies other than military suppliers are recognizing the importance of tolerancing their drawings with GD&T.

Traditional tolerancing methods have been in use since the mid-1800s. These methods do a good job of dimensioning and tolerancing the size of features and are still used in that capacity today, but they do a poor job of locating and orienting features of size. GD&T is used extensively for tolerancing size, shape, form, orientation, and location of features. Tolerancing with GD&T has a number of advantages over conventional tolerancing methods; three dramatic advantages are illustrated in this chapter.

The purpose of this introductory chapter is to provide an understanding of what GD&T is and why it was developed, when to use it, and what advantages it has over conventional tolerancing methods. With a knowledge of this subject, technical practitioners will be more likely to understand tolerancing in general. With this new skill, engineers will have a greater understanding of how parts assemble, do a better job of communicating design requirements, and ultimately be able to make a greater contribution to their companies' bottom line.

Chapter Objectives

After completing this chapter, the learner will be able to:

- *Define* GD&T
- *Explain* when to use GD&T
- *Identify* three advantages of GD&T over coordinate tolerancing

What Is GD&T?

GD&T is a symbolic language. It is used to specify the size, shape, form, orientation, and location of features on a part. Features toleranced with GD&T reflect the actual relationship between mating parts. Drawings with properly applied geometric tolerancing provide the best opportunity for uniform interpretation and cost-effective assembly. GD&T was created to ensure the proper assembly of mating parts, to improve quality, and to reduce cost.

GD&T is a design tool. Before designers can apply geometric tolerancing properly, they must carefully consider the fit and function of each feature of every part. GD&T, in effect, serves as a checklist to remind the designer to consider all aspects of each feature. GD&T allows the designer to specify the maximum available tolerance and, consequently, design the most economical parts. Properly applied geometric dimensioning and tolerancing ensures that every part will assemble every time.

GD&T communicates design requirements. This tolerancing scheme identifies all applicable datum features, that is, reference surfaces, and the features being controlled to these datum features. A properly toleranced drawing not only is a picture that communicates the shape and size of the part but also tells a story that explains the tolerance relationships between features.

When Should GD&T Be Used?

Many designers ask, when should I use GD&T? Because GD&T was designed to position features of size, the simplest answer is to locate all features of size with GD&T controls. Designers should tolerance parts with GD&T when:

- Drawing delineation and interpretation need to be the same
- Features are critical to function or interchangeability
- It is important to stop scrapping perfectly good parts
- It is important to reduce drawing changes
- Automated equipment is used
- Functional gaging is required
- It is important to increase productivity
- Companies want across-the-board savings

Advantages of GD&T over Coordinate Dimensioning and Tolerancing

Since the middle of the nineteenth century, industry has been using the plus or minus tolerancing system for tolerancing drawings. This system has several limitations. The plus or minus tolerancing system generates rectangular tolerance zones. A rectangular tolerance zone, such as the example in Fig. 1-1, is a boundary within which the axis of a feature that is in tolerance must lie. Rectangular tolerance zones do not have a uniform distance from the center to the outer edges. In Fig. 1-1, from left to right and top to bottom, the tolerance is ± .005; across the diagonals, the tolerance is ± .007. Therefore, when designers tolerance features with a plus or minus .005 tolerance, they must tolerance the mating parts to accept a plus or minus .007 tolerance, which exists across the diagonals of the tolerance zones.

With the plus or minus tolerancing system, features of size can be specified only at the *regardless of feature size* condition. *Regardless of feature size* means that the location tolerance remains the same, ± .005, no matter what size the feature happens to be within its size tolerance. If a hole, like the one in Fig. 1-1, increases in size, it actually has more location tolerance, but, with the plus or minus tolerancing system, there is no way to capture that additional tolerance.

Datum features are usually not specified where the plus or minus tolerancing system is used. Consequently, machinists and inspectors don't know which datum features apply or in what order they apply. In Fig. 1-1, measurements are taken from the lower and left sides of the part. The fact that measurements are taken from these sides indicates that they are datum features. However, since these datum features are not specified, they are called *implied datum features*. Where datum features are implied, the designer has not indicated which datum feature is more important and has not specified whether a third datum feature is included. It would be

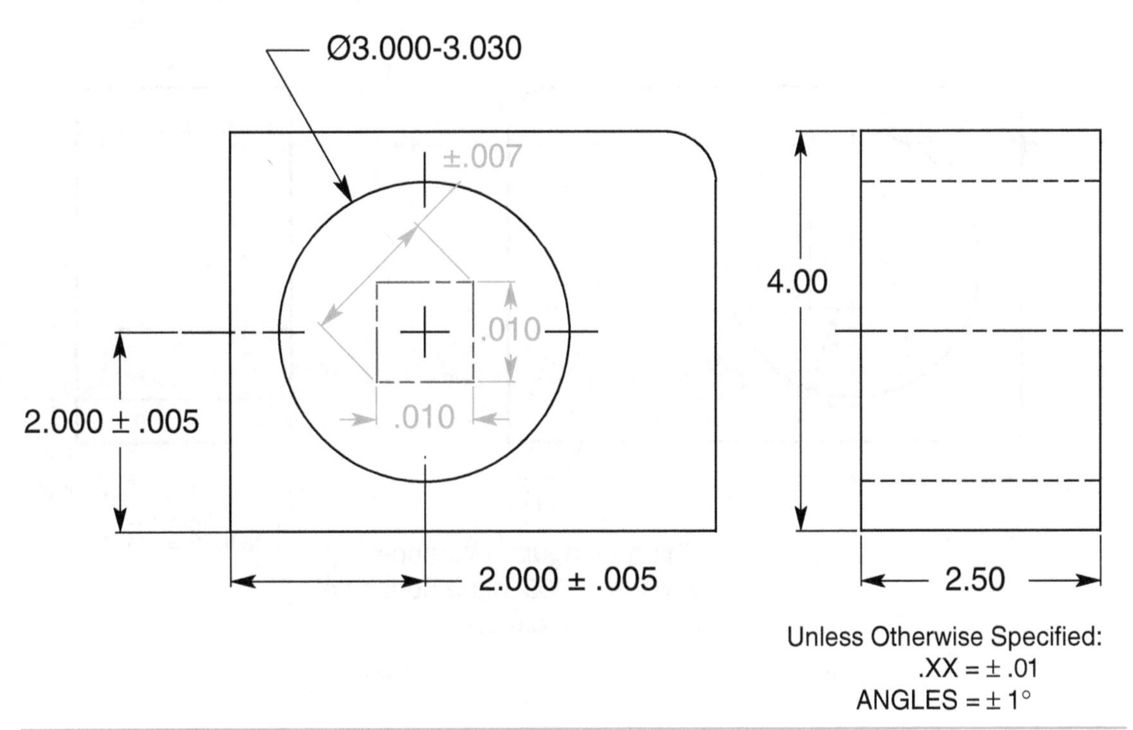

FIGURE 1-1 The traditional plus or minus tolerancing system. The axis of the 3-inch-diameter hole, to be in tolerance, must fall inside of the .010 square tolerance zone.

logical to assume that a third datum feature does exist because the datum reference frame consists of three mutually perpendicular planes, even though a third datum feature is not implied. When locating features with GD&T, there are three important advantages over the coordinate tolerancing system:

- The cylindrical tolerance zone
- The maximum material condition modifier
- Datum features specified in order of precedence

The Cylindrical Tolerance Zone

The cylindrical tolerance zone is located and oriented to a specified datum reference frame. In Fig. 1-2, the tolerance zone is oriented perpendicular to datum plane A and located with basic dimensions to datum planes B and C. There are no tolerances directly associated with a basic dimension; consequently, basic dimensions eliminate undesirable tolerance stack-up. Because the cylindrical tolerance zone is established at a basic 90° angle to datum plane A and extends through the entire length of the feature, it easily controls the orientation of the axis.

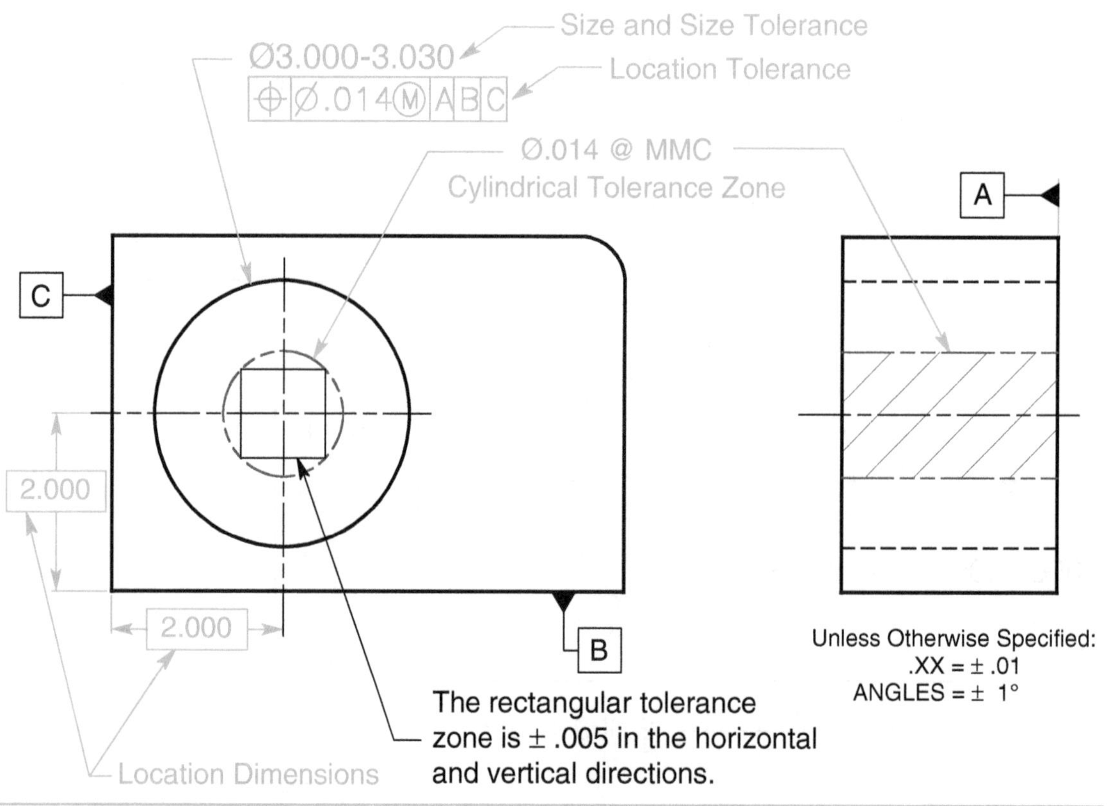

FIGURE 1-2 The cylindrical tolerance zone compared with the rectangular tolerance zone.

Unlike the rectangular tolerance zone, the cylindrical tolerance zone defines a uniform distance from true position, the theoretically perfect center of the hole, to the tolerance zone boundary. When a .014-diameter cylindrical tolerance zone is specified about true position, there is a

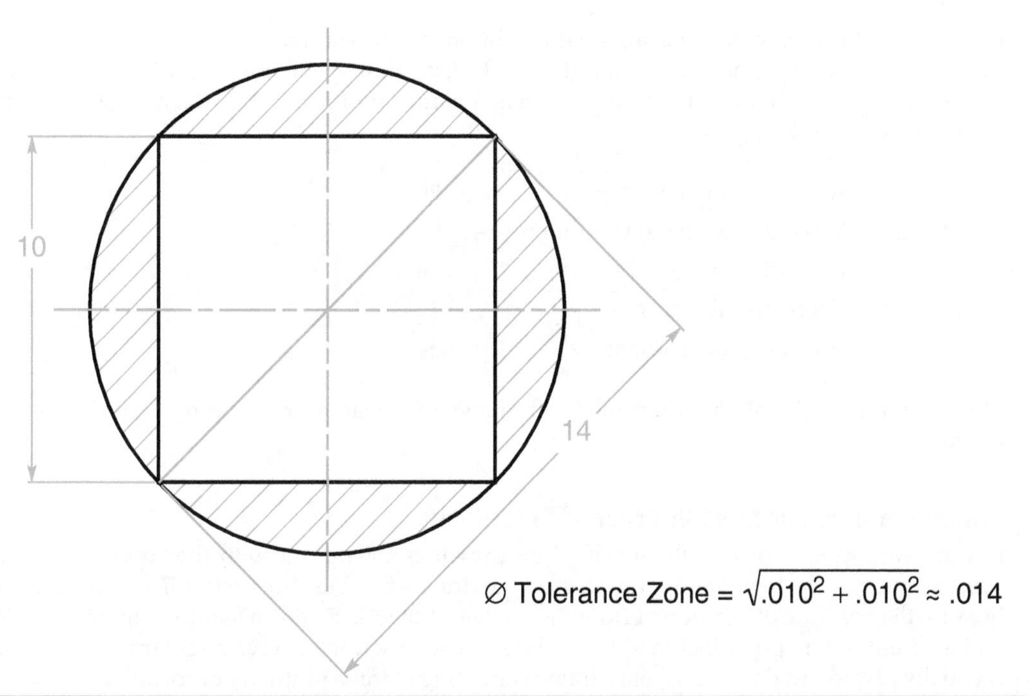

FIGURE 1-3 A cylindrical tolerance zone provides a uniform distance from the axis to the tolerance zone boundary.

tolerance if .007 from true position in all directions. A cylindrical tolerance zone circumscribed about a square tolerance zone, such as the one in Fig. 1-3, has 57% more area than the square tolerance.

The Maximum Material Condition Modifier

The maximum material condition symbol (circle M) in the feature control frame is a modifier. It specifies that as the hole in Fig. 1-2 increases in size, a bonus tolerance is added to the tolerance stated in the feature control frame.

The limit tolerance in Fig. 1-4 indicates that the hole size can be as small as 3.000 (maximum material condition) and as large as 3.030 (least material condition) in diameter. The geometric tolerance specifies that the hole be positioned with a cylindrical tolerance zone of .014 in diameter when the hole is produced at its maximum material condition size. The tolerance zone is oriented perpendicular to datum plane A and located with basic dimensions to datum planes B and C. Since the .014-diameter tolerance is specified with a maximum material condition modifier, circle M, a bonus tolerance is available. As the hole size in Fig. 1-2 departs from maximum

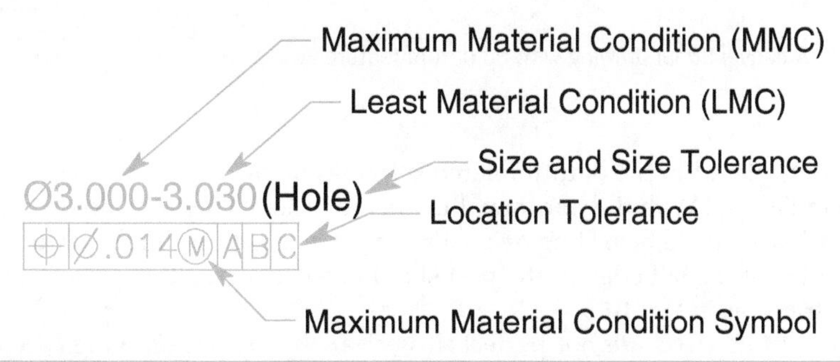

FIGURE 1-4 The size, size tolerance, and feature control frame for the hole in Fig. 1-2.

material condition toward least material condition, additional location tolerance, called *bonus tolerance,* is allowed in the exact amount of such departure. If the hole specified by the feature control frame in Fig. 1-4 is actually produced at a diameter of 3.020, the total available tolerance is a diameter of .034.

	Actual Mating Envelope	3.020
Minus	Maximum Material Condition	−3.000
	Bonus Tolerance	.020
Plus	Geometric Tolerance	+.014
	Total Positional Tolerance	.034

The maximum material condition modifier allows the designer to capture all of the available tolerance.

Datum Features Specified in Order of Precedence

Datum features are not usually specified on drawings toleranced with the coordinate dimensioning system. The lower and left edges on the drawing in Fig. 1-5 are implied datum features because the holes are dimensioned from these edges. But which datum feature is more important, and is a third datum plane included in the datum reference frame? A rectangular part such as this is usually placed in a datum reference frame consisting of three mutually perpendicular intersecting planes. When datum features are not specified, machinists and inspectors are forced to make assumptions that could be very costly.

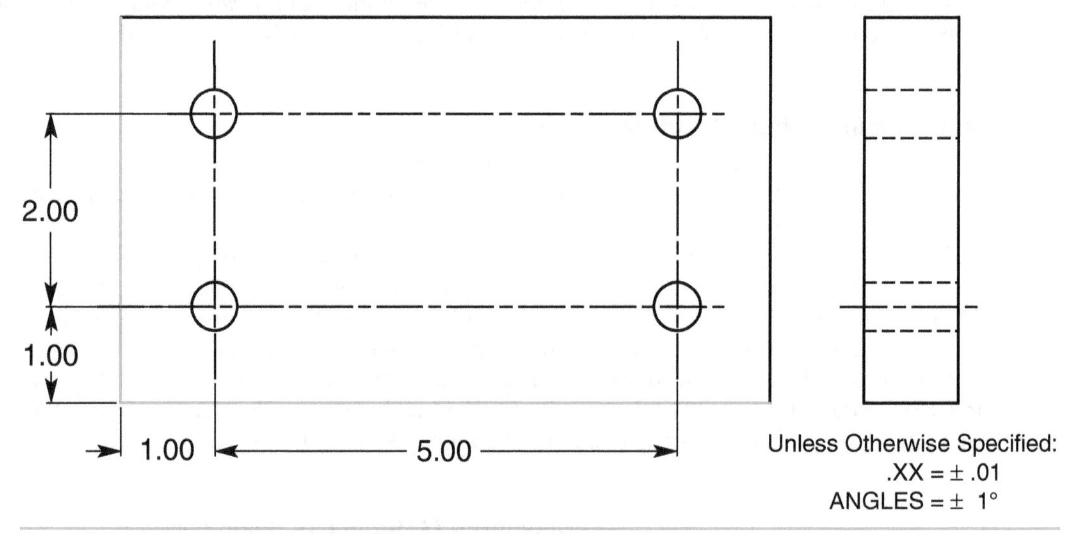

FIGURE 1-5 A conventional drawing with no datum features specified.

The parts placed in the datum reference frames in Fig. 1-6 shows two interpretations of the drawing in Fig. 1-5. With the traditional method of tolerancing, it is not clear whether the lower edge of the part should be resting against the horizontal surface of the datum reference frame as in Fig. 1-6*A* or if the left edge of the part should be contacting the vertical surface of the datum reference frame as in Fig. 1-6*B*.

Manufactured parts are not perfect. It is clear that, when drawings are dimensioned with traditional tolerancing methods, a considerable amount of information is left to the machinists'

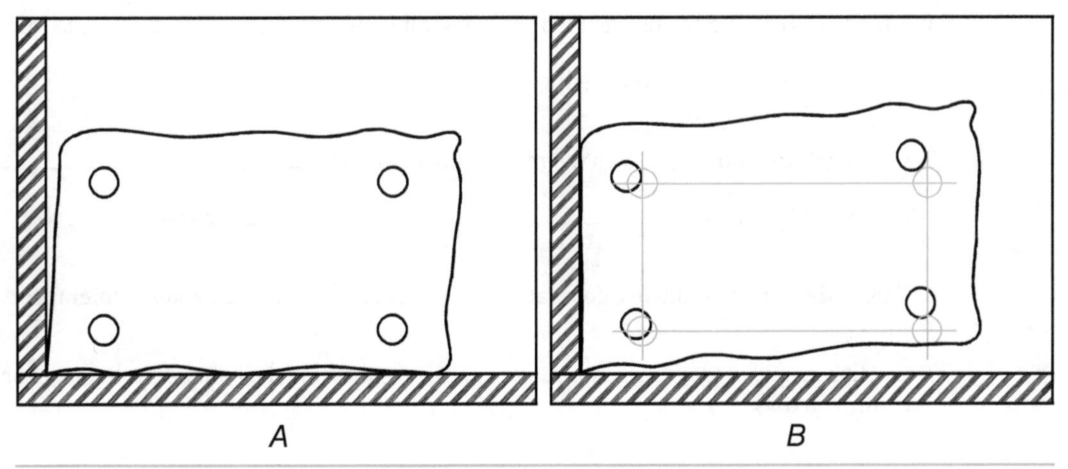

FIGURE 1-6 Possible datum feature interpretations of the drawing in Fig. 1-5.

and inspectors' judgment. If a part is to be inspected the same way every time, the drawing must specify how the part is to fit in the datum reference frame. Each datum feature must be specified in the feature control frame in its proper order of precedence.

Summary

- GD&T is a symbolic language used to specify the size, shape, form, orientation, and location of features on a part.
- GD&T was created to ensure the proper assembly of mating parts, to improve quality, and to reduce cost.
- GD&T is a design tool.
- GD&T communicates design requirements.
- This text is based on the standard *Dimensioning and Tolerancing* ASME Y14.5–2018.
- The cylindrical tolerance zone defines a uniform distance from true position to the tolerance zone boundary.
- The maximum material condition symbol in the feature control frame is a modifier that allows a bonus tolerance.
- All of the datum features must be specified in order of precedence.

Chapter Review

1. ________________________________ is the current, authoritative reference document that specifies the proper application of GD&T.

2. GD&T is a symbolic language used to specify the ______________, ______________, ______________, ______________, and ______________ of features on a part.

3. Features toleranced with GD&T reflect the ______________________________ between mating parts.

4. GD&T was created to ensure the proper assembly of ____________________, to improve ____________________, and to reduce ____________________.

5. Geometric tolerancing allows the maximum available ____________________ and, consequently, the most ____________________ parts.

6. Plus and minus tolerancing generates a ____________________ shaped tolerance zone.

7. ____________________ generates a cylindrical shaped tolerance zone to control an axis.

8. If the distance across a square tolerance zone is ± .005, or a total of .010, what is the approximate distance across the diagonal? ____________________

9. The ____________________ defines a uniform distance from true position to the tolerance zone boundary.

10. Bonus tolerance equals the difference between the actual mating envelope size and the ____________________.

11. While processing, a rectangular part usually rests against a ____________________ consisting of three mutually perpendicular intersecting planes.

CHAPTER 2

Dimensioning and Tolerancing Fundamentals

Many people know how to design parts and make drawings, yet they lack the basic knowledge to produce engineering drawings that conform to industry standards. Nonconforming drawings can be confusing, cause misunderstanding, and produce unacceptable parts. This chapter will familiarize the reader with some of the lesser-known but important standards-based dimensioning and tolerancing practices. All of the drawings in this book are dimensioned and toleranced with the inch system of measurement because most drawings produced in the United States are dimensioned with this system. Metric dimensioning is shown for illustration purposes only.

Chapter Objectives

After completing this chapter, the learner will be able to:

- *Identify* fundamental drawing rules
- *Demonstrate* the proper way to specify units of measurement
- *Demonstrate* the proper way to specify dimensions and tolerances
- *Interpret* limits
- *Explain* the need for dimensioning and tolerancing on CAD/CAM database models

Fundamental Drawing Rules

Dimensioning and tolerancing must clearly define engineering intent and shall conform to the following rules:

1. Each feature must be toleranced. Those dimensions specifically identified as reference, maximum, minimum, or stock do not require the application of a tolerance.
2. Dimensioning and tolerancing must be complete so that there is full understanding of the characteristics of each feature. Values may be expressed in an engineering drawing or in a CAD product definition data set specified in ASME Y14.41. Neither scaling nor assumption of a distance or a size is permitted.
3. Each necessary dimension of an end product must be shown or defined by model data. No more dimensions than those necessary for complete definition shall be given. The use of reference dimensions on a drawing should be minimized.

4. Dimensions must be selected and arranged to suit the function and mating relationship of a part and must not be subject to more than one interpretation.
5. The drawing should define a part without specifying manufacturing methods.
6. Nonmandatory processing dimensions must be identified by an appropriate note, such as NONMANDATORY (MFG DATA).
7. Dimensions should be arranged to provide required information for optimum readability.
8. Dimensions in orthographic views should be shown in true profile views and refer to visible outlines. When dimensions are shown in models, the dimension must be applied in a manner that shows the true value.
9. Wires, cables, sheets, rods, and other materials manufactured to gage or code numbers must be specified by linear dimensions indicating the diameter or thickness. Gage or code numbers may be shown in parentheses following the dimension.
10. An implied 90° angle always applies where centerlines and lines depicting features are shown on orthographic views at right angles and no angle is specified.
11. An implied 90° basic angle always applies where centerlines of features or surfaces shown at right angles on an orthographic view are located or defined by basic dimensions and no angle is specified.
12. A zero basic dimension always applies where axes, center planes, or surfaces are shown coincident on orthographic views and geometric tolerances establish the relationship between the features.
13. Unless otherwise specified, all dimensions and tolerances are applicable at 68°F (20°C). Compensation may be made for measurements made at other temperatures.
14. Unless otherwise specified, all dimensions and tolerances apply in a free-state condition, except for restrained nonrigid parts.
15. Unless otherwise specified, all tolerances and datum features apply for the full depth, length, and width of the feature.
16. Dimensions and tolerances apply only at the drawing level where they are specified. A dimension specified for a given feature on one level of a drawing is not mandatory for that feature at any other level.
17. Unless otherwise specified by a drawing/model note or reference to a separate document, the as-designed dimension value does not establish a functional or manufacturing target.
18. Where a coordinate system is shown on the orthographic views or in the model, it must be right-handed unless otherwise specified. Each axis must be labeled and the positive direction shown.
19. Unless otherwise specified, elements of a surface include surface texture and flaws. All elements of a surface must be within the applicable specified tolerance zone boundaries.

Units of Linear Measurement

Units of linear measurement are typically expressed in either the inch system or the metric system. The system of measurement used on the drawing must be specified in a note, usually in the title block. A typical note reads, UNLESS OTHERWISE SPECIFIED, ALL

DIMENSIONS ARE IN INCHES (or MILLIMETERS, as applicable). Some drawings have both inch and metric systems of measurement on them. On drawings dimensioned with the inch system where some dimensions are expressed in millimeters, the millimeter values are followed by the millimeter symbol, mm. On drawings dimensioned with the millimeter system where some dimensions are expressed in inches, the inch values are followed by the inch symbol, IN.

Specifying Linear Dimensions

Where specifying decimal inch dimensions on drawings as described in Table 2-1:

- A zero *is never placed* before the decimal point for values less than 1 inch. Some designers routinely place zeros before the decimal point. This practice is incorrect and confusing for the reader who knows the proper convention.
- A dimension is specified with the same number of decimal places as its tolerance even if zeros need to be added to the right of the decimal point.

	Decimal Inch Dimensions		Millimeter Dimensions	
	Correct	Incorrect	Correct	Incorrect
1.	**.25**	0.25	**0.25**	.25
2.	**4.500 ± .005**	4.5 ± .005	**4.5**	4.500
3.			**4**	4.000

TABLE 2-1 Decimal Inch and Millimeter Dimensions

Where specifying millimeter dimensions on drawings as described in Table 2-1:

- A zero *is placed* before the decimal point for values less than 1 millimeter.
- Zeros *are not added* to the right of the decimal point where dimensions are a whole number plus some decimal fraction of a millimeter. (This practice differs where tolerances are written bilaterally or as limits. See "Specifying Linear Tolerances" below.)
- Neither a decimal point nor a zero is shown where the dimension is a whole number.

Specifying Linear Tolerances

There are three types of direct tolerancing methods:

- Limit tolerancing
- Plus and minus tolerancing
- Geometric tolerancing directly applied to features

Where using limit dimensioning, the high limit or largest value is placed above the lower limit. If the tolerance is written on a single line, the lower limit precedes the higher limit separated by a dash. With plus and minus dimensioning, the dimension is followed by a plus and minus sign and the required tolerance.

	Inch Tolerances		Millimeter Tolerances	
	Correct	**Incorrect**	**Correct**	**Incorrect**
1.	**+.000** **.250** **−.005**	0 .250 −.005	**0** **45** **−0.05**	+.00 45 −.05
2.	**+.025** **.250** **−.010**	+.025 .25 −.01	**+0.25** **45** **−0.10**	+.25 45 −.1
3.	**.500 ± .005**	.5 ± .005	**45 ± 0.05**	45.00 ± 0.05
4.	**.500** **.495**	.5 .495	**45.25** **45.00**	45.25 45

TABLE 2-2 Inch and Millimeter Tolerances

Where inch tolerances are used on drawings (Table 2-2):

- Where a unilateral tolerance is specified and either the plus or the minus limit is zero, its dimension and zero value must have the same number of decimal places as the other limit, and the zero value must have the opposite sign of the nonzero value.
- Where unequal bilateral tolerancing is specified, both the dimension and the tolerance values have the same number of decimal places. Zeros are added where necessary.
- Where bilateral tolerancing is used, both the plus and minus value and the dimension value have the same number of decimal places. Zeros are added where necessary.
- Where limit dimensioning and tolerancing is used, both values shall have the same number of decimal places, even if zeros must to be added after the decimal point.

Where millimeter tolerances are used on drawings (Table 2-2):

- Where a unilateral tolerance is specified and either the plus or the minus limit is zero, a single zero is shown, and no plus or minus sign is used.
- Where unequal bilateral tolerancing is specified, both tolerance values have the same number of decimal places. Zeros are added where necessary. The dimension value is not required to have the same number of decimal places as the tolerance values.
- Where bilateral tolerancing is used, the dimension value is not required to have the same number of decimal places as the tolerance values.
- Where limit dimensioning and tolerancing is used, both values must have the same number of decimal places, even if zeros must to be added after the decimal point.

Where basic inch dimensions are used, the basic dimension values are not required to have the same number of decimal places as the associated tolerances shown in Fig. 2-1. Where basic

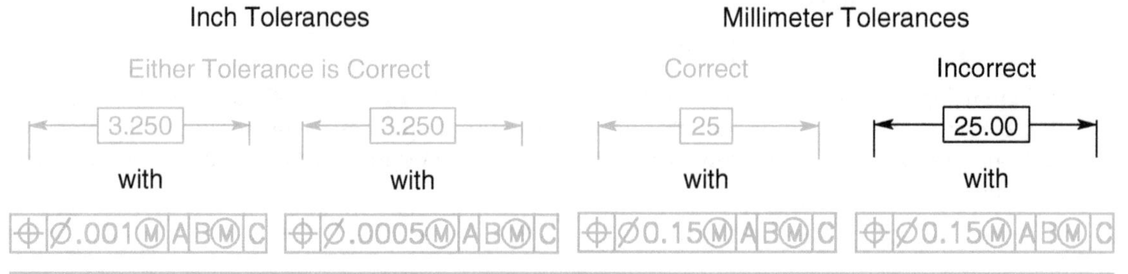

FIGURE 2-1 Basic dimensions and geometric tolerances need not have the same number of decimal places.

metric dimensions are used, the basic dimension values are specified with the practices shown in Table 2-1 for millimeter dimensioning.

Interpreting Dimensional Limits

All dimensional limits are absolute as shown in Table 2-3. Regardless of the number of decimal places, dimensional limit values are used as if an infinite number of zeros followed the last digit after the decimal point.

4.0	means	4.000...0
4.2	means	4.200...0
4.25	means	4.250...0

TABLE 2-3 Dimensional Limits Are Absolute

Specifying Angular Dimensions

Angular dimensions are specified in either of two conventions:

- Degrees and decimal parts of a degree (44.72°)
- Degrees (°), minutes (′), and seconds (″)

Where only minutes or seconds are specified, the number of minutes or seconds shall be preceded by zero degrees (0° 10′) or zero degrees and zero minutes (0° 0′ 30″). Where only degrees are assigned, the value is followed with the degree symbol (44.72°). Where decimal values less than 1° are specified, a zero shall precede the decimal value (0.55°) as shown in Fig. 2-2.

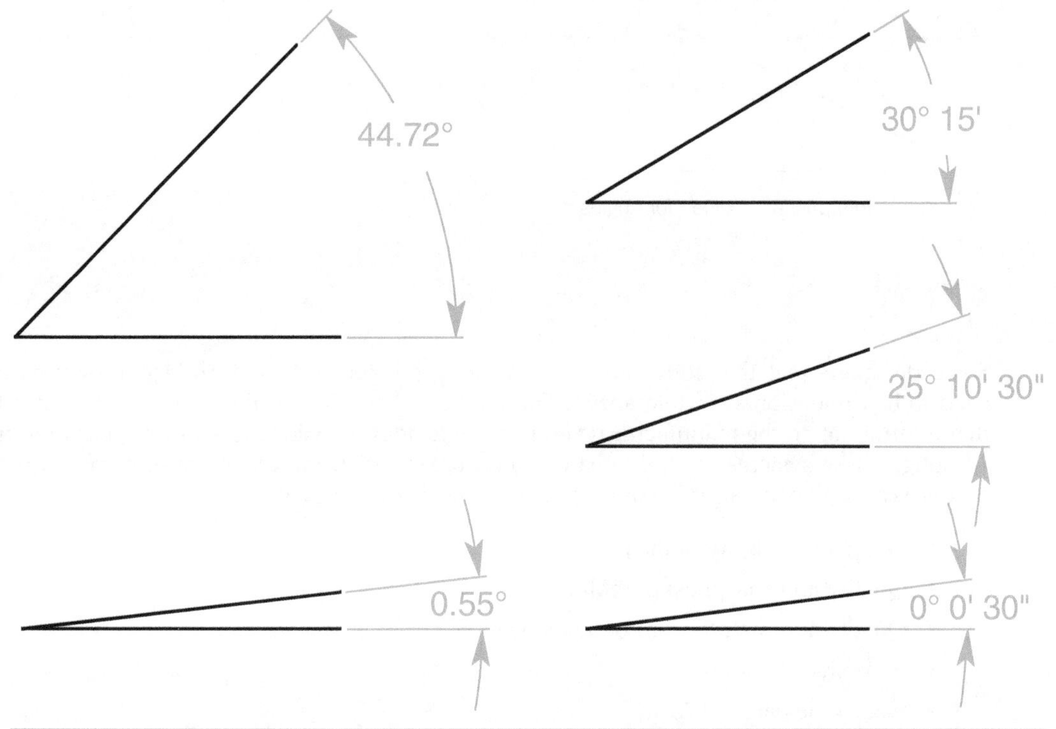

FIGURE 2-2 Angular dimensions may be expressed with degrees and a decimal part of a degree or degrees, minutes, and seconds.

Features appearing to be 90° in a view are, in fact, at an implied dimension of 90°. Unless otherwise specified, the tolerance for an implied 90° angle is the same as the tolerance for any other angle on the field of the drawing governed by a general note or the general, angular title block tolerance.

Two dimensions, 90° angles and zero dimensions, are not placed on the field of the drawing. A zero distance, such as the distance between two axes of coaxial features, must be toleranced separately. The title block tolerance does not control the orientation or location relationship between individual features.

Specifying Angular Tolerances

Where specifying angle tolerances on drawings in terms of degrees and decimal fractions of a degree, as shown in Fig. 2-3, the angle and the plus and minus tolerance values are written with the same number of decimal places. A decimal tolerance less than 1° is preceded by a zero. Where specifying angle tolerances in terms of degrees and minutes, the angle and the plus and minus tolerance values are written in degrees and minutes, even if the number of degrees is zero.

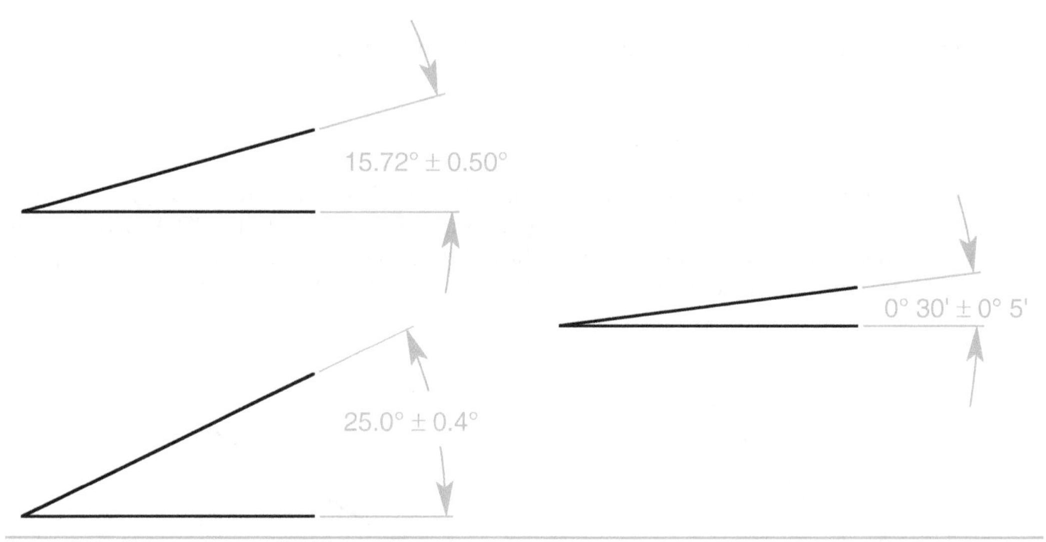

FIGURE 2-3 Tolerances specified for angles.

Dimensioning and Tolerancing for CAD/CAM Database Models

Some designers feel that solid model drawings produced with CAD/CAM programs don't need to be dimensioned or toleranced. The method of producing the design and transmitting that information to the manufacturing equipment is not the major cause of irregularity in parts. Although these systems may eliminate some human error, the major cause of part variation occurs as a result of a variety of other sources, such as the following:

- Setup and stability of the part
- Quality and sharpness of tooling
- Quality and maintenance of machine tools
- Excessive clamping
- Size of the part

- The material the part is made from
- Heat treating
- Plating

None of these problems are addressed with the use of solid modeling programs. To quote *Dimensioning and Tolerancing ASME Y14.5–2018,*

> Dimensioning and tolerancing must be complete so there is full understanding of the characteristics of each feature. Values may be expressed in an engineering drawing or in a CAD product definition data set specified in ASME Y14.41.

Drawings are inspection tools. If features on orthographic drawings are not dimensioned and toleranced, they cannot be inspected. The most effective way to communicate design intent is through the proper use of geometric dimensioning and tolerancing.

Summary

- Units of linear measurement are typically expressed either in the inch system or in the metric system, and that system must be specified on the drawing.
- A zero is never placed before the decimal point for values less than 1 inch. Inch dimensions are specified with the same number of decimal places as their tolerances even if zeros need to be added to the right of the decimal point.
- There are three types of direct tolerancing methods: limit tolerancing, plus and minus tolerancing, and geometric tolerancing.
- Where a unilateral inch tolerance is specified and either the plus or the minus limit is zero, its dimension and zero value must have the same number of decimal places as the other limit, and the tolerance will include the appropriate plus or minus sign.
- Where bilateral inch tolerancing is specified, both the dimension and tolerance values have the same number of decimal places.
- Angular dimensions may be expressed with degrees and a decimal part of a degree or degrees, minutes, and seconds.
- Where specifying angle tolerances on drawings, the angle dimension and the plus and minus tolerance values are expressed with the same number of places after the decimal point.
- Regardless of the number of decimal places, dimensional limits are used as if an infinite number of zeros followed the last digit after the decimal point.
- If CAD/CAM database models do not include tolerances, they must be communicated outside of the database on a referenced document.

Chapter Review

1. Each feature must be ______________________. Those dimensions specifically identified as reference, maximum, minimum, or stock do not require the application of a tolerance.

2. Dimensioning and tolerancing must be ______________________ so that there is a full understanding of the ______________________________________ of each feature.

3. Dimensions must be selected and arranged to suite the function and mating relationship of a part and must not be subject to more than one ______________________.

4. The drawing should define a part without specifying ______________________ methods.

5. An ______________________ always applies where centerlines and lines depicting features are shown on orthographic views at right angles and no angle is specified.

6. An ______________________ always applies where centerlines of features or surfaces shown at right angles on an orthographic view are located or defined by basic dimensions and no angle is specified.

7. Unless otherwise specified, all dimensions and tolerances are applicable at ______________ Measurements made at other temperatures may be adjusted mathematically.

8. Unless otherwise specified, all dimensions and tolerances apply in a ______________ ______________ condition except for constrained nonrigid parts.

9. Unless otherwise specified, all tolerances and datum features apply for the full ______________________ and ______________________ of the feature.

10. Dimensions and tolerances apply only at the ______________________ where they are specified.

11. Units of linear measurement are typically expressed either in the ______________ system or in the ______________________ system.

12. Where specifying decimal inch dimensions, a ______________________ is never placed before the decimal point for values less than 1 inch.

13. What are the three types of direct tolerancing methods?

____

14. Where inch tolerances are used, a dimension is specified with the same number of decimal places as its ______________________.

15. Where decimal inch unilateral tolerance is specified and either the plus or the minus limit is zero, its dimension and zero value must have the ______________ ______________________ as the other limit, and the zero value must have the ______________________________ of the nonzero value.

16. For decimal inch tolerances, where bilateral tolerancing or limit dimensioning and tolerancing is used, both values ________________________.

17. Dimensional limits are used as if an ________________________ followed the last digit after the decimal point.

18. Angular dimensions may be expressed with ________________________ or with ________________________.

19. What two dimensions are not placed on the field of the drawing?

____

20. When dimensioning and tolerancing CAD/CAM database models, dimensioning and tolerancing must be complete so that there is full understanding of the characteristics of each feature. Values may be expressed in an engineering drawing or in a

____.

CHAPTER 3

Symbols, Terms, and Rules

Symbols, terms, and rules are the basics of geometric dimensioning and tolerancing (GD&T). They are the alphabet, the definitions, and the syntax of this language. Imagine trying to read a book or write a composition without knowing the alphabet, without a good vocabulary, and without a working knowledge of how a sentence is constructed. The GD&T practitioner must be very familiar with these concepts and know how to use them. It is best to commit them to memory; a little memorization up front will save time and reduce frustration in the future.

Chapter Objectives

After completing this chapter, the learner will be able to:

- *List* the 12 geometric characteristic symbols
- *Identify* the datum feature symbol
- *Explain* the elements of the feature control frame
- *Identify* the other symbols used with GD&T
- *Define* 15 critical terms
- *Explain* the 2 general rules

Symbols

Geometric Characteristic Symbols

Geometric characteristic symbols are the essence of this graphic language. It is important not only to know each symbol but also to know how to apply these symbols on drawings. The 12 geometric characteristic symbols, shown in Fig. 3-1, are divided into five categories:

- Form
- Profile
- Orientation
- Location
- Runout

It is important to learn these symbols in their respective categories because many characteristics that apply to one geometric control also apply to other geometric controls in the same category. For example, it is not appropriate to specify datum features for any of the form controls.

Pertains to	Type of Tolerance	Geometric Characteristics	Symbol
Individual Feature Only	Form	STRAIGHTNESS	⏤
		FLATNESS	⏥
		CIRCULARITY	○
		CYLINDRICITY	⌭
Individual Feature or Related Features	Profile	PROFILE OF A LINE	⌒
		PROFILE OF A SURFACE	⌓
Related Features	Orientation	PERPENDICULARITY	⊥
		PARALLELISM	//
		ANGULARITY	∠
	Location	POSITION	⌖
	Runout	CIRCULAR RUNOUT	↗
		TOTAL RUNOUT	⌰

FIGURE 3-1 Geometric characteristic symbols.

Notice that form controls apply only to individual features. In other words, form controls are not related to other features. Features toleranced with orientation, location, or runout controls are related to datum features. Profile controls may relate features to a datum feature(s) or they may apply without datum features as required.

The Datum Feature Symbol

The datum feature symbol is used to identify physical features of a part as datum features. The datum feature symbol consists of a capital letter enclosed in a square box. The square box is connected to a leader directed to the datum feature ending in a triangle. The triangle may be solid or open. The datum feature identifying letters may be any letter of the alphabet

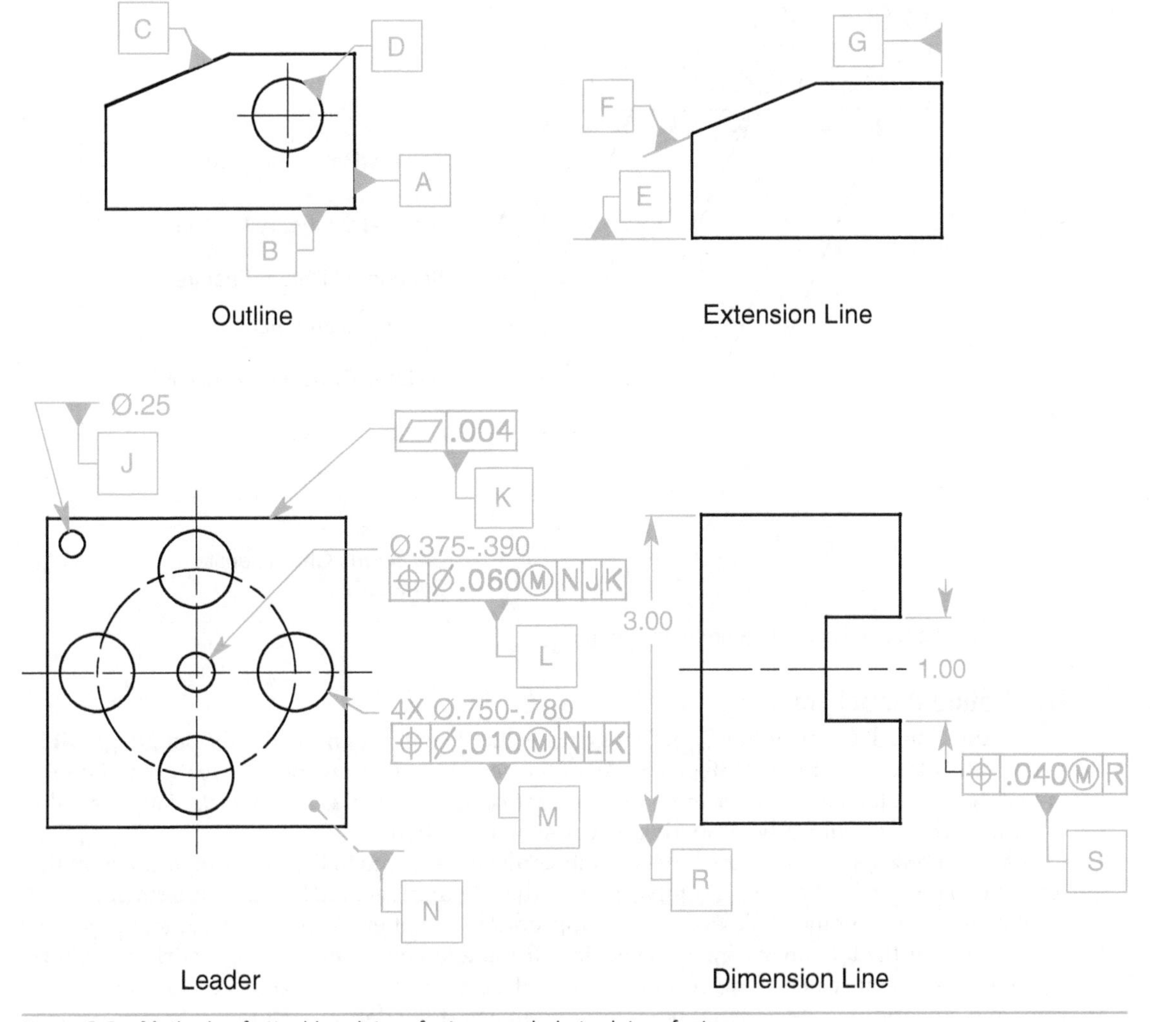

FIGURE 3-2 Methods of attaching datum feature symbols to datum features.

except I, O, and Q. Multiple letters, such as AA through AZ, BA through BZ, and so on, may be used if a large number of datum features is required. Datum feature symbols must not be attached to centerlines, center planes, or axes (see Fig. 4-5). They may be directed to outlines or extension lines of features, such as datum feature symbols A through G, as shown on the top two drawings in Fig. 3-2. Datum feature symbols may also be attached to leaders, feature control frames, or dimension lines, as shown in the lower two drawings in Fig. 3-2. If only a leader is used, the datum feature symbol is attached to the tail, such as datum features J and N. The dashed leader from datum feature N to the part specifies the far side of the part. This leader terminates in a dot indicating a surface. That is, datum feature symbol N identifies the back side of the part as datum feature N. Also, a datum feature symbol may be attached to a feature control frame directed to the datum feature with a leader, such as datum features K, L, and M. If the datum feature symbol is placed in line with a dimension line of a feature of size or on a feature control frame associated with a feature of size is the datum feature. For example, in Fig. 3-2, datum feature R is the 3.00-inch feature of size between the top and bottom surfaces, and datum feature S is the 1.00-inch slot.

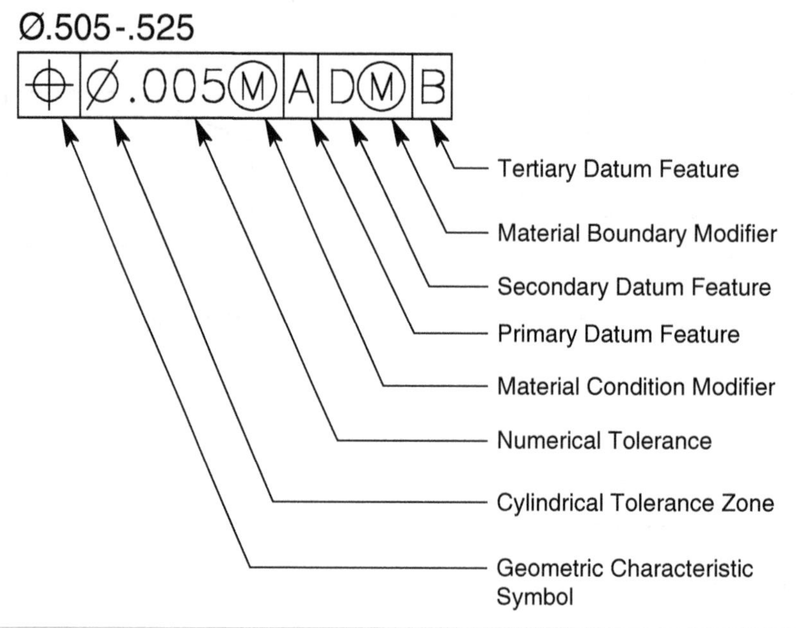

FIGURE 3-3 The feature control frame explained.

The Feature Control Frame

The feature control frame in the GD&T language is like a sentence in the English language—it is a complete tolerancing thought (Fig. 3-3). All of the geometric tolerancing for a feature or pattern of features is contained in one or more feature control frames. Just as in any other language, the feature control frame must be properly and completely written.

One of the 12 geometric characteristic symbols always appears in the first compartment of the feature control frame. The second compartment is the tolerance section. In this compartment, there is, of course, the tolerance followed by any appropriate modifiers. A diameter symbol precedes the tolerance if the tolerance zone is cylindrical. If the tolerance zone is not cylindrical, nothing precedes the tolerance. The final section of the feature control frame is reserved for datum features and any appropriate material boundary modifiers. If the datum feature is a feature of size, a material boundary applies; if no material boundary modifier is specified for a datum feature of size, regardless of material boundary automatically applies. Datum features are arranged in order of precedence or importance. The first letter to appear in the datum section of the feature control frame, the primary datum feature, is the most important datum feature. The second letter, the secondary datum feature, is the next most important datum feature, and the third letter, the tertiary datum feature, is the least important. Datum features need not be specified in alphabetical order.

Reading the Feature Control Frame

Those who can read a feature control frame understand the significance of each symbol and how it relates to the tolerance of the feature. Each numbered statement corresponds with the appropriately numbered item under the feature control frames shown in Figs. 3-4 through 3-7.

1. The geometric characteristic symbol
2. The controlled element of the feature (determined by the geometry of the feature shown on the drawing, not in the feature control frame)

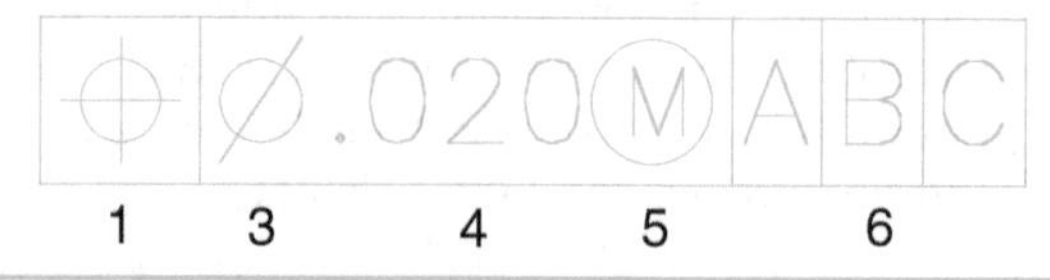

FIGURE 3-4 Reading the feature control frame.

3. The tolerance zone shape
4. The tolerance zone size
5. Any tolerance modifier(s)
6. The datum feature(s)

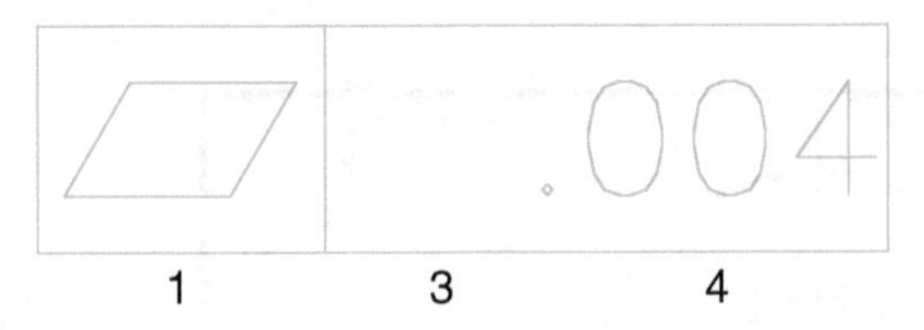

FIGURE 3-5 Reading the flatness of a surface control.

1. The **flatness** tolerance requires that
2. All elements on the controlled surface
3. Must lie within a tolerance zone between two parallel planes
4. .004 apart

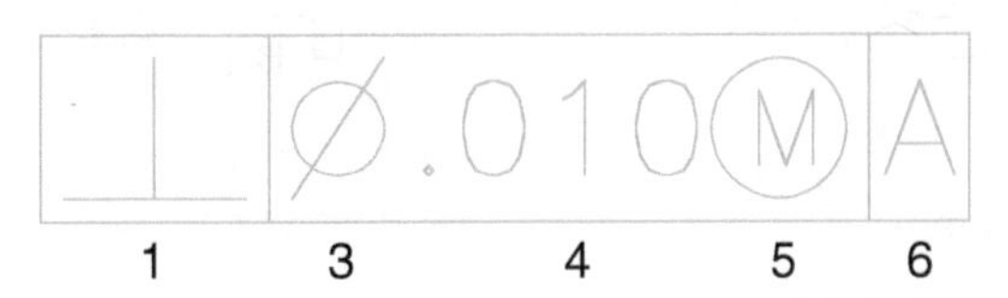

FIGURE 3-6 Reading the perpendicularity of an axis control.

1. The **perpendicularity** tolerance requires that
2. The axis of the controlled feature
3. Must lie within a cylindrical tolerance zone
4. .010 in diameter
5. At maximum material condition (MMC)
6. Oriented with a basic 90° angle to datum feature A

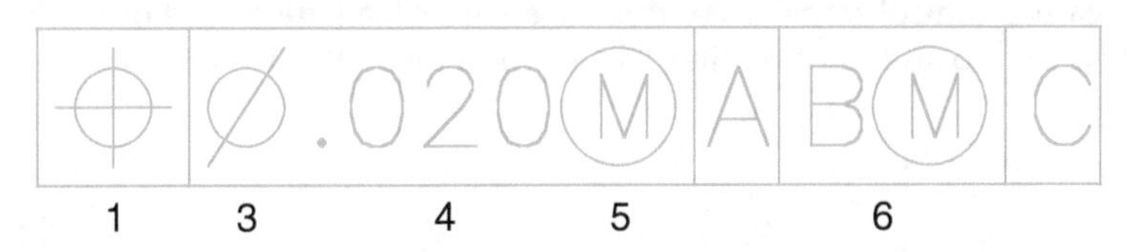

FIGURE 3-7 Reading the position of an axis control.

1. The **position** tolerance requires that
2. The axis of the controlled feature
3. Must lie within a cylindrical tolerance zone
4. .020 in diameter
5. At MMC
6. Oriented and located with basic dimensions to a datum reference frame established by datum feature A, datum feature B at its maximum material boundary, and datum feature C

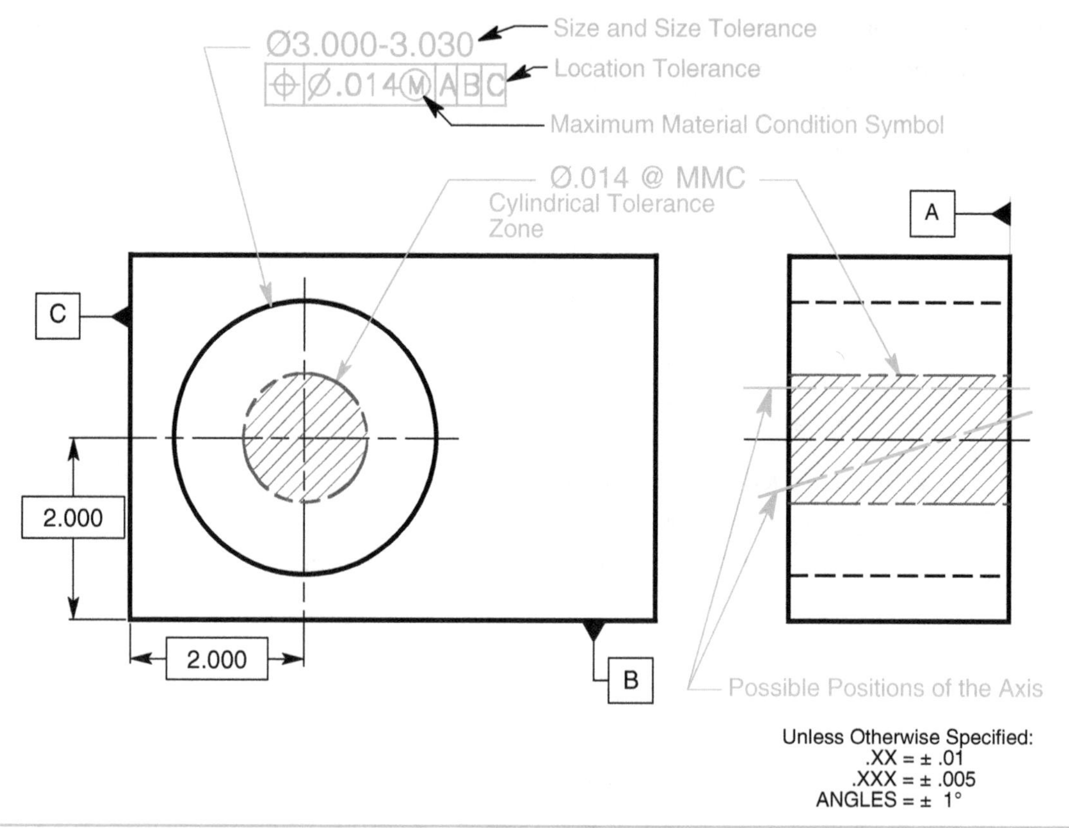

FIGURE 3-8 A feature control frame specifying the tolerance zone size, shape, and relationship to its datum features.

The feature control frame in Fig. 3-8 can be read as follows. The position control requires that the axis of the hole must lie within a cylindrical tolerance zone of .014 in diameter at MMC (circle M). The tolerance zone is oriented and located with basic dimensions perpendicular to datum feature A, located up from datum feature B, and over from datum feature C. If the hole diameter is produced at its MMC size, Ø3.000, the diameter of the tolerance zone is .014. If the hole diameter is produced at Ø3.020, the diameter of the tolerance zone is .034, as shown in Chap. 1.

Attaching a Feature Control Frame to a Feature

Feature control frames may be attached to features with extension lines, dimension lines, or leaders. Where a feature control frame is controlling a surface, a side or end of the feature control frame may be attached to an extension line, as shown in Fig. 3-9*A*. Even a corner of the feature control

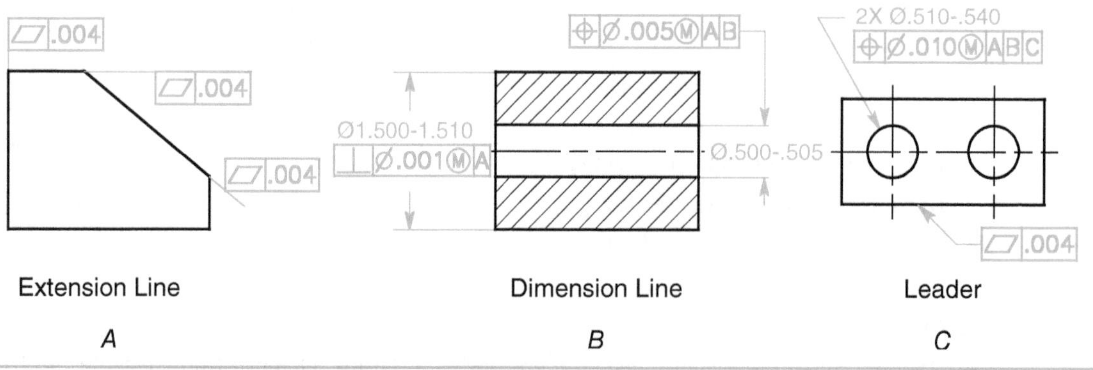

FIGURE 3-9 Feature control frames attached to features.

frame may be attached to an extension line extending from a surface at an angle to the horizontal plane. Where a feature control frame is controlling the orientation or location of a feature of size, the feature control frame may be placed beneath a dimension or attached to an extension of a dimension line, as shown in Fig. 3-9*B*. Finally, a feature control frame may be attached to a leader directed to a feature surface or placed beneath a dimension directed with a leader to a feature of size, such as the hole pattern shown in Fig. 3-9*C*.

A composite feature control frame consists of one geometric characteristic symbol followed by two tolerance and datum feature sections, as shown in Fig. 3-10*A*. The lower segment is a refinement of the upper segment. The multiple single-segment feature control frames in Fig. 3-10*B* consist of two complete feature control frames, one below the other, which may have different datum references as shown. The lower segment is a refinement of the upper segment. In Fig. 3-10*C*, a single feature control frame may have one or more feature control frames refining the tolerance of specific feature characteristics. These controls will be discussed in more detail in later chapters.

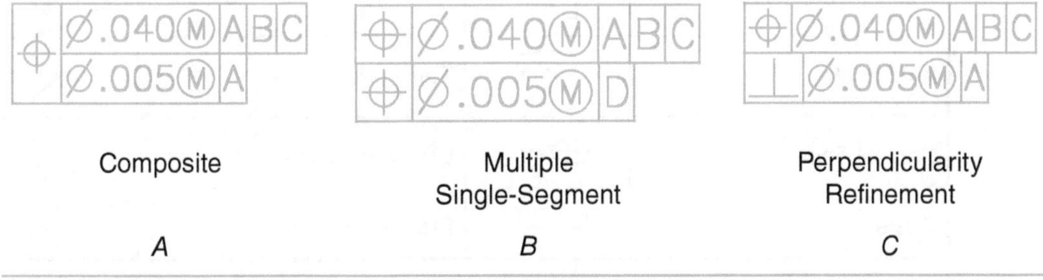

FIGURE 3-10 Feature control frames with tolerance refinements.

Other Symbols Used with Geometric Tolerancing

A number of other symbols used with geometric dimensioning and tolerancing are listed in Fig. 3-11. They are discussed in more detail below and in subsequent chapters.

All Around	○	Free State	Ⓕ
All Over	◎	Independency	Ⓘ
Between	↔	Projected Tolerance Zone	Ⓟ
Number of Places	X	Tangent Plane	Ⓣ
Continuous Feature	⟨CF⟩	Unequally Disposed Profile	Ⓤ
Counterbore	⌴	Spotface	⌴SF
Countersink	⌵	Radius	R
Depth/Deep	↧	Radius, Controlled	CR

FIGURE 3-11 Other symbols used on prints.

Diameter	Ø	Spherical Radius	SR
Dimension, Basic	1.000 (boxed)	Spherical Diameter	SØ
Dimension, Reference	(60)	Square	□
Dimension Not To Scale	15 (underlined)	Statistical Tolerance	⟨ST⟩
Dimension Origin	◄—⊕	Datum Target	Ø.500 / A1
Datum Translation	▷	Movable Datum Target	A1
Arc Length	110 (with arc above)	Target Point	X
Conical Taper	(conical taper symbol)	Dynamic Profile	△
Slope	(slope symbol)	From - To	→

FIGURE 3-11 *(Continued)*

The **all around** and **between** symbols are used with the profile control, as shown in Fig. 3-12. Where a small circle is placed at the joint of the leader, a profile tolerance is specified all around the surface of the part. The between symbol, represented by a double-arrow line segment in Fig. 3-12, indicates that the tolerance applies between points X and Z on the portion of the profile where the leader is pointing.

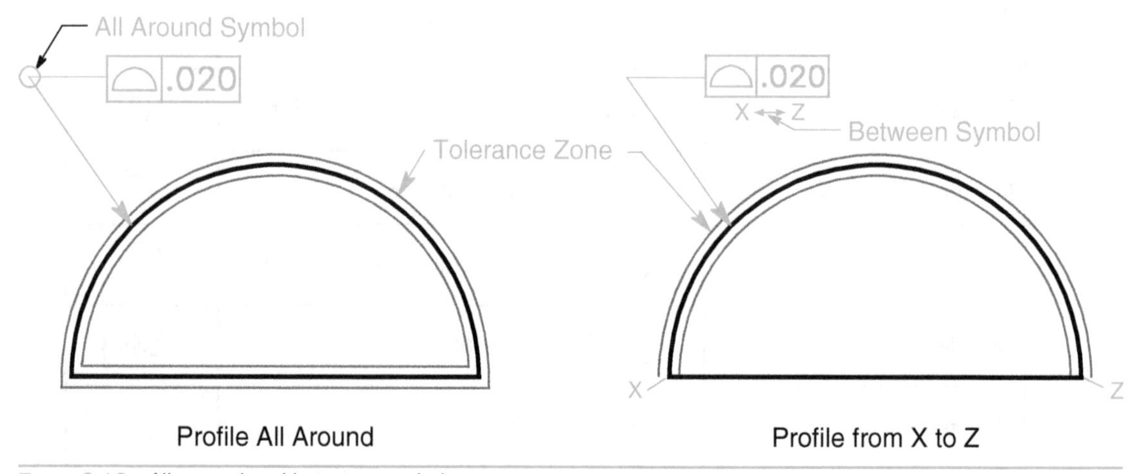

FIGURE 3-12 All around and between symbols.

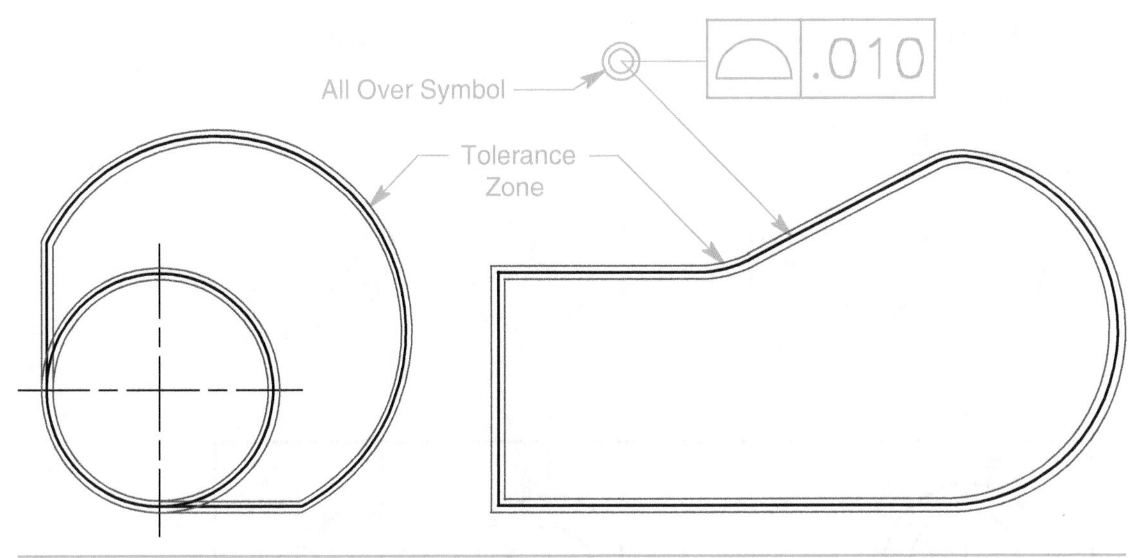

FIGURE 3-13 All over symbol.

The **all over** symbol, also used with the profile control, consists of two small concentric circles placed at the joint of the leader connecting the feature control frame to the feature. Where the all over symbol is specified, the profile applies all over the three-dimensional surface of the part, as shown in Fig. 3-13.

The **continuous feature** symbol specifies that a group of two or more interrupted regular features of size are to be considered one single feature of size, as shown in Fig. 3-14.

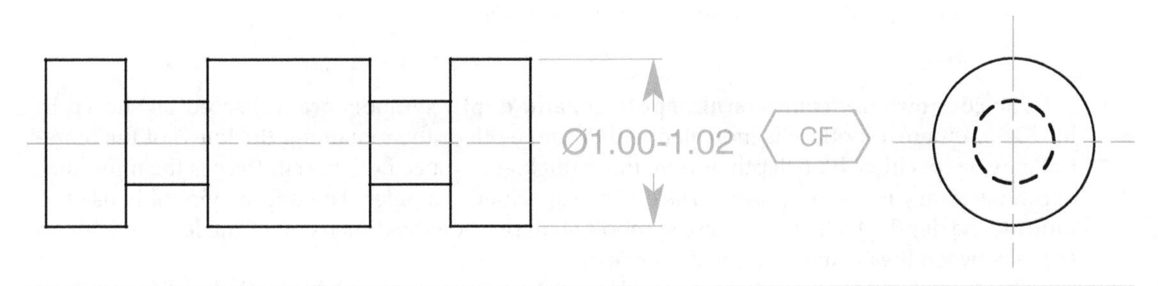

FIGURE 3-14 Continuous feature symbol.

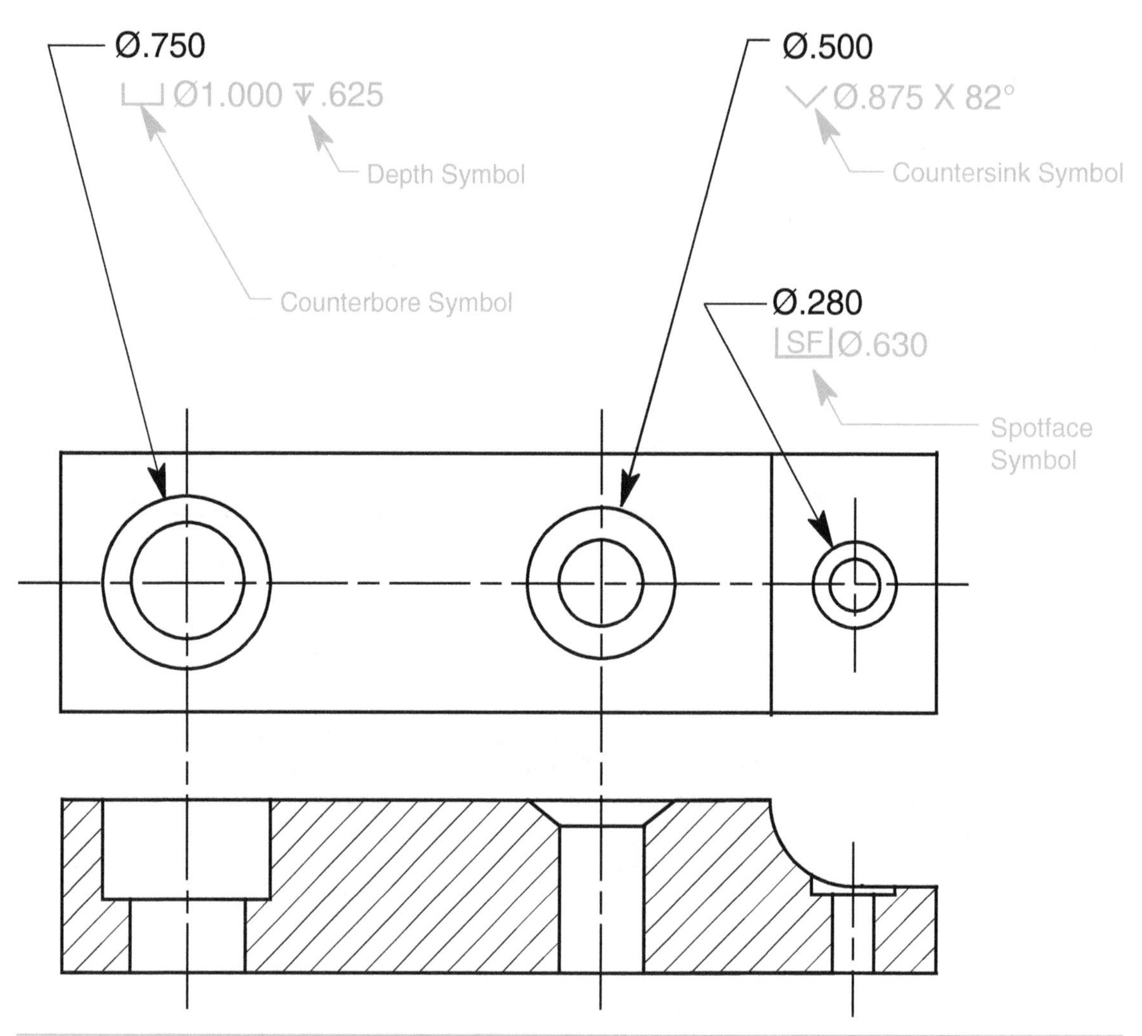

FIGURE 3-15 Counterbore, countersink, spotface, and depth symbols.

The **counterbore**, **countersink**, **spotface**, and **depth** symbols are indicated as shown in Fig. 3-15. Where a spotface is indicated, either the depth or the remaining thickness of the material may be specified. If no depth or remaining thickness is specified, the spotface is the minimum depth necessary to clean up the surface of the specified diameter. The depth symbol is used to indicate the depth of a feature. These symbols shall precede the dimension of the feature with no space between the symbol and the dimension.

The **diameter**, **spherical diameter**, **radius**, **spherical radius**, and **controlled radius** symbols are shown in Fig. 3-11. These symbols shall precede the value of a dimension or tolerance given as a diameter or radius, as applicable. The symbol and value must not be separated by a space.

The **number of times** symbol, X, may be used to indicate the number of times the specified value applies. The X is preceded by the number of occurrences with no space between the number and the X. The X is followed by a space between the X and the dimension, as shown in Fig. 3-16. An X may be used to indicate "by" between coordinate dimensions, as shown for the chamfer in Fig. 3-16. In this case, the X shall be preceded and followed by a space.

The **basic dimension** has a box around the numerical value. The title block tolerance does not apply to basic dimensions. The tolerance associated with a basic dimension usually appears in a feature control frame or a note.

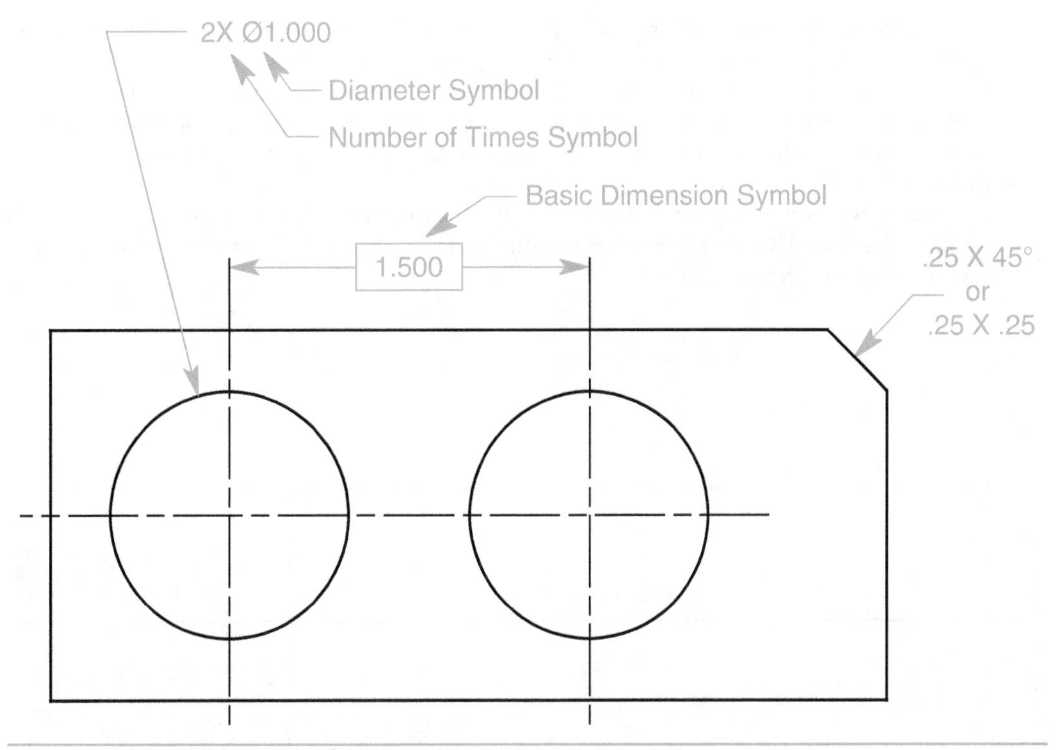

FIGURE 3-16 Diameter, number of times, and basic dimension symbols.

A *radius* is a straight line connecting the center and the periphery of an arc, a circle, or a sphere.

The **radius** symbol, R, shown in Fig. 3-17, defines a tolerance zone bounded by a maximum radius arc and a minimum radius arc that are tangent to the adjacent surfaces. The surface of the toleranced radius must lie within this tolerance zone.

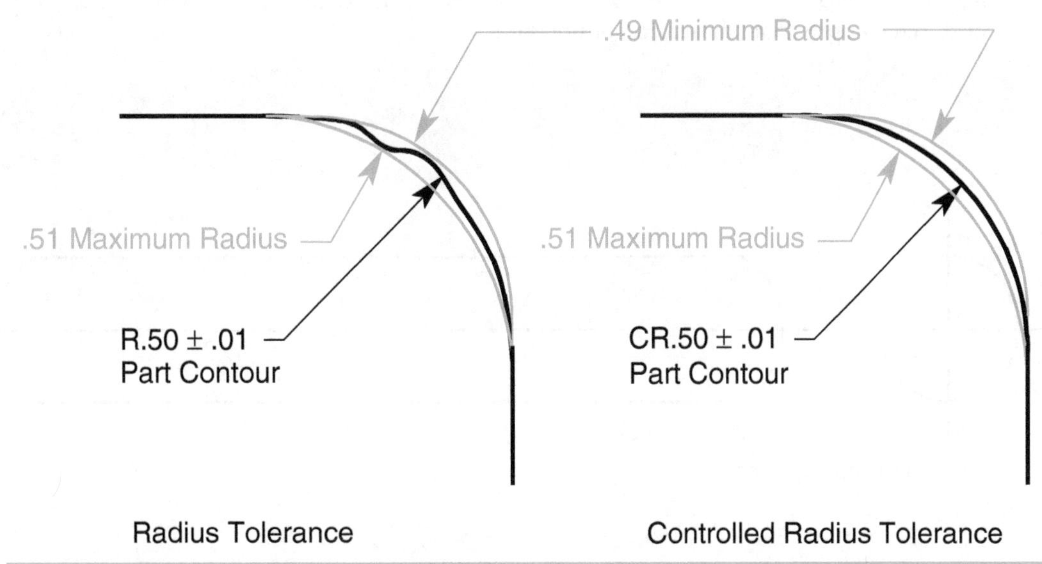

FIGURE 3-17 Radius and controlled radius tolerances.

The **controlled radius** symbol, CR, also defines a tolerance zone bounded by maximum and minimum radii arcs that are tangent to the adjacent surfaces. However, the surface of the controlled radius not only must lie within this tolerance zone but also must be a fair (smooth) curve with no reversals. In addition, at no point on the radius can the curve be greater than the maximum limit or smaller than the minimum limit arcs. Additional requirements may be specified in a note.

The **dimension origin** symbol indicates that the measurement of a feature starts at the origin, which is the end of the dimension line within the circle. Figure 3-18 shows several ways to specify the dimension origin symbol.

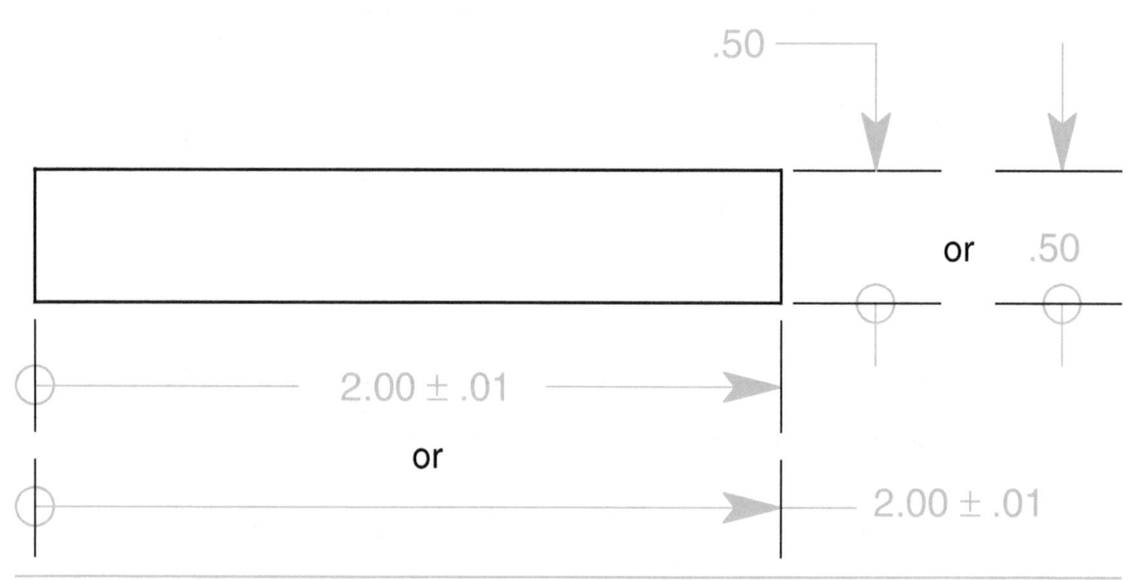

FIGURE 3-18 Dimension origin symbol.

The **independency** symbol, circle I, shown in Fig. 3-19, indicates that perfect form of a regular feature of size at MMC (or least material condition) is not required. The symbol must be placed next to the appropriate dimension or notation.

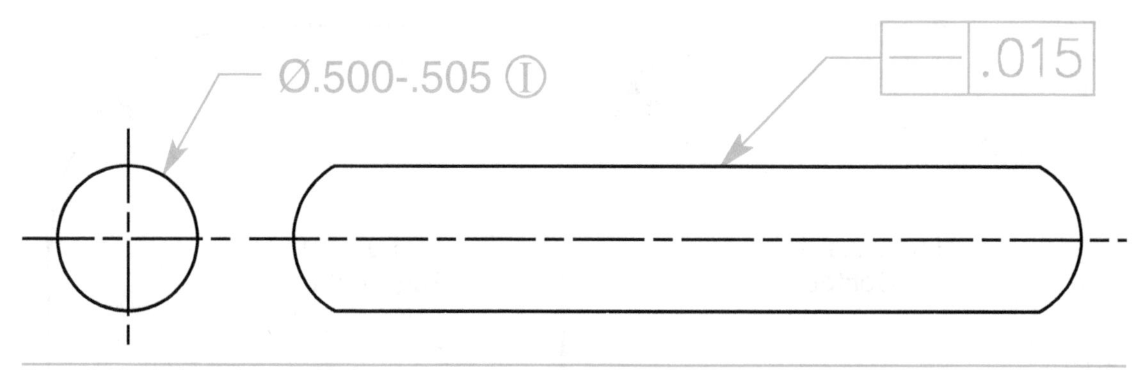

FIGURE 3-19 Independency symbol.

The **free state** symbol, circle F, specifies that where tolerances are to be inspected in a restrained condition and an additional tolerance is to be inspected in the free state, the circle F is used to clarify that situation. The **projected tolerance zone** symbol, circle P, specifies that the tolerance zone is to be projected into the mating part. The **tangent plane** symbol, circle T, specifies that if a precision plane contacting the high points of a surface falls within the specified tolerance zone, the surface is in tolerance.

The **statistical tolerance** symbol indicates that the geometric tolerance is based on statistical tolerancing. The statistical tolerance symbol may also be applied to a size tolerance. Use of the statistical tolerance symbol requires that the toleranced feature be charted and that the control chart be in statistical control. The four modifiers mentioned earlier are placed in the feature control frame after the tolerance and any material condition symbols, as shown in Fig. 3-20.

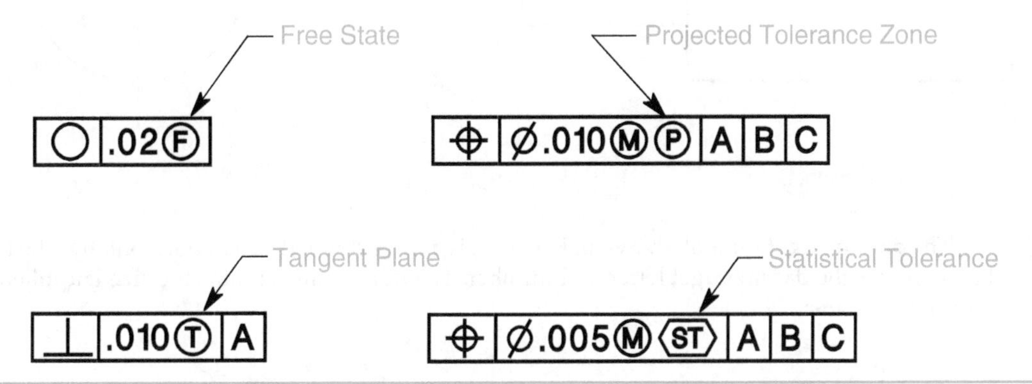

FIGURE 3-20 Free state, projected tolerance zone, tangent plane, and statistical tolerance symbols.

The **unequally disposed profile** symbol, circle U, indicates that the profile tolerance is unilaterally or unequally disposed about the true profile. This symbol shall be placed in the feature control frame following the tolerance value, as shown in Fig. 3-21. The tolerance that would allow additional material added to the true profile is placed after the circle U.

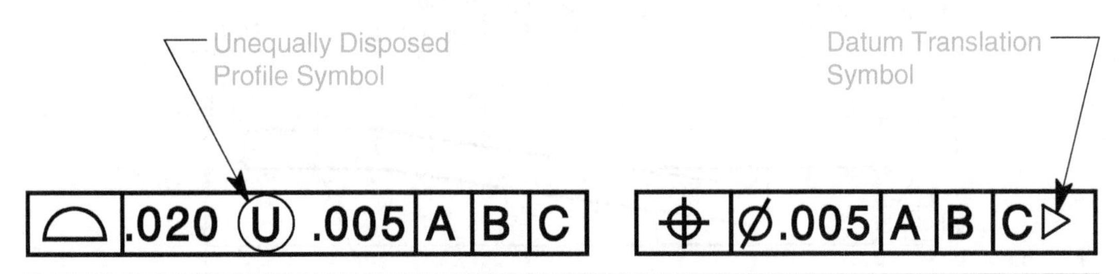

FIGURE 3-21 Unequally disposed profile and datum translation symbols.

The **datum translation** symbol, shown in Fig. 3-21, indicates that a datum feature simulator is not fixed at its basic location and must be free to translate to fully engage the feature.

The **square** symbol preceding a dimension specifies that the toleranced feature is square and that the dimension applies in both directions, as shown in Fig. 3-22. The square symbol applies to square features in the same way a diameter symbol applies to circular features. The symbol and the value are not separated by a space.

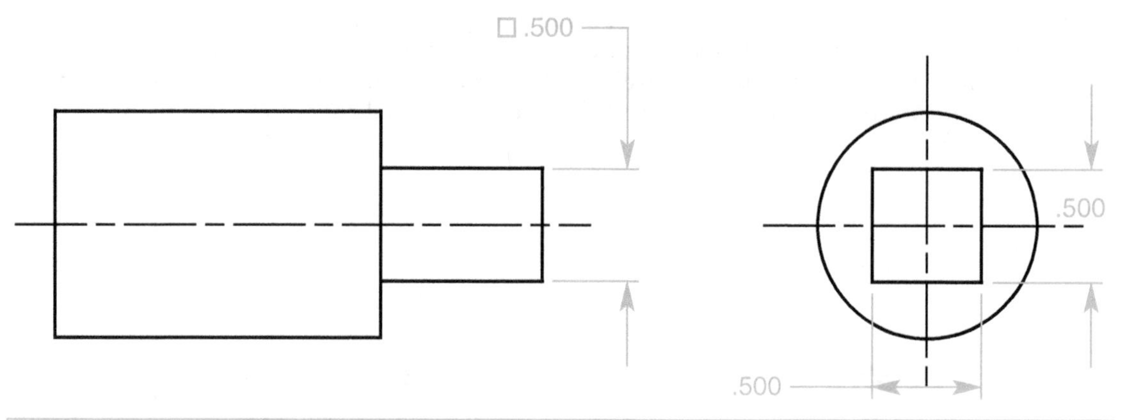

FIGURE 3-22 Square symbol.

The **datum target** symbol, shown in Fig. 3-23, is a circle divided in half horizontally. The lower half contains the datum target letter and number. The upper half contains the size if applied to a datum target area.

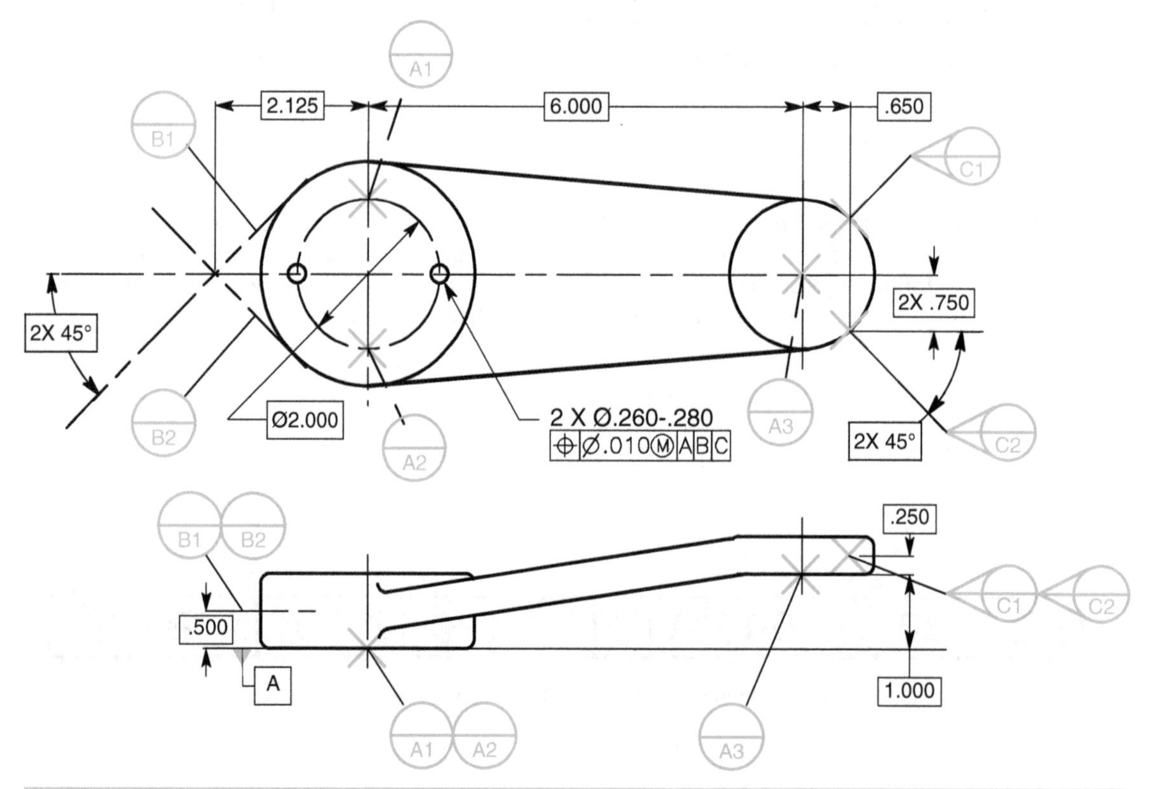

FIGURE 3-23 Datum target symbol, datum target point, and movable datum target symbol.

The **datum target point** represented by an X is located with basic dimensions and is attached to a datum target symbol with a radial line, as shown for datum targets A1, A2, and A3.

The **movable datum target** symbol is similar to a datum target symbol but is not fixed to its basic location and is allowed to translate. The movable datum target symbol is attached to the datum target with a radial line, as shown for datum targets C1 and C2 (see Fig. 3-23).

Conical taper, shown in Fig. 3-24, is defined as the ratio of the difference between two diameters, perpendicular to the axis of a cone, divided by the length between the two diameters:

$$\text{Taper} = (D - d)/L$$

D is the larger diameter, **d** is the smaller diameter, and **L** is the length between the two diameters.

Slope is defined as the ratio of the difference in heights at both ends of an inclined surface, measured at right angles above a base line, and divided by the length between the two heights (see Fig. 3-24):

$$\text{Slope} = (H - h)/L$$

H is the larger height, **h** is the smaller height, and **L** is the length between the two heights.

A **reference dimension** is a numerical value without a tolerance, used only for general information, and may not be used for manufacturing or inspection. The reference dimension is indicated by placing parentheses around the numerical value, as shown in Fig. 3-24.

The **arc length** symbol, the **dimension not to scale** symbol, the **dynamic profile tolerance zone modifier** symbol, and the **"from-to"** symbol, shown in Fig. 3-11, are not illustrated. The arc length symbol indicates that a linear dimension is used to measure an arc along its curved outline. The dimension not to scale symbol is a thick straight line placed beneath an altered dimension that does not agree with the scale of the drawing. The dynamic profile tolerance zone modifier symbol is the symbolic means for specifying the refinement of the form independent of size of a considered feature that is controlled by a profile tolerance. The from-to symbol is the symbolic means to indicate that a specification transitions from one location to a second location. The from-to symbol is similar to the between symbol except that it specifies a particular direction.

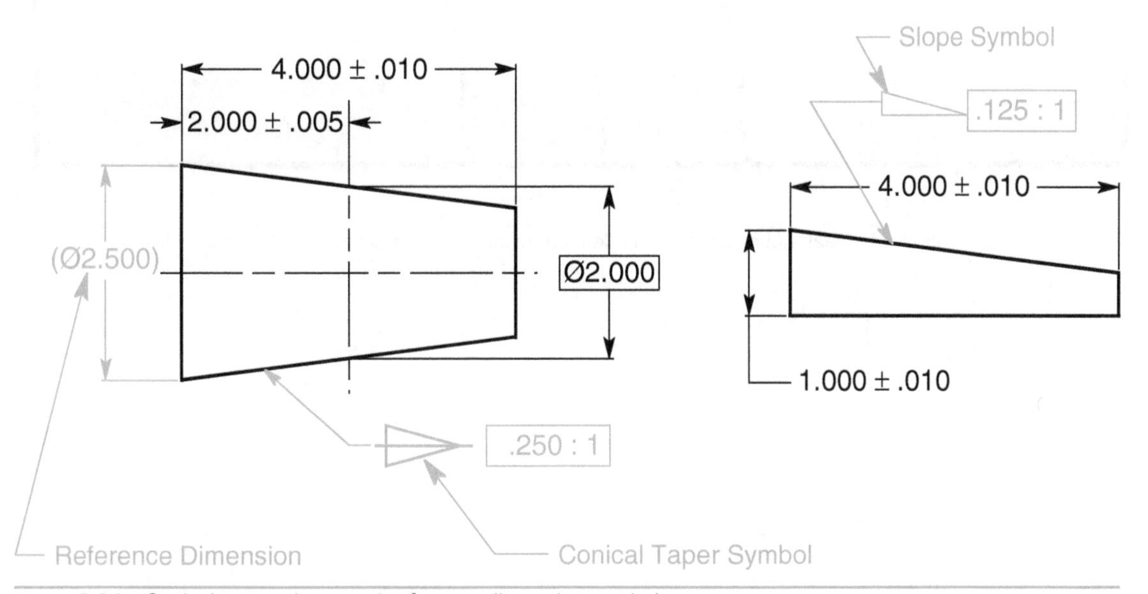

FIGURE 3-24 Conical taper, slope, and reference dimension symbols.

Terms

The names and definitions of many GD&T terms have very specific meanings. In some cases, they are quite different from general English usage. To be able to function in this language, it is important for each GD&T practitioner to be very familiar with these terms.

1. **Actual Mating Envelope**
   The actual mating envelope is a similar perfect feature(s) counterpart of smallest size that can be contracted about an external feature(s) or largest size that can be expanded within an internal feature(s) so that it coincides with the surface(s) at the highest points. Two types of actual mating envelopes are described below.

   - **Unrelated Actual Mating Envelope**
     An unrelated actual mating envelope is a similar perfect feature(s) counterpart contracted about an external feature(s) or expanded within an internal feature(s) and not constrained to any datum feature(s), as shown in Fig. 3-25.

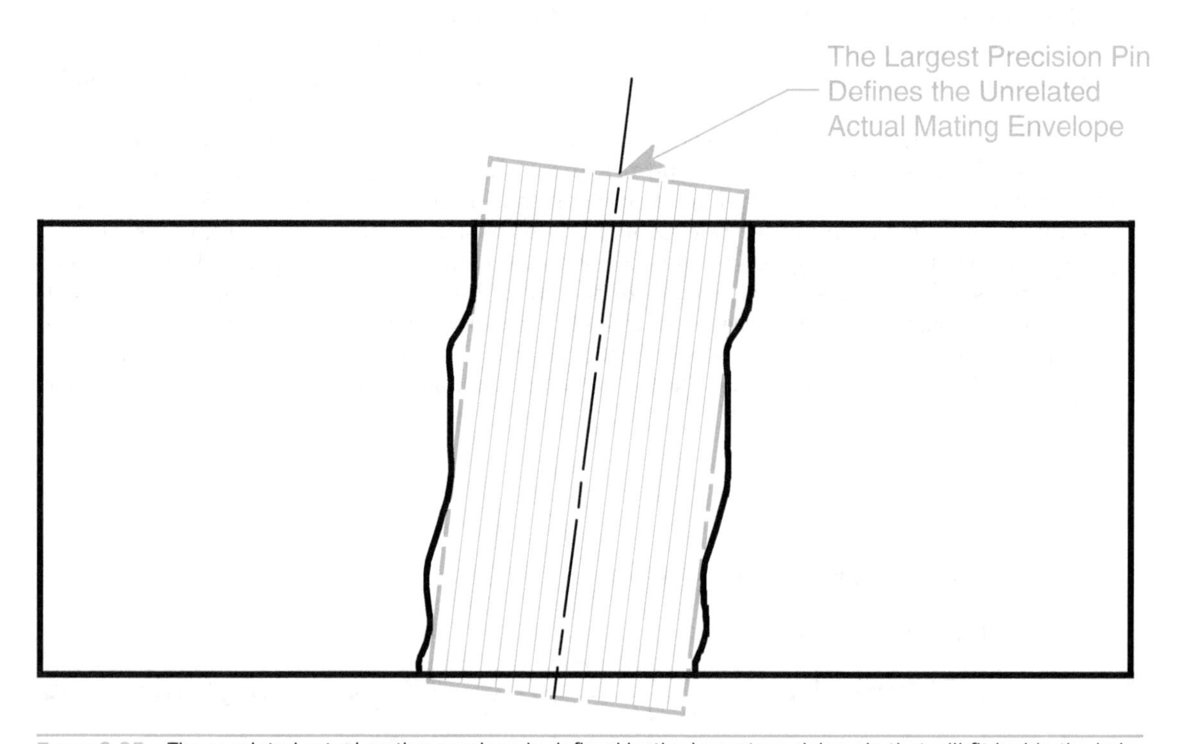

FIGURE 3-25 The unrelated actual mating envelope is defined by the largest precision pin that will fit inside the hole.

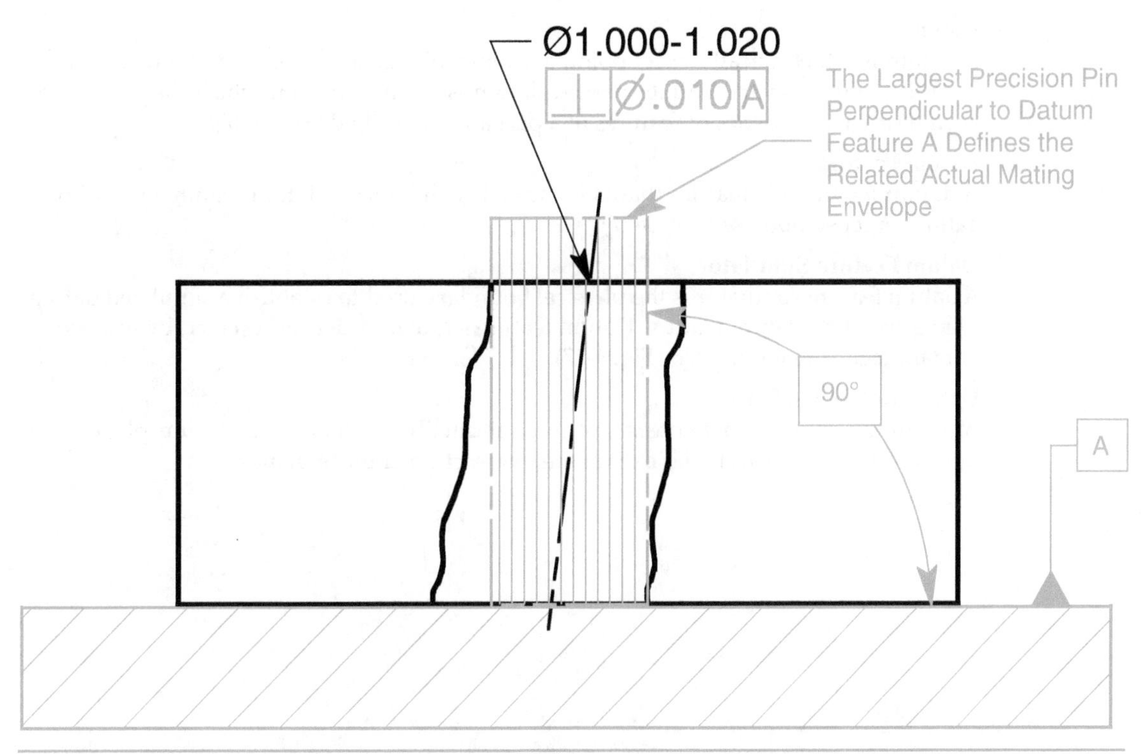

FIGURE 3-26 The related actual mating envelope is defined by the largest precision pin, perpendicular to datum feature A, that will fit inside the hole.

- **Related Actual Mating Envelope**
  A related actual mating envelope is a similar perfect feature(s) counterpart contracted about an external feature(s) or expanded within an internal feature(s) while constrained in orientation, in location, or in both orientation and location to the applicable datum feature(s). If the actual mating envelope of the hole in Fig. 3-26 is 1.010 in diameter and is out of perpendicularity by Ø.005, the largest perpendicular pin that will fit inside this hole would be Ø1.005, the related actual mating envelope.

2. **Basic Dimension**
   A basic dimension is a theoretically exact dimension. Basic dimensions are used to define or position tolerance zones. Title block tolerances do not apply to basic dimensions. The tolerance associated with a basic dimension usually appears in a feature control frame or a note.

3. **Datum**
   A datum is a theoretically exact point, axis, line, plane, or combination thereof derived from the true geometric counterpart. A datum is the origin from which the location or geometric characteristics of features of a part are established (Fig. 3-27).
4. **Datum Feature**
   A datum feature is a feature that is identified with either a datum feature symbol or a datum target symbol (see Fig. 3-27).
5. **Datum Feature Simulator**
   A datum feature simulator is the physical boundary used to establish a simulated datum from a specified datum feature. Datum feature simulators are represented by inspection or manufacturing tooling (see Fig. 3-27).
6. **Datum Reference Frame**
   A datum reference frame consists of three mutually perpendicular datum planes and three mutually perpendicular axes at the intersection of those planes.

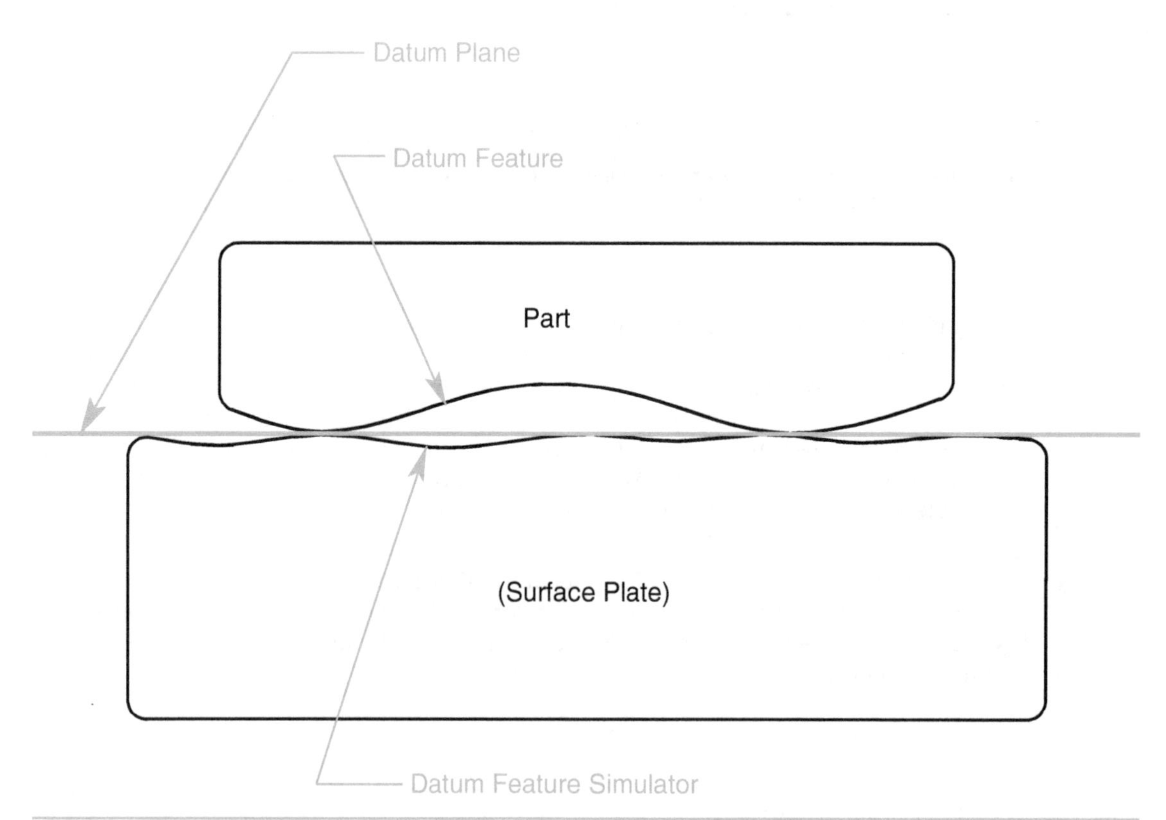

FIGURE 3-27 The relationship between a theoretical datum plane, a physical datum feature simulator, and a datum feature.

7. **Feature**
   A feature is a physical portion of a part, such as a surface, pin, hole, tab, slot, or its representation on drawings, models, or digital data files.

8. **Feature of Size**
   A feature of size encompasses two types:

   - **Regular Feature of Size**
     A regular feature of size is a feature that is associated with a directly toleranced dimension and takes one of the following forms:

     a) A cylindrical surface
     b) A set of two opposed parallel surfaces
     c) A spherical surface
     d) A circular element
     e) A set of two opposed parallel line elements

     Cylindrical surfaces and two opposed parallel surfaces are the most common features of size.

   - **Irregular Feature of Size**
     The two types of irregular features of size are the following:

     a) A directly toleranced feature or collection of features that may contain or be contained by an unrelated actual mating envelope that is a sphere, cylinder, or pair of parallel planes.
     b) A directly toleranced feature or collection of features that may contain or be contained by an unrelated actual mating envelope other than a sphere, cylinder, or pair of parallel planes.

9. **Limits of Size**

   - **Maximum Material Condition (MMC)**
     The MMC is the condition in which a feature of size contains the maximum amount of material within the stated limits of size, such as the minimum hole diameter and the maximum shaft diameter.

   - **Least Material Condition (LMC)**
     The LMC is the condition in which a feature of size contains the least amount of material within the stated limits of size, such as the maximum hole diameter and the minimum shaft diameter.

10. **Material Condition Modifiers**
A material condition modifier is specified in a feature control frame, associated with the geometric tolerance of a feature of size or a datum feature of size. The material condition modifiers are shown in Fig. 3-28 and Table 3-1.

- **Maximum Material Condition Modifier**
A circle M, specified in a feature control frame following the geometric tolerance of a feature of size, is called the MMC modifier. It indicates that the tolerance applies at the MMC size of the feature and that a bonus tolerance is available as the size of the feature departs from MMC toward LMC.

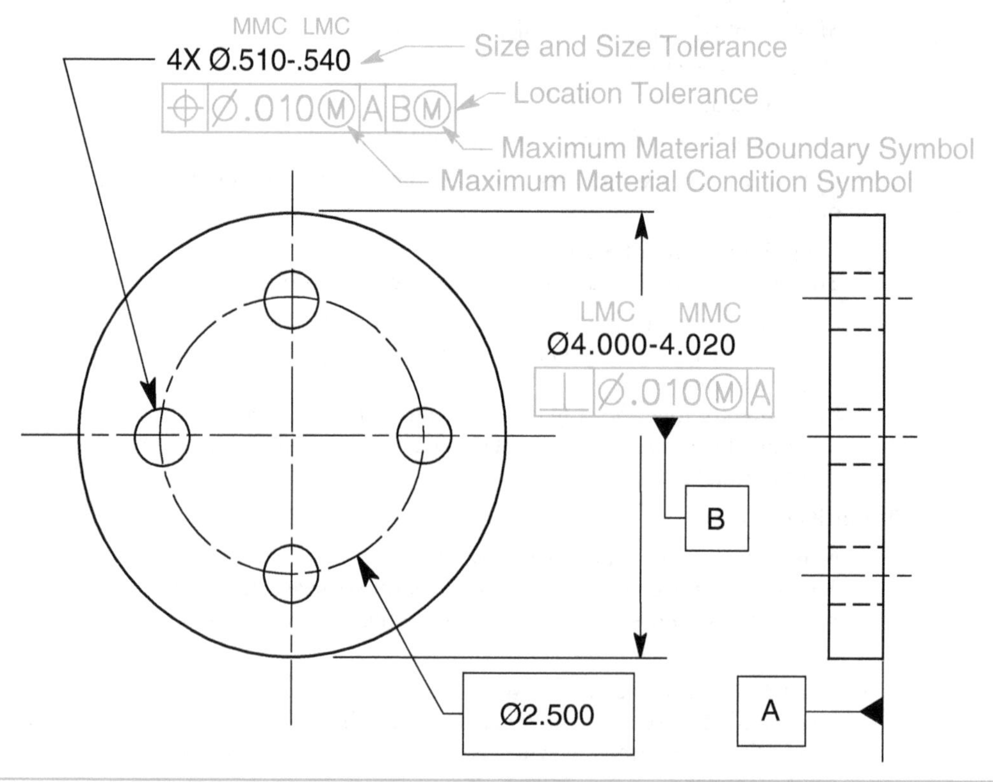

FIGURE 3-28 MMC and LMC limits of size and material condition modifiers.

Material Condition Modifier	Abbreviation	Symbol
Maximum Material Condition (when applied to a tolerance value)	MMC	Ⓜ
Maximum Material Boundary (when applied to a datum reference)	MMB	Ⓜ
Least Material Condition (when applied to a tolerance value)	LMC	Ⓛ
Least Material Boundary (when applied to a datum reference)	LMB	Ⓛ
Regardless of Feature Size (when applied to a tolerance value)	RFS	None
Regardless of Material Boundary (when applied to a datum reference)	RMB	None

TABLE 3-1 Material Condition Modifiers, Abbreviations, and Symbols

Bonus tolerance is the positive difference or the absolute value between the actual mating envelope and MMC. The following formulas are used to calculate the bonus tolerance and total positional tolerance at MMC (Fig. 3-29 and Table 3-2):

Actual Mating Envelope – MMC = Bonus

Bonus + Geometric Tolerance = Total Positional Tolerance

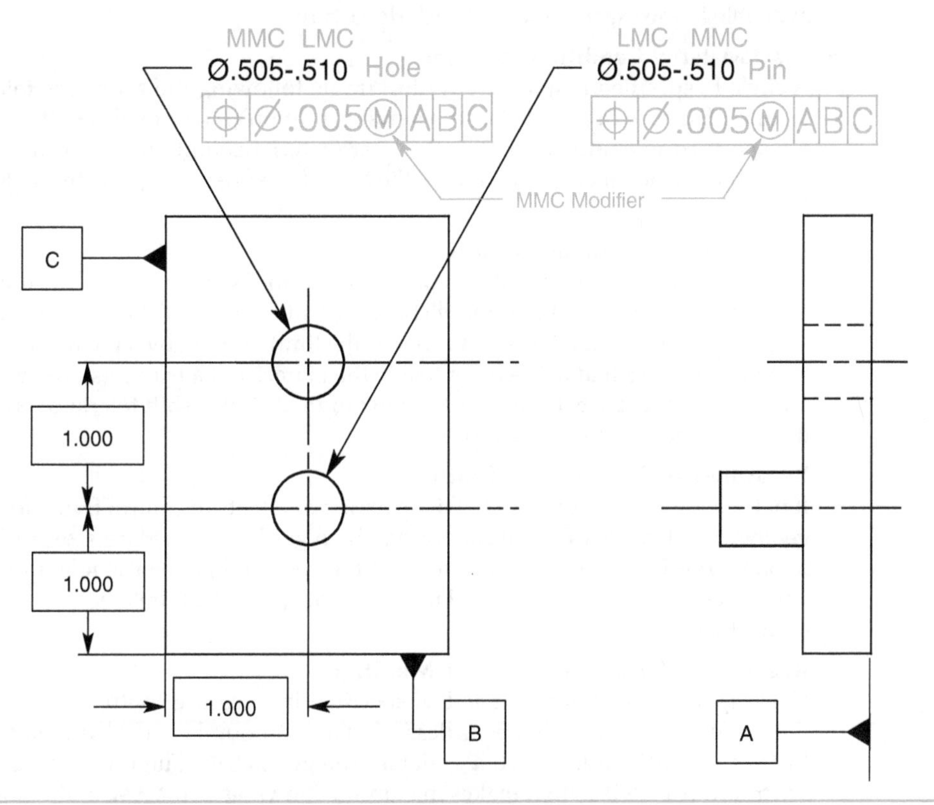

FIGURE 3-29 A drawing with a hole and a pin for calculating bonus tolerances at MMC.

	Internal Feature (Hole)				
	Actual Mating Envelope	**MMC**	**Bonus**	**Geometric Tolerance**	**Total Positional Tolerance**
MMC	.505	.505	.000	.005	.005
	.506	.505	.001	.005	.006
	.507	.505	.002	.005	.007
	.508	.505	.003	.005	.008
	.509	.505	.004	.005	.009
LMC	.510	.505	.005	.005	.010

	External Feature (Pin)				
	Actual Mating Envelope	**MMC**	**Bonus**	**Geometric Tolerance**	**Total Positional Tolerance**
MMC	.510	.510	.000	.005	.005
	.509	.510	.001	.005	.006
	.508	.510	.002	.005	.007
	.507	.510	.003	.005	.008
	.506	.510	.004	.005	.009
LMC	.505	.510	.005	.005	.010

TABLE 3-2 The Increase in Bonus and Total Tolerance as the Actual Mating Envelope Departs from MMC Toward LMC

- **Maximum Material Boundary Modifier**
  A circle M, specified in a feature control frame following a datum feature of size, is called the maximum material boundary (MMB) modifier. The MMB modifier indicates that a datum feature simulator is produced at the limit defined by a tolerance or combination of tolerances that exists on or outside the material of a feature(s). As the size of the datum feature departs from the MMB toward the LMB, a shift tolerance is available in the exact amount of such departure.
- **Least Material Condition Modifier**
  A circle L, specified in a feature control frame following the geometric tolerance of a feature of size, is called the least material condition (LMC) modifier. It indicates that the specified tolerance applies at the least material condition size of the feature and that a bonus tolerance is available as the size of the feature departs from LMC toward MMC.
- **Least Material Boundary Modifier**
  A circle L, specified in a feature control frame following a datum feature of size, is called the least material boundary (LMB) modifier. The LMB modifier indicates that a datum feature simulator is produced at the limit defined by a tolerance or combination of tolerances that exists on or inside the material of a feature(s). As the size of the datum feature departs from the LMB toward the MMB, a shift tolerance is available in the exact amount of such departure.
- **Regardless of Feature Size Modifier**
  If neither a circle M nor a circle L is specified in a feature control frame following the geometric tolerance of a feature of size, the regardless of feature size (RFS) modifier applies. The RFS modifier indicates that the specified geometric tolerance applies at any increment of size of the actual mating envelope of the unrelated feature of size. No bonus tolerance is available.
- **Regardless of Material Boundary Modifier**
  If neither a circle M nor a circle L is specified in a feature control frame following a datum feature of size, the regardless of material boundary (RMB) modifier applies. RMB is a condition in which a variable true geometric counterpart progresses from MMB toward LMB until it makes maximum allowable contact with the extremities of a datum feature(s) to establish a datum. No shift tolerance is available.

11. **Resultant Condition**
    The resultant condition of a feature of size specified with an MMC modifier is the single worst-case boundary generated by the collective effects of the LMC limit of size, the specified geometric tolerance, and the size tolerance. The size tolerance is the available bonus tolerance when the feature is produced at its LMC size. Features specified with an LMC modifier also have a resultant condition.

    Resultant condition calculations for features toleranced with an MMC modifier:

External Features (Pin)	**Internal Features (Hole)**
LMC	LMC
Minus Geometric Tolerance @ MMC	*Plus* Geometric Tolerance @ MMC
Minus The Size Tolerance (Bonus)	*Plus* The Size Tolerance (Bonus)
Resultant Condition	Resultant Condition

12. **True Position**
    True position is the theoretically exact location of a feature of size, as established by basic dimensions. Tolerance zones are located at true position.
13. **True Profile**
    True profile is the theoretically exact profile on a drawing defined by basic dimensions or a digital data file. Tolerance zones are located about the true profile.
14. **Virtual Condition**
    The virtual condition of a feature of size specified with an MMC modifier is a constant boundary generated by the collective effects of the considered feature's MMC limit of size and the specified geometric tolerance. Features specified with an LMC modifier also have a virtual condition.
    Virtual condition calculations for features toleranced with an MMC modifier:

External Features (Pin)	Internal Features (Hole)
MMC	MMC
Plus Geometric Tolerance @ MMC	*Minus* Geometric Tolerance @ MMC
Virtual Condition	Virtual Condition

15. **Worst-Case Boundary**
    The worst-case boundary of a feature is a general term that describes the smallest or largest boundary generated by the collective effects of the MMC or LMC of a feature and any applicable geometric tolerance.
    - **Inner Boundary Specified at MMC**
      The worst-case inner boundary is the virtual condition of an internal feature and the resultant condition of an external feature.
    - **Outer Boundary Specified at MMC**
      The worst-case outer boundary is the resultant condition of an internal feature and the virtual condition of an external feature. Features specified with an LMC modifier also have worst-case boundaries.

Rules

There are two rules that apply to drawings in general and to GD&T in particular. They govern specific characteristics of features on a drawing. It is important for each geometric dimensioning and tolerancing practitioner to know these rules and to know how to apply them.

Rule #1: Limits of Size Prescribe Variations of Form

(a) Rule #1 states that for an individual regular feature of size, no element of the feature shall extend beyond an envelope that is a boundary of perfect form at MMC. This boundary is the true geometric form represented by the drawing views or model. No elements of the produced surface shall violate this boundary unless the requirement for perfect form at MMC has been removed. There are several methods of overriding Rule #1 explained below.

(b) Where the actual local size of a regular feature of size has departed from MMC toward LMC, a local variation in form is allowed equal to the amount of such departure.

(c) There is no default requirement for a boundary of perfect form at LMC. Thus, a regular feature of size produced at its LMC limit of size is permitted to vary from true form to the maximum variation allowed by its tolerance of size.

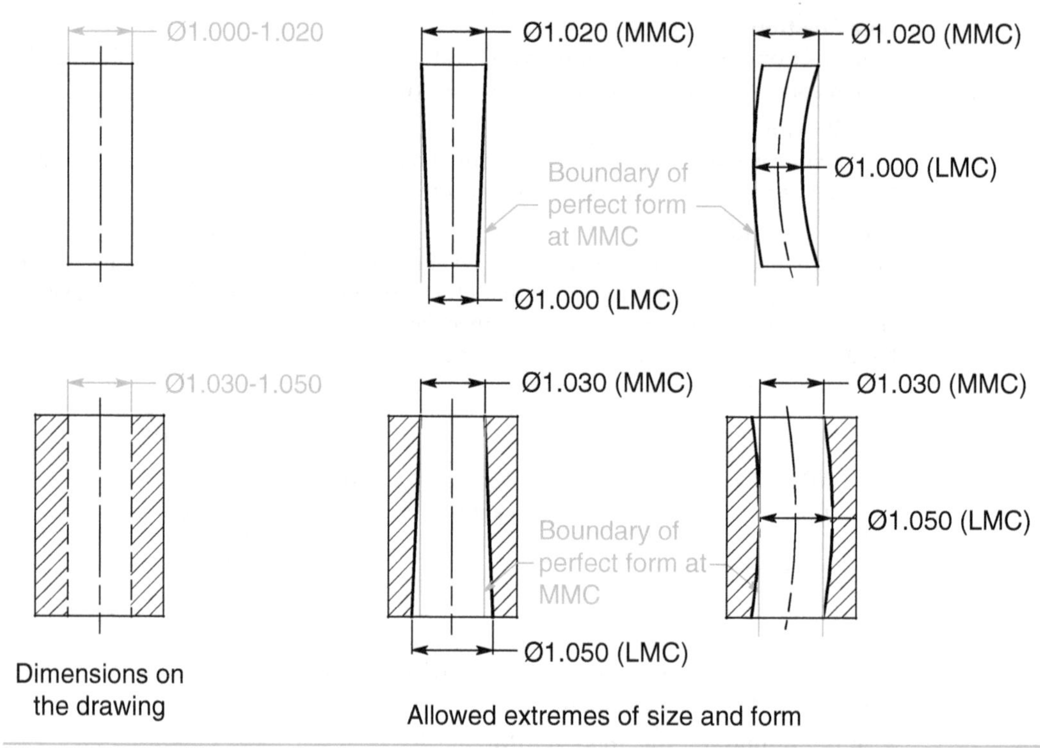

FIGURE 3-30 Rule #1—examples of size and form variations allowed by the size tolerance.

In Fig. 3-30, the MMC of the pin is 1.020. The pin may in no way fall outside this MMC boundary or envelope of perfect form. That is, if the pin is produced at a diameter of 1.020 at each and every cross section, it must not be bowed or out of circularity in any way. If the pin is produced at a diameter of 1.010, it may be out of straightness and/or out of circularity by a total of .010. If the pin is produced at a diameter of 1.000, its LMC, it may vary the full .020 tolerance from perfect form.

Rule #1 does not apply to stock, to features of size toleranced with a form control, or to features subject to free-state variation in the unrestrained condition. Where the word *stock* is specified on a drawing, it indicates bar, plate, sheet, and so on as it comes from the supplier. Stock items are manufactured to industry or government standards and are not controlled by Rule #1. Stock is used as is, unless otherwise specified by a geometric tolerance or a note. Rule #1 does not apply when a form tolerance of straightness or flatness is applied to a feature of size. Finally, Rule #1 does not apply to parts that are flexible and are to be measured in their free state.

Where perfect form at MMC is not required, the independency symbol, circle I, may be placed next to the appropriate dimension or notation. However, a supplementary form tolerance(s) may be required to limit excessive variations of form, as shown in Fig. 3-31.

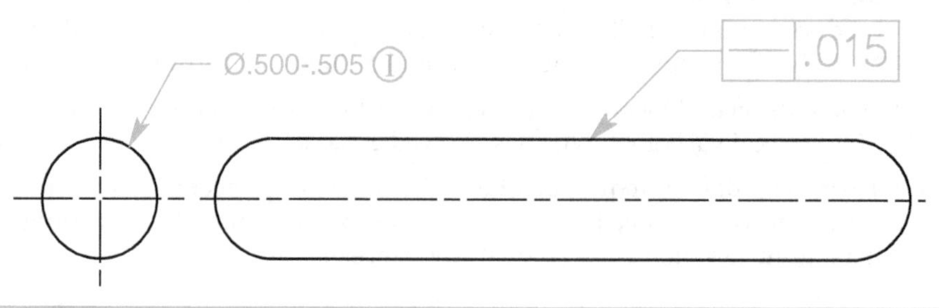

FIGURE 3-31 Independency symbol.

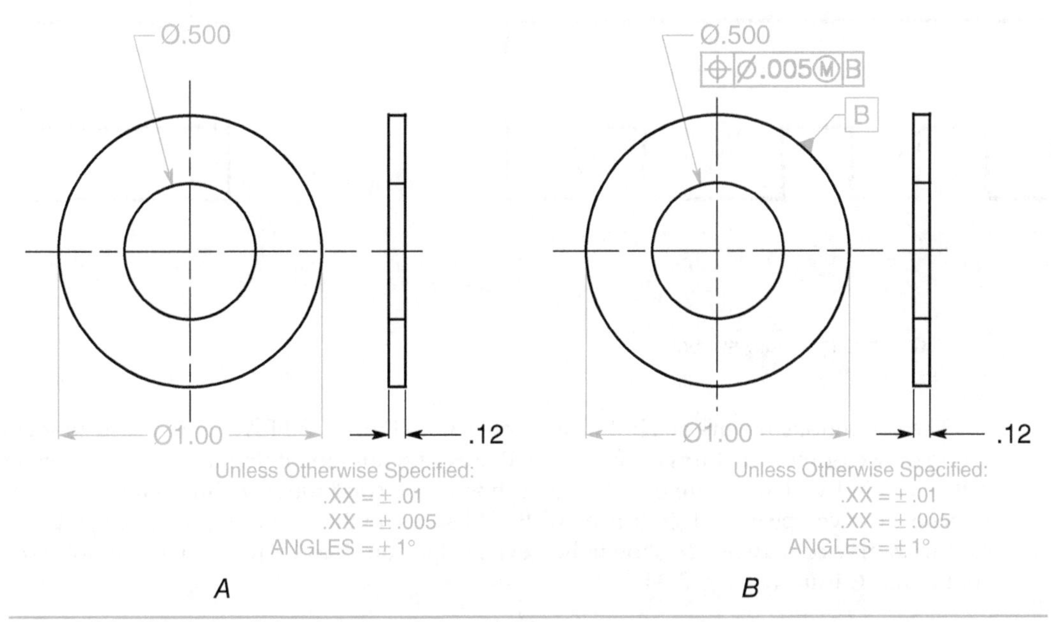

FIGURE 3-32 The limits of size do not control coaxiality.

The relationship between individual features is not controlled by the limits of size. If features on a drawing are shown coaxial, or symmetrical to each other and not toleranced for location or orientation, the drawing is incomplete. Fig. 3-32*A* is incomplete because there is no coaxiality control between the inside and the outside diameters. Fig. 3-32*B* shows one way of specifying the coaxiality of the inside and outside diameters.

As shown by the view in Fig. 3-33, the perpendicularity between features of size is not controlled by the size tolerance. There is a misconception that the corners of a rectangle must be perfectly square if the sides are produced at MMC. If no orientation tolerance is specified, perpendicularity is controlled not by the size tolerance but by the angularity tolerance. The right angles of the rectangle in Fig. 3-33 may fall between 89° and 91° as specified by the angular tolerance in the title block.

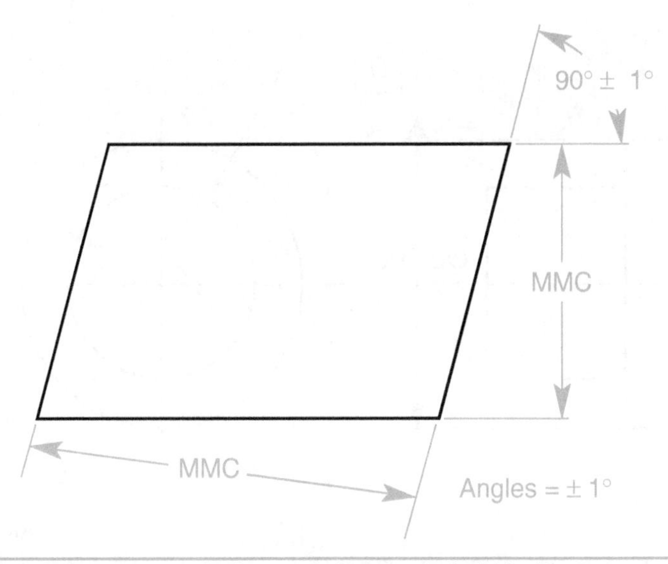

FIGURE 3-33 Angularity tolerance controls the angularity between individual features.

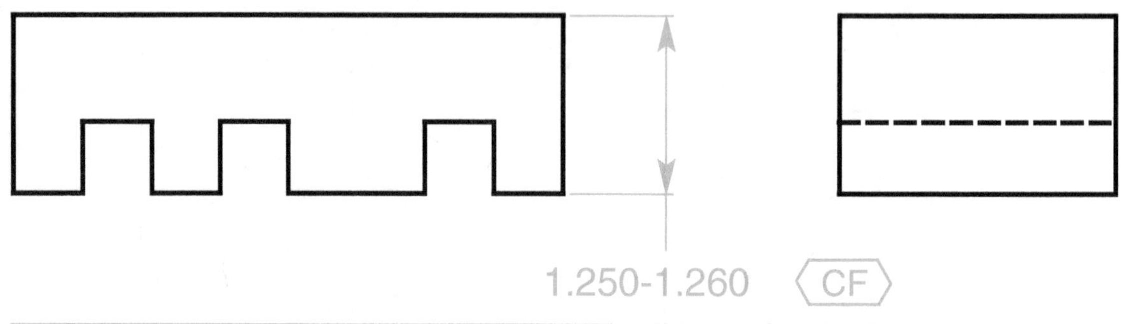

FIGURE 3-34 Continuous feature symbol.

The continuous feature symbol or the note CONTINUOUS FEATURE is used to identify a group of two or more features of size where there is a requirement that they be treated geometrically as an individual feature of size. No portion of the continuous feature is allowed to extend outside the envelope of perfect form at MMC. When using the continuous feature symbol, extension lines may or may not be shown; however, extension lines by themselves do not indicate a continuous feature. See Fig. 3-34.

Rule #2: Applicability of Modifiers in Feature Control Frames

MMC, RFS, or LMC applies to each geometric tolerance specified for a feature of size, and MMB, RMB, or LMB applies to each datum feature of size.

Rule #2 states that in feature control frames, the RFS modifier automatically applies to individual tolerances of features of size, and the RMB modifier automatically applies to datum features of size where no modifying symbol is specified. MMC/MMB and LMC/LMB must be specified for tolerances and features of size where they are required.

In Fig. 3-35, both the feature being controlled and the datum feature are features of size. The feature control frame labeled *A* has no material condition modifiers. Consequently, the coaxiality tolerance in feature control frame *A* applies at RFS, and datum feature D applies at RMB. If the controlled feature is toleranced with feature control frame *A*, the tolerance is .005 no matter

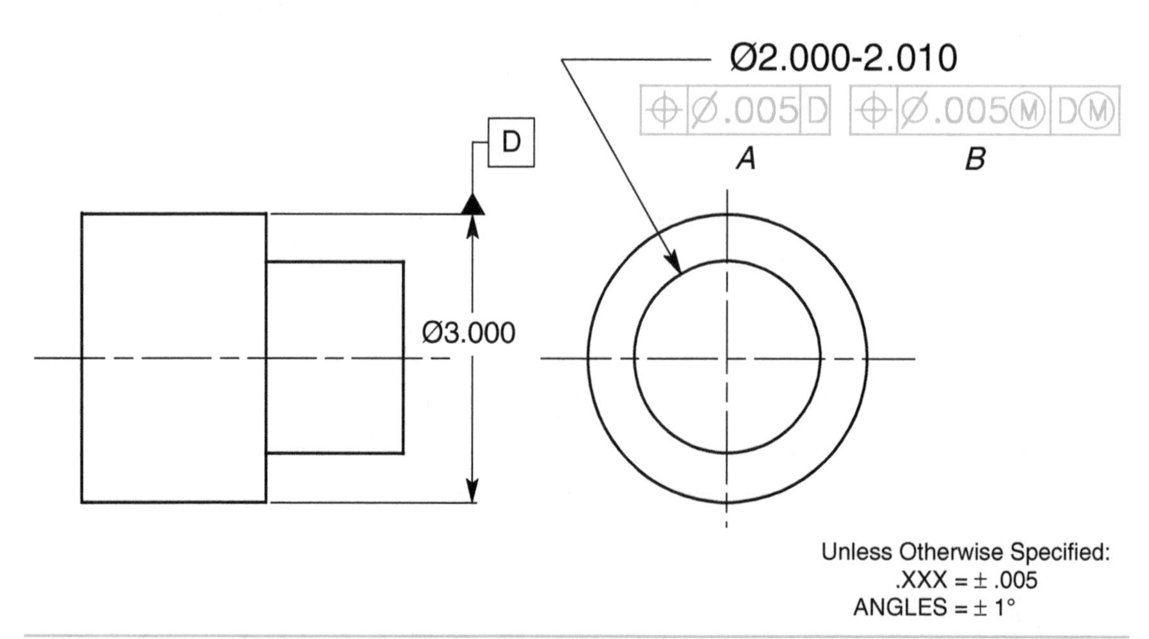

FIGURE 3-35 Feature control frame A is specified at RFS and RMB. Feature control frame B is specified at MMC and MMB.

what diameter the cylinder happens to be between 2.000 and 2.010. Datum feature D must make physical contact with the inspection equipment during inspection, and no shift tolerance applies to datum feature D. For the feature control frame labeled *B*, MMC applies to the tolerance, and MMB applies to datum feature D. MMC allows a bonus tolerance for the controlled feature, and MMB allows a shift tolerance to apply to datum feature D. Material condition modifiers will be discussed in more detail in Chap. 7.

The Pitch Diameter Rule

The discussion about pitch diameter was once called the pitch diameter rule. Although it is no longer considered a rule, it is important to mention here.

Each tolerance of orientation or position and datum reference specified for a screw thread applies to the axis of the pitch cylinder. Exceptions to this rule may be specified by placing a note, such as MAJOR DIA or MINOR DIA, beneath the feature control frame or beneath or adjacent to the datum feature symbol.

Each tolerance of orientation or position and datum reference specified for features other than screw threads, such as gears and splines, must designate the specific feature as MAJOR DIA, PITCH DIA, or MINOR DIA, at which each applies. A note is placed beneath the feature control frame or beneath or adjacent to the datum feature symbol.

Summary

- There are 12 geometric characteristic symbols. They are divided into five categories: form, profile, orientation, location, and runout.
- The datum feature symbol consists of a capital letter enclosed in a square box. It is connected to a leader directed to the datum feature ending in a triangle.
- The datum feature symbol is used to identify physical features of a part as datum features. It must not be attached to centerlines, center planes, or axes.
- Datum feature symbols placed in line with a dimension line or on a feature control frame associated with a feature of size identify the feature of size as the datum feature.
- The feature control frame is the sentence of the GD&T language.
- Feature control frames may be attached to features with extension lines, dimension lines, or leaders.
- A number of other symbols used with geometric dimensioning and tolerancing are listed in Fig. 3-11. The reader should be able to recognize each of these symbols.
- The names and definitions of many geometric dimensioning and tolerancing concepts have very specific meanings. To be able to properly read and apply GD&T, it is important to be very familiar with these terms.
- A circle M specified in a feature control frame following the geometric tolerance of a feature of size is known as the MMC modifier. It indicates that the specified tolerance applies at the MMC size of the feature and that a bonus tolerance is available.
- A circle L specified in a feature control frame following the geometric tolerance of a feature of size is known as the LMC modifier. It indicates that the specified tolerance applies at the least material condition size of the feature and that a bonus tolerance is available.
- If neither a circle M nor a circle L is specified in a feature control frame following the geometric tolerance of a feature of size, the RFS modifier applies, and no bonus tolerance is available.
- A circle M specified in a feature control frame following a datum feature of size indicates that the MMB modifier applies and that a shift tolerance is available.

- A circle L specified in a feature control frame following a datum feature of size indicates that the LMB modifier applies and that a shift tolerance is available.
- Rule #1 states that for an individual regular feature of size, where only a tolerance of size is specified, the limits of size prescribe the extent to which variations in its geometric form, as well as its size, are allowed.
- Rule #2 states that in feature control frames, the RFS modifier automatically applies to individual tolerances of features of size and the RMB modifier automatically applies to datum features of size. MMC/MMB and LMC/LMB must be specified for features of size where they are required.
- The pitch diameter (former rule) states that each geometric tolerance or datum reference specified for a screw thread applies to the axis of the thread derived from the pitch cylinder.

Chapter Review

1. What type of geometric tolerances has no datum features? ____________________

2. What is the name of the symbol used to identify physical features of a part as a datum feature and must not be applied to centerlines, center planes, or axes?

   ____

3. Datum feature identifying letters may be any letter of the alphabet except?

   ____

4. If the datum feature symbol is placed in line with a dimension line or on a feature control frame associated with a feature of size, the datum feature is what kind of feature?

   ____

5. One of the 12 geometric characteristic symbols always appears in the ____________

   ____________ compartment of the feature control frame.

6. The second compartment of the feature control frame is the ____________________ section.

7. The tolerance is preceded by a diameter symbol only if the tolerance zone is ________

   ____________________.

8. Datum features are arranged in order of ____________________.

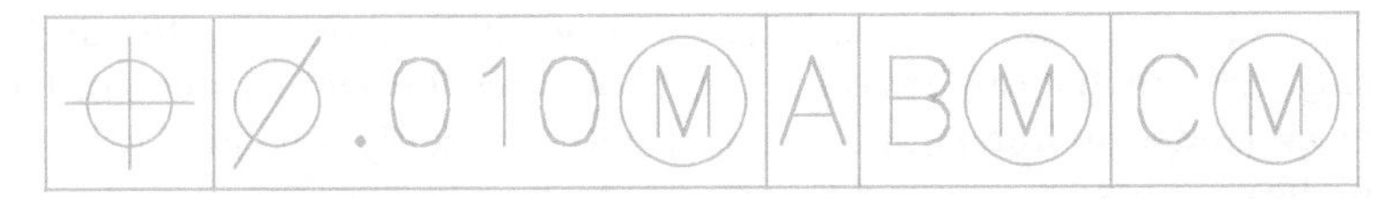

FIGURE 3-36 Position feature control frame: Question 9.

9. Read the feature control frame in Fig. 3-36.

   1. ______________________________

   2. ______________________________

   3. ______________________________

   4. ______________________________

   5. ______________________________

   6. ______________________________

   ____

10. The all around and between symbols are used with what control?

    ____

11. The all over symbol consists of two small ______________ placed at the joint of the leader connecting the feature control frame to the feature.

12. The ______________ symbol specifies that a group of two or more interrupted features of size are to be considered one single feature of size.

13. If no depth or remaining thickness is specified, the spotface is the ______________ depth necessary to clean up the surface of the specified diameter.

14. The ______________ symbol indicates that perfect form of a feature of size at MMC or LMC is not required.

15. The ______________ symbol indicates that the profile tolerance is unilateral or unequally disposed about the true profile.

16. The ______________ symbol indicates that a datum feature simulator is not fixed and is free to translate within the specified geometric tolerance.

17. The ______________________________ is a similar, perfect, feature(s) counterpart of smallest size that can be contracted about an external feature(s) or largest size that can be expanded within an internal feature(s) so that it coincides with the surface(s) at its highest points.

18. A theoretically exact dimension is called a ______________________________.

19. What is the theoretically exact point, axis, line, plane, or combination thereof derived from the theoretical datum feature simulator called? ______________________________

20. A ______________________________ is a feature that is identified with either a datum feature symbol or a datum target symbol.

21. A ______________________________ is the physical boundary used to establish a simulated datum from a specified datum feature.

22. A ______________________________ consists of three mutually perpendicular intersecting datum planes and three mutually perpendicular axes at the intersection of those planes.

23. What is the name of a physical portion of a part, such as a surface, pin, hole, tab, or slot?

____

24. A regular feature of size is a feature that is associated with a directly toleranced dimension and takes one of the following forms:

a) A ____

b) A ____

c) A ____

d) A ____

e) A ____

25. Write the names and draw the geometric characteristic symbols where indicated in Fig. 3-37.

Pertains to	Type of Tolerance	Geometric Characteristics	Symbol
Individual Feature Only	Form	________ ________ ________ ________	
Individual Feature or Related Features	Profile	________ ________	
Related Features	Orientation	________ ________ ________	
	Location	________	
	Runout	________ ________	

FIGURE 3-37 Geometric characteristic symbols: Question 25.

26. Draw the indicated geometric tolerancing symbols in the spaces on Fig. 3-38.

Name	Symbol	Name	Symbol
All Around		Free State	
All Over		Independency	
Between		Projected Tolerance Zone	
Number of Places		Tangent Plane	
Continuous Feature		Unequally Disposed Profile	
Counterbore		Spotface	
Countersink		Radius	
Depth/Deep		Radius, Controlled	
Diameter		Spherical Radius	
Dimension, Basic	1.000	Spherical Diameter	
Dimension, Reference	60	Square	
Dimension Not To Scale	15	Statistical Tolerance	
Dimension Origin		Datum Target	
Datum Translation		Movable Datum Target	
Arc Length		Target Point	
Conical Taper		Dynamic Profile	
Slope		From - To	

FIGURE 3-38 Geometric tolerancing symbols: Question 26.

27. The ______________________ is the condition in which a feature of size contains the maximum amount of material within the stated limits of size.

28. The ______________________ is the condition in which a feature of size contains the least amount of material within the stated limits of size.

29. What kind of feature always applies at MMC/MMB, LMC/LMB, or RFS/RMB?

____

30. The MMC modifier specifies that the tolerance applies at the __________ __________

__________ size of the feature.

31. The MMC modifier indicates that the tolerance applies at the MMC size of the feature and that a ______________________ tolerance is available as the size of the feature departs from MMC toward LMC.

32. ______________________ tolerance is the positive difference or the absolute value between the actual mating envelope and MMC.

33. The total positional tolerance equals the sum of the ______________________

tolerance and the ______________________ tolerance.

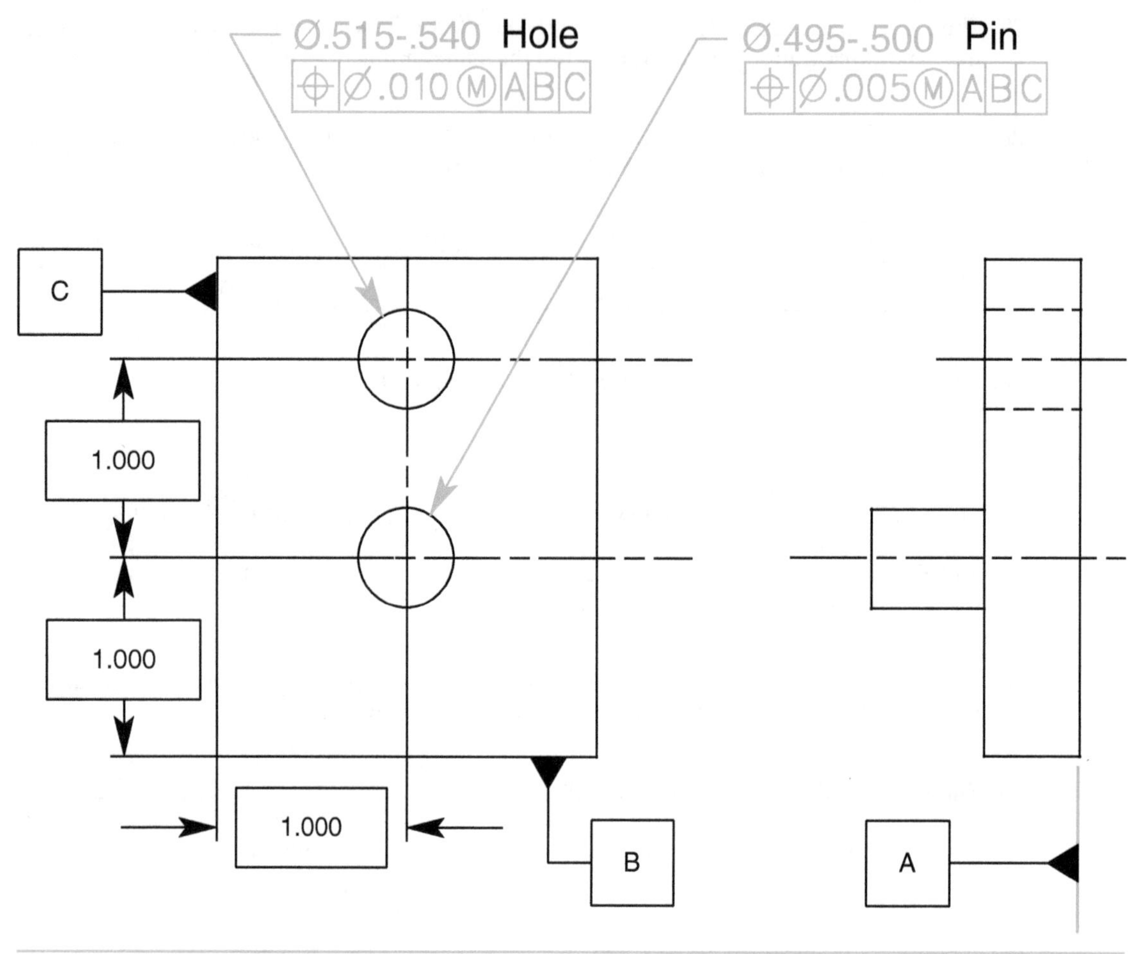

FIGURE 3-39 MMC exercise: Questions 34 through 41.

Refer to Fig. 3-39 to answer Questions 34 through 41.

	Hole	Pin
34. What is the MMC?		
35. What is the LMC?		
36. What is the geometric tolerance?		
37. What material condition modifier is specified?		
38. What datum feature(s) control(s) perpendicularity?		
39. What datum feature(s) control(s) location?		

40. Using the drawing in Fig. 3-39, complete Table 3-3.

	Internal Feature (Hole)				
	Actual Mating Envelope	MMC	Bonus	Geometric Tolerance	Total Positional Tolerance
MMC	.515				
	.520				
	.525				
	.530				
	.535				
LMC	.540				

TABLE 3-3 Total Positional Tolerance for Holes

41. Using the drawing in Fig. 3-39, complete Table 3-4.

	External Feature (Pin)				
	Actual Mating Envelope	MMC	Bonus	Geometric Tolerance	Total Positional Tolerance
MMC	.500				
	.499				
	.498				
	.497				
	.496				
LMC	.495				

TABLE 3-4 Total Positional Tolerance for Pins

42. A ____________________________ is specified in a feature control frame when it is associated with the geometric tolerance of a feature of size or a datum feature of size.

43. The ____________________________ modifier indicates that the specified geometric tolerance applies at any increment of size of the actual mating envelope of the unrelated feature of size.

44. The ____________________________ of a feature of size specified with an MMC modifier is the single worst-case boundary generated by the collective effects of the LMC limit of size, the specified geometric tolerance, and the size tolerance.

45. ____________________________ is the theoretically exact location of a feature of size, as established by basic dimensions.

46. ____________________________ is the theoretically exact profile on a drawing defined by basic dimensions or a digital data file.

47. The ________________________________ of a feature of size specified with an MMC modifier is a constant boundary generated by the collective effects of the considered feature's MMC limit of size and the specified geometric tolerance.

48. For an individual regular feature of size, no element of the feature shall extend beyond the MMC boundary (envelope) of perfect form.

    This statement is the essence of ________________________________.

49. The local form tolerance increases as the actual local size of the feature departs from

    MMC toward ________________________________.

50. If features on a drawing are shown coaxial, or symmetrical to each other and not

    controlled for ________________________ or ________________________, the drawing is incomplete.

51. If there is no orientation control specified for a rectangle on a drawing, the perpendicularity

    is controlled not by the size ________________________________ but by the

    ________________________________ tolerance.

52. Rule #2 states that ________________________________ automatically applies to

    individual tolerances of feature of size and ________________________________ automatically applies to datum features of size.

53. Each tolerance of orientation or position and datum reference specified for a screw thread

    applies to the axis of the ________________________________.

54. Each geometric tolerance or datum reference specified for gears and splines must

    designate the specific feature at which each applies, such as ________________________

    ________________________________.

Problems

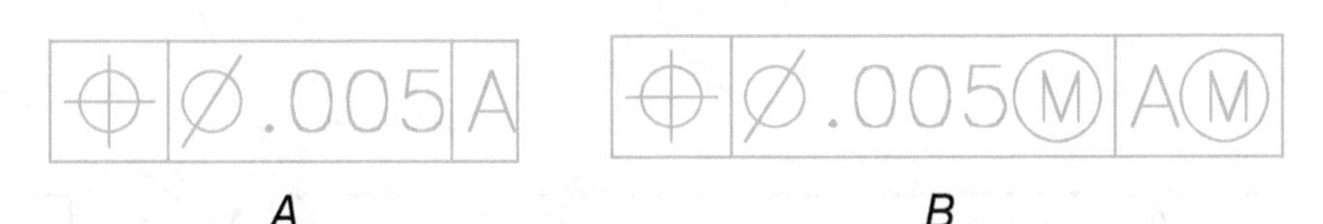

FIGURE 3-40 Feature control frames with material condition symbols: Prob. 1.

1. Read the complete tolerance in each feature control frame in Fig. 3-40 and write it below (datum feature A is a feature of size).

A. 1. ______________________________

2. ______________________________

3. ______________________________

4. ______________________________

5. ______________________________

6. ______________________________

B. 1. ______________________________

2. ______________________________

3. ______________________________

4. ______________________________

5. ______________________________

6. ______________________________

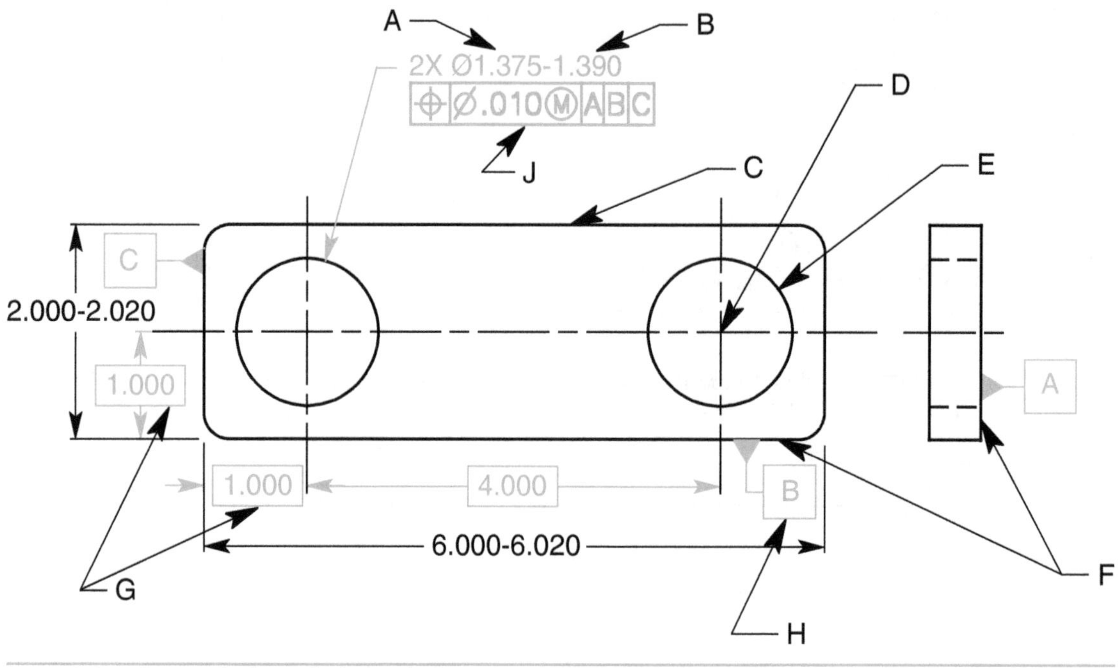

FIGURE 3-41 Geometric dimensioning and tolerancing terms: Prob. 2.

2. Place each letter of the items on the drawing in Fig. 3-41 next to the most correct term below.

______ Feature	______ Basic dimensions	______ Feature control frame
______ MMC	______ Datum feature	______ True position
______ LMC	______ Feature of size	______ Datum feature symbol

CHAPTER 4
Datums

Datums are the references or the starting points for the location and orientation of toleranced features. They are essential for appropriate and complete tolerancing of a part. Datum geometries can become very complicated when they are features of size, compound datums, or features of an unusual shape. Geometric dimensioning and tolerancing provides the framework necessary for dealing with these complex datum features. The simpler plus and minus tolerancing system ignores these complexities, meaning that plus and minus toleranced drawings cannot adequately tolerance feature of sizes. As a result, many plus and minus toleranced drawings are subject to more than one interpretation.

Chapter Objectives

After completing this chapter, the learner will be able to:

- *Define* a datum
- *Explain* how datums apply
- *Explain* how a part is immobilized
- *Demonstrate* how a datum reference frame provides origin and direction
- *Select* datum features
- *Demonstrate* how datum features are identified
- *Demonstrate* how to specify an inclined datum feature
- *Explain* how datum planes are associated with a cylindrical part
- *Explain* how datum features are established
- *Demonstrate* the application of irregular features of size as datum features
- *Explain* how common datum features are used to create a single datum feature
- *Demonstrate* how to specify partial datum features
- *Explain* the use of datum targets

Definition of a Datum

A datum is a theoretically perfect point, axis, line, plane, or combination thereof. Datums establish the origin from which the location or geometric characteristics of features of a part are established. These points, axes, lines, and planes exist within a structure of three mutually perpendicular intersecting datum planes known as a datum reference frame, as shown in Fig. 4-1.

Application of Datums

Datums are considered to be absolutely perfect, which makes them imaginary. Measurements cannot be made from theoretical surfaces. Therefore, datums are assumed to exist in and be simulated by datum feature simulators. A datum feature simulator is merely processing equipment, such as surface plates, gages, machine tables, and vises. Processing equipment is not perfect but is made

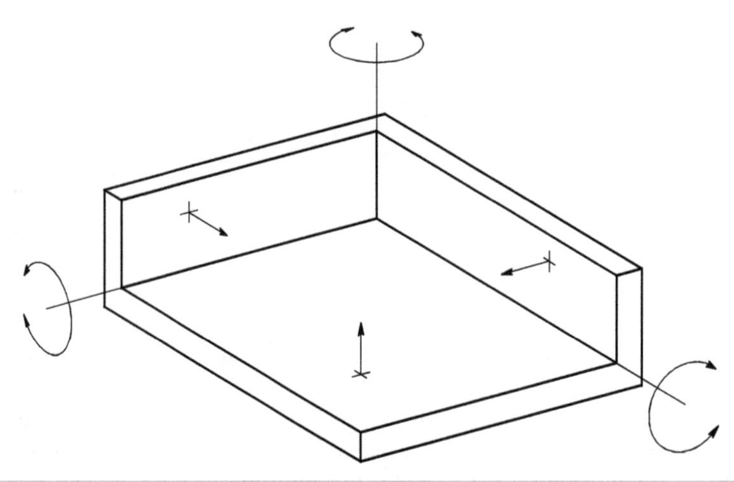

FIGURE 4-1 A datum reference frame consists of three mutually perpendicular datum planes and three mutually perpendicular axes at the intersection of those planes.

sufficiently accurately to simulate datums. The three mutually perpendicular planes of a datum reference frame provide origin and direction for measurements from simulated datums to features.

Immobilization of a Part

Parts are thought to have six degrees of freedom, three degrees of translational freedom, and three degrees of rotational freedom. A part can move back and forth in the X direction, in and out in the Y direction, and up and down in the Z direction and rotate around the X-axis, around the Y-axis, and around the Z-axis.

A part is oriented and immobilized relative to the three mutually perpendicular intersecting datum planes of the datum reference frame in a selected order of precedence, as shown in Fig. 4-2. In order to properly place an imperfect, rectangular part in a datum reference frame, the primary datum feature sits flat on one of the planes with a minimum of three points of contact, not in a straight line. The secondary datum feature is pushed up against a second plane with a minimum of two points of contact. Finally, the part is slid along the first two planes until it contacts the third plane with a minimum of one point of contact. The primary datum feature on the part contacting the simulated datum reference frame eliminates three degrees of freedom, translation in the Z direction, and rotation around the X-axis and the Y-axis. The secondary datum feature on the part contacting the simulated datum reference frame eliminates two degrees of freedom, translation in the Y direction, and rotation around the Z-axis. The tertiary datum feature on the part

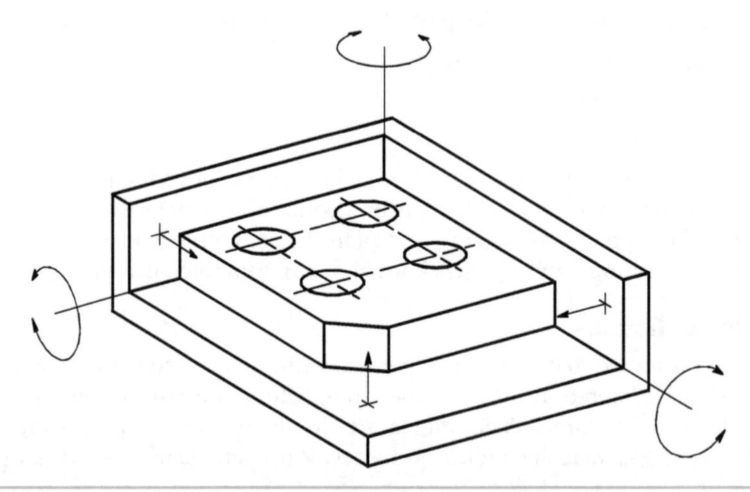

FIGURE 4-2 Immobilizing a part within the three mutually perpendicular intersecting datum planes of a simulated datum reference frame.

FIGURE 4-3 The order of precedence of datum features is determined by the order in which they appear from left to right in the feature control frame.

contacting the simulated datum reference frame eliminates one degree of freedom, that is, it eliminates translation in the X direction.

Datum features are specified in order of precedence as they appear from left to right in the feature control frame; they need not be in alphabetical order. Datum feature A in the feature control frame in Fig. 4-3 is the primary datum feature, datum feature B is the secondary datum feature, and datum feature C is the tertiary datum feature because this is the order in which they appear in the feature control frame.

A Datum Reference Frame Provides Origin and Direction

Figure 4-4 shows an imperfect part placed in a relatively perfect simulated datum reference frame. The back of the part is identified as datum feature A, which is specified in the feature control frame as the primary datum feature. In this example, the primary datum feature must contact the primary datum reference plane with a minimum of three points of contact not in a straight line; as a result, the primary datum feature controls orientation. Datum feature A eliminates rotation about the X-axis and the Y-axis. Datum feature B eliminates rotation about the Z-axis. Datum features B and C are the lower and left edges of the part and are identified as the secondary and tertiary datum features, respectively. Dimensions are measured from—and are perpendicular to—the almost perfect datum reference frame, not the imperfect datum features of the part.

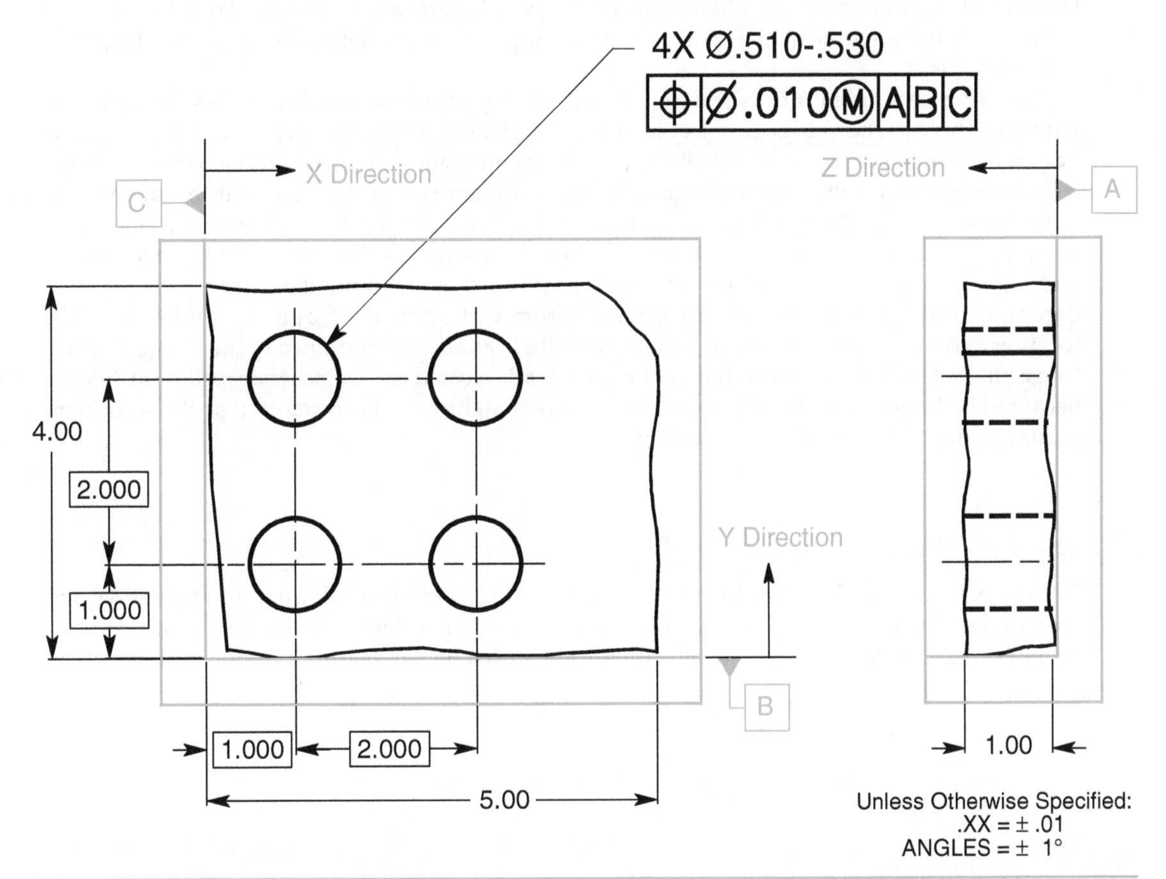

FIGURE 4-4 The planes of a datum reference frame are simulated in a mutually perpendicular relationship to provide direction as well as origin for related dimensions.

The selection of secondary and tertiary datum features depends on characteristics of these features, such as feature size and whether they are mating surfaces. However, if the two features are the same size, do not mate with other features, and are essentially equal in every respect, then either one of them could be the secondary datum feature. Even though selecting a secondary datum feature over a tertiary datum feature may be arbitrary, one datum feature must precede the other, keeping in mind that all applicable datum features must be specified. Specification of datum features in order of precedence allows the part to be placed in the datum reference frame the same way every time.

Variations of form that fall within the size tolerance may occur on the datum feature. If variations on datum features fall within the size tolerance but exceed design requirements, they should be controlled with a form tolerance.

Datum Feature Selection

A datum feature is selected on the basis of its functional relationship to the toleranced feature(s) and the requirements of the design. When selecting datum features, the designer should consider the following characteristics:

- Functional surfaces
- Mating surfaces
- Readily accessible surfaces
- Surfaces of sufficient size to allow repeatable measurements

Datum features must be easily identifiable on the part. If parts are symmetrical or have identical features making identification of datum features impossible, the datum features must be physically identified.

Selecting datum features is the first step in dimensioning a part. Figure 4-4 shows a part with four holes. The designer selected the back of the part as the primary datum feature, datum feature A, because the back of the part mates with another part, and the parts are bolted together with four bolts. Datum feature A makes a good primary datum feature for the four holes because the primary datum feature controls orientation, and it is desirable to have bolt holes perpendicular to mating surfaces. The hole locations are dimensioned from the bottom and left edges of the part. Datum feature B is specified as the secondary datum feature, and datum feature C is specified as the tertiary datum feature in the feature control frame. The locating datum features are selected because of their relative importance to the controlled features: the holes. The bottom edge of the part was selected as the secondary datum feature because it is larger than the left edge. The left edge might have been selected as the secondary datum feature if it were a mating surface.

Datum Feature Identification

Datum features must be identified with datum feature symbols or datum target symbols and specified in a feature control frame. That is, if a feature is oriented or located relative to one or more surfaces, each surface must be identified with a datum feature symbol, and the letters in the datum feature symbols must appear in a feature control frame in their proper order of precedence or importance. Datum features may be designated with any letter of the alphabet except I, O, and Q. The datum feature symbol is used to identify physical features of a part as datum features. Datum feature symbols must not be applied to centerlines, center planes, or axes.

Datum feature symbols B and C attached to the center planes in Fig. 4-5 are ambiguous. It is not clear whether the outside edges, one of the hole patterns, or the slots are the features that determine these center planes. The other datum feature symbols in Fig. 4-5 are attached to actual features and are acceptable as datum features. The center planes can then be determined from actual features on the part.

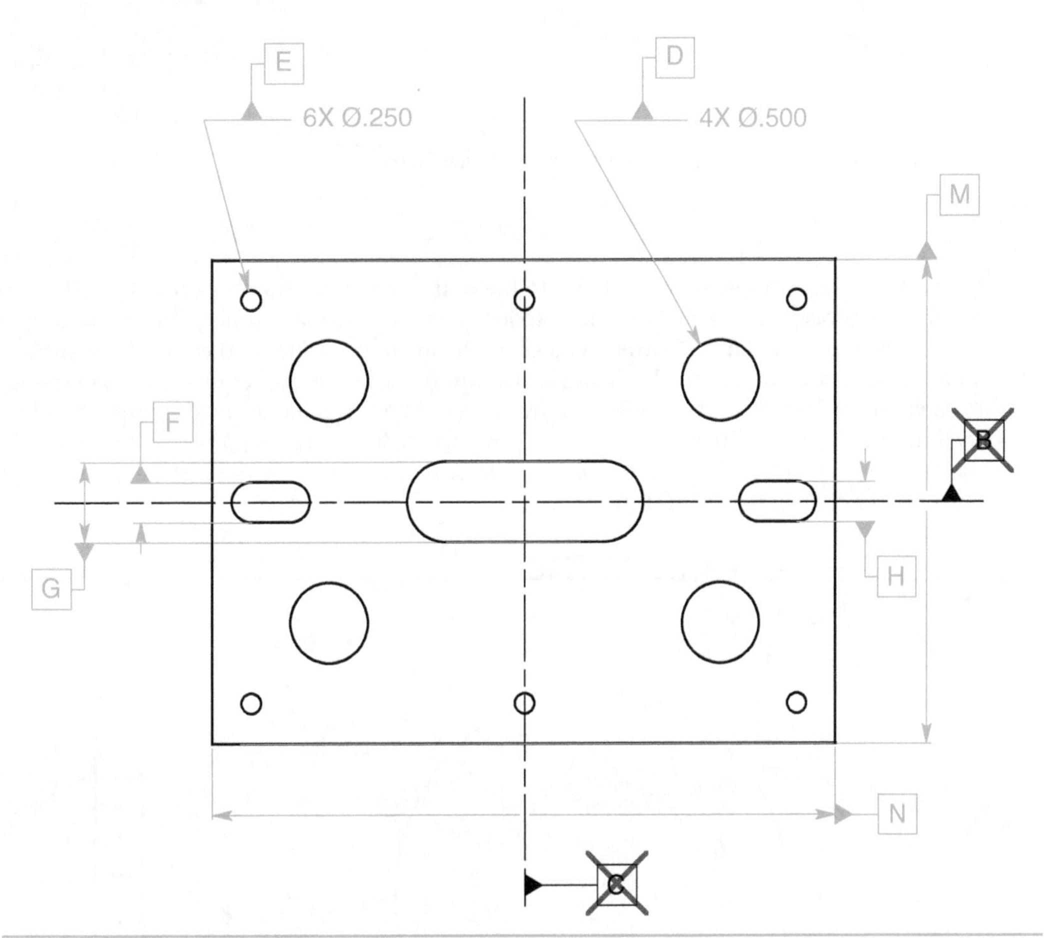

FIGURE 4-5 Datum feature symbols shall not be applied to imaginary planes or lines.

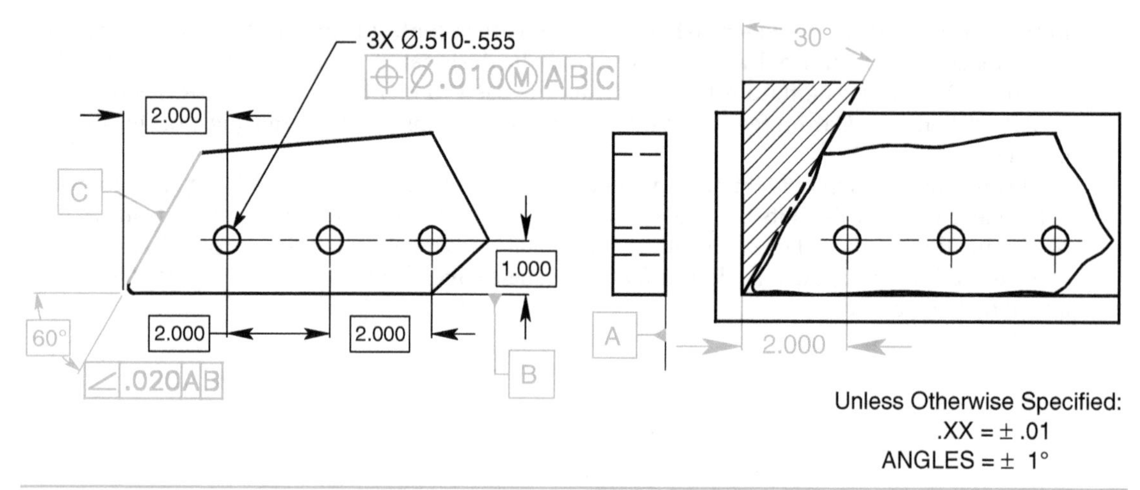

FIGURE 4-6 Datum features at an angle to the datum reference frame.

Inclined Datum Features

If a surface is at an angle other than 90° to the datum reference frame, especially if the corner is rounded or broken off, it may be difficult to locate features to that surface. One method, shown in Fig. 4-6, is to place a datum feature symbol on the inclined surface and control that surface with an angularity tolerance and a basic angle. Datum features are not required to be perpendicular to each other. Only the datum reference frame is defined as three mutually perpendicular intersecting datum planes. To inspect this part, a precision 30° wedge is placed in a datum reference frame. The part is then placed in the datum reference frame with datum feature C making at least one point of contact with the 30° wedge.

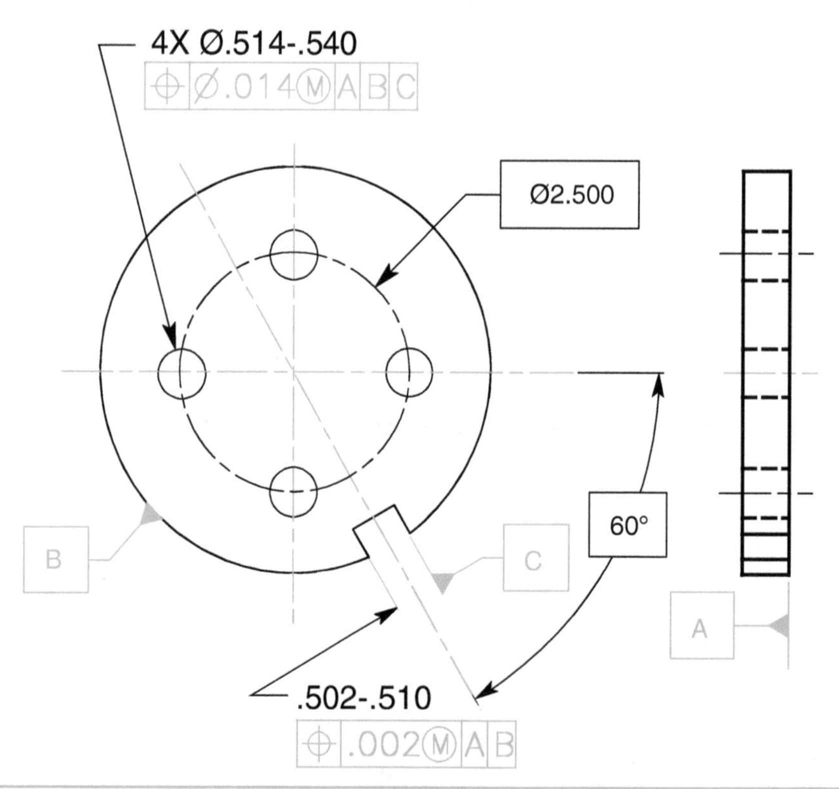

FIGURE 4-7 A pattern of features located to a cylindrical datum feature and clocked to a third datum feature.

Cylindrical Datum Features

Cylindrical parts might have an inside or outside diameter as a datum feature. A cylindrical datum feature is always associated with two theoretical planes meeting at right angles at its datum axis. The part in Fig. 4-7 may be mounted in a centering device, such as a chuck or a V-block, so that the center planes intersecting the datum axis can be determined. Another datum feature, datum feature C, may be established to control rotational orientation or clocking of the hole pattern around the datum axis.

Establishing Datum Features

Two kinds of features may be specified as datum features:

- Plane flat surfaces
- Features of size

Plane Flat Surfaces Specified as Datum Features

Where plane flat surfaces are specified as datum features, such as datum features A, B, and C specified in feature control frame number 1 in Fig. 4-8, the corresponding datum features are simulated by the plane surfaces of a datum reference frame. Plane flat surface features on a rectangular shaped part make the most convenient datum features. Unfortunately, many parts are not rectangular, and designers are often forced to select datum features that are features of size, such as cylinders.

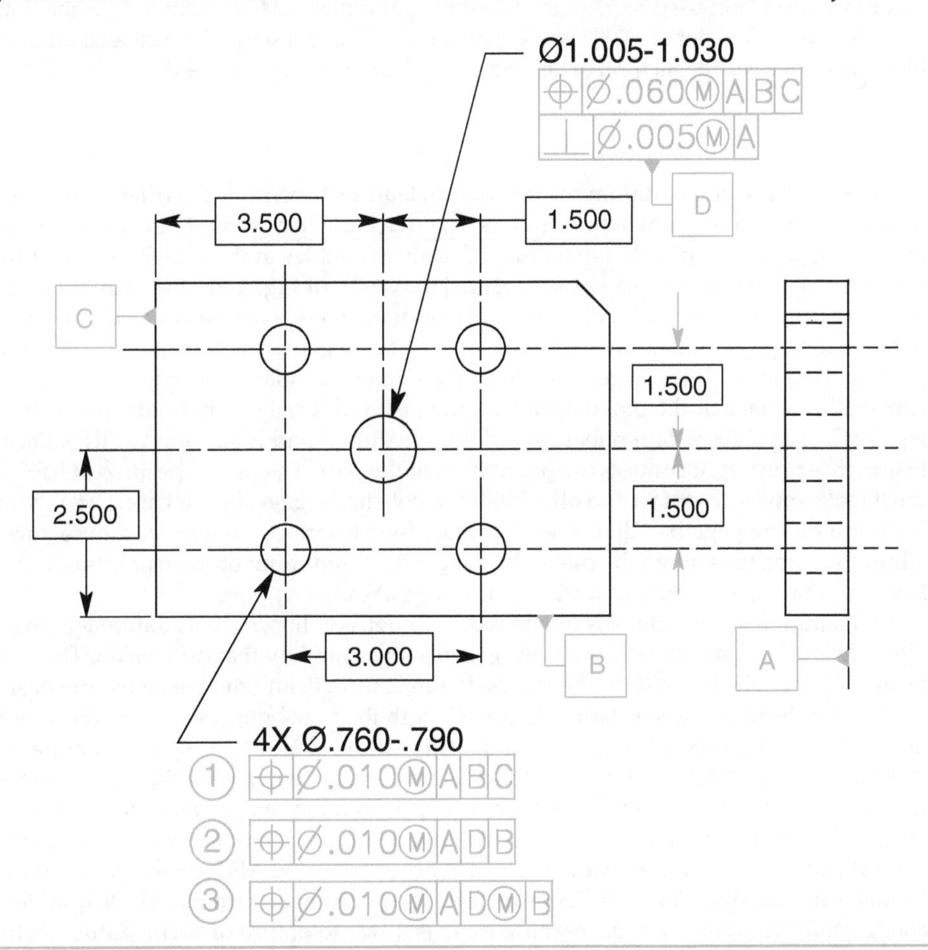

FIGURE 4-8 Datum reference frames established with datum planes A, B, and C and a feature of size, datum feature D.

Datum Features of Size at RMB

Features of size such as datum feature D in Fig. 4-8 are specified with one of the material condition modifiers: regardless of material boundary (RMB), maximum material boundary (MMB), or least material boundary (LMB). If a datum feature of size is specified at RMB, then processing equipment, such as gages, chucks, and mandrels, must make physical contact with the datum feature. This means that when gaging the four-hole pattern to the center hole—datum feature D, specified at RMB in feature control frame number 2—the inspector must use the largest gage pin that fits through the hole, datum feature D, in order to make physical contact with the hole. The pin must first be perpendicular to datum feature A. Then the four-hole pattern is located to the axis of the gage pin. If other inspection techniques are used, such as a coordinate measuring machine, the four-hole pattern is located to the axis of the largest gage pin that fits through the datum feature D hole and is perpendicular to datum feature A.

Datum Features of Size at MMB

If a datum feature of size is specified at MMB, feature control frame number 3, the size of the mating feature on the processing equipment has a constant boundary. The constant boundary pin is specified at the maximum material condition (MMC) or the MMC virtual condition of the datum feature. Where a datum feature of size is toleranced with a geometric tolerance and is referenced in a feature control frame at MMB, circle M, the resulting MMB for the datum feature is equal to its virtual condition with respect to the preceding datum feature. The virtual condition for datum feature D with respect to datum feature A is a diameter of 1.000. Therefore, the simulated datum feature D pin used to gage the four-hole pattern is 1.000 in diameter. As the datum feature departs from a diameter of 1.000 toward a diameter of 1.030, a shift tolerance exists about datum feature D in the amount of such departure. See Chap. 7 for a more complete discussion of shift tolerance.

Plane Flat Surfaces versus Features of Size

In Fig. 4-9*A*, the primary datum feature, datum feature A, controls the orientation of the part and must maintain a minimum of three points of contact, not in a straight line, with the top surface of the mating gage. The pin, datum feature B, easily assembles in the Ø2.000-inch mating hole with a possible shift tolerance since it is specified at MMB. In Fig. 4-9*B*, the primary datum feature, datum feature A, also must maintain a minimum of three points of contact with the top surface of the mating gage, but datum feature B, specified at regardless of feature size (RFS), must make physical contact with the gage. Therefore, the size of the hole in the gage must be adjustable to contact the surface of the pin, datum feature B, even if it only contacts the pin at two points. In Fig. 4-9*C*, the primary datum feature is datum feature B and is specified at RFS. Because datum feature B is primary, it controls the orientation of the part. The pin is specified at RFS; therefore, it must make physical contact and align with the hole in the gage. In this case, not only must datum feature B on the gage be adjustable to contact the surface of the pin, datum feature B, but the adjustable gage must align the pin to the gage with a minimum of three points of contact. Datum feature A may contact the top surface of the gage at only one point.

If a datum feature symbol is in line with a dimension line, such as datum features B and C in Fig. 4-10, the datum feature is the feature of size measured by that dimension. The 7-inch feature of size between the left and right edges is datum feature B, and the 5-inch feature of size between the top and bottom edges is datum feature C. Both the four-hole pattern and the 3-inch-diameter hole are located on the center planes of datum features B and C, as specified in the feature control frames. It is understood that the four-hole pattern is centered on the center planes of datum features B and C, and no dimensions are required from the center planes to the pattern. The axis of the 3-inch-diameter hole is also centered on the intersection of the center planes of datum features B and C. Since datum features B and C are specified at MMB, circle M, a shift tolerance is available in each direction as each datum feature of size departs from MMC toward least material condition (LMC). For example, because there is a size tolerance of ±.010, datum feature B could be as small as 6.990 and as large as 7.010. Suppose datum feature B actually measures 7.002 wide. In this case, datum feature B is .008 smaller than MMC (7.010–7.002). Both the center hole and

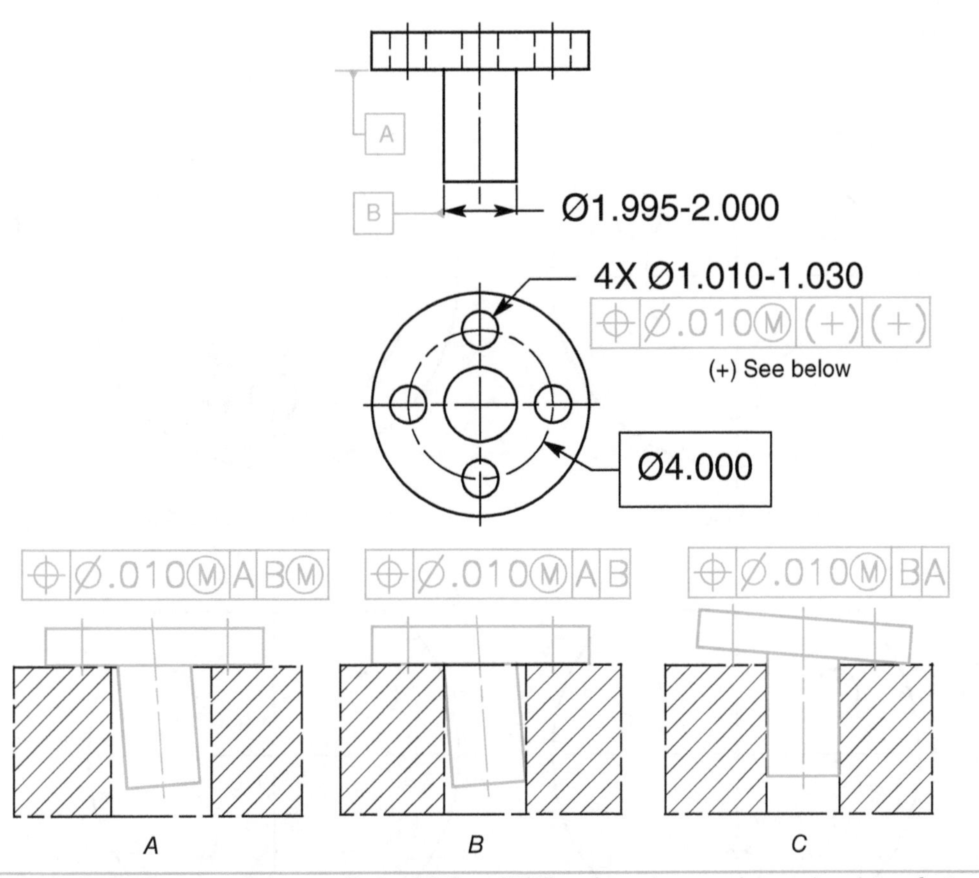

FIGURE 4-9 Datum features of size specified at MMB and RMB and as primary and secondary datum features.

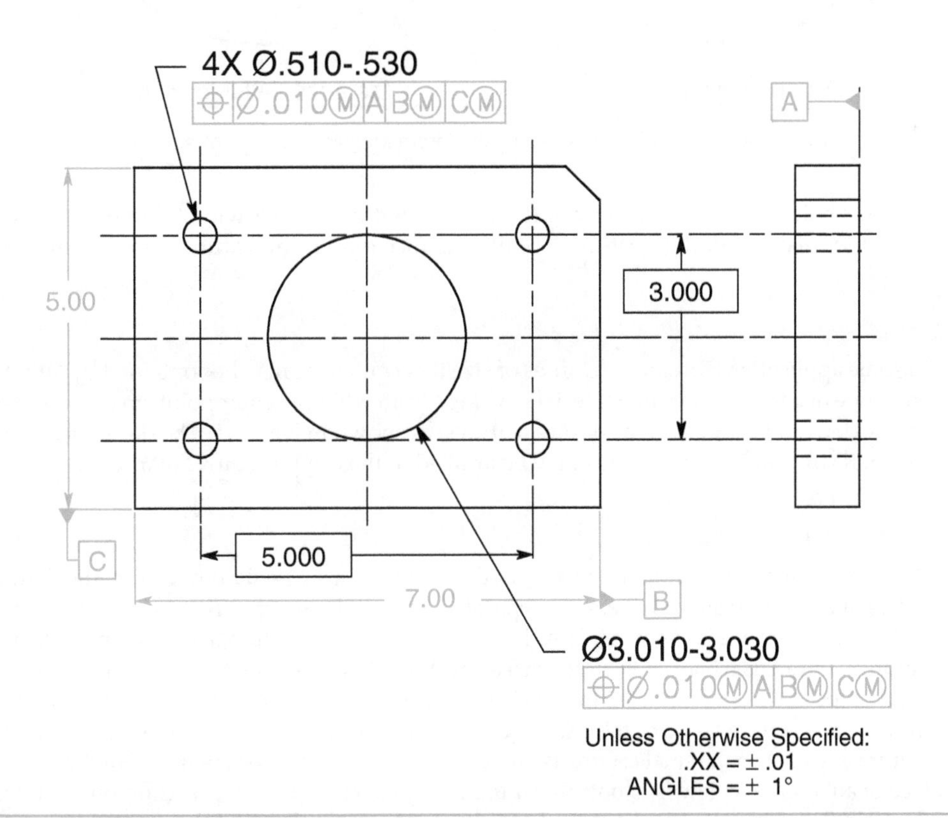

FIGURE 4-10 Features controlled to datum features of size.

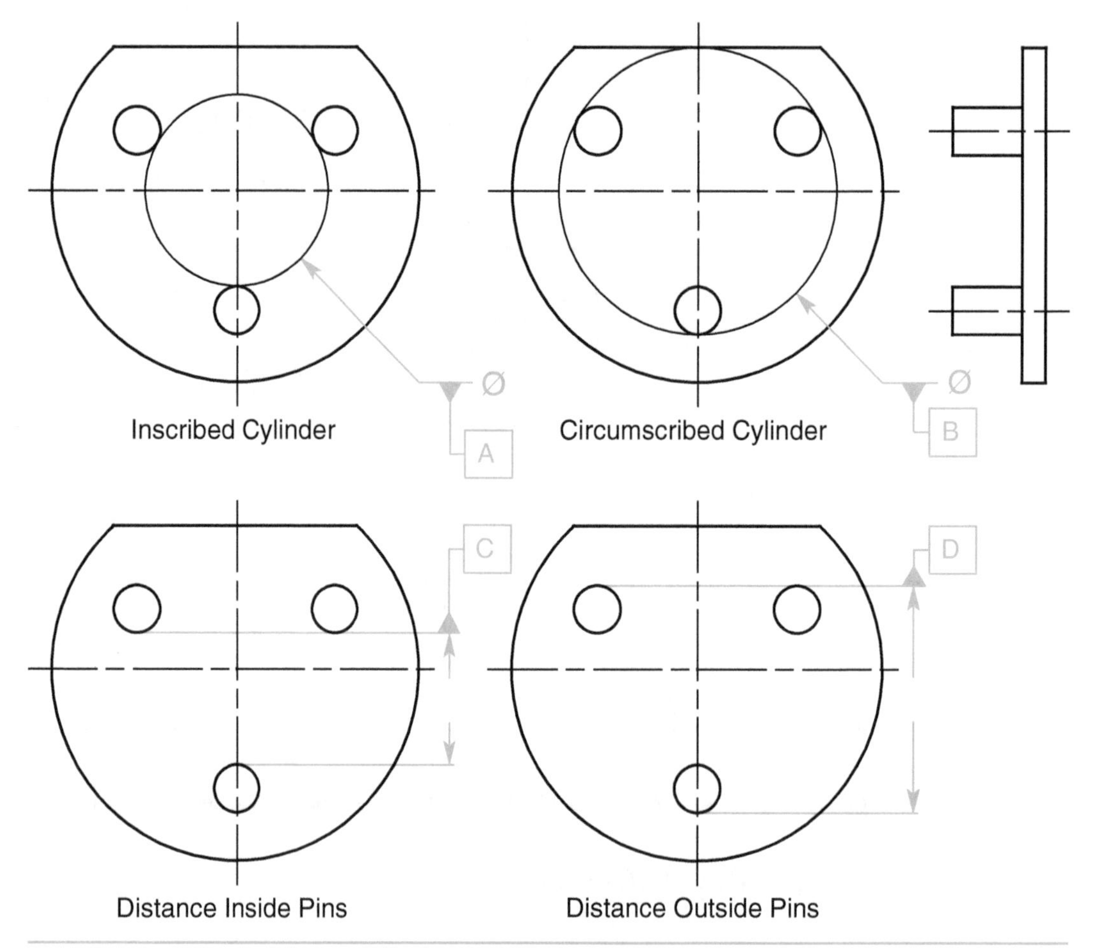

FIGURE 4-11 Possible datum features established from irregular features of size.

the four-hole pattern can shift from left to right within an .008-wide tolerance zone centered on the center plane of datum feature B. Shift tolerance for datum feature C applies in the same way.

Irregular Datum Features of Size

In some applications, irregular features of size that contain or may be contained by an actual mating envelope or actual minimum material envelope from which a center point, an axis, or a center plane can be derived may be used as datum features, as shown in Fig. 4-11. The datum feature modifiers, MMB, RMB, and LMB principles may be applied to these types of irregular features of size.

Common Datum Features

Where more than one datum feature is used to establish a single datum feature, the datum reference letters and appropriate modifiers are separated by a dash and specified in one compartment of the feature control frame, as shown in Fig. 4-12. Together, the two datum features constitute one common datum feature axis where neither datum feature A nor datum feature B is more important. Where a cylinder is specified as a datum feature, such as datum feature A or datum feature B, the entire surface of the feature is considered to be the datum feature. Theoretically, the entire surface of a cylindrical datum feature is to contact the smallest precision sleeve that will fit over the cylinder. Similarly, the entire surface of an internal cylindrical datum feature is to contact the largest precision pin that will fit inside the hole. This almost never happens since inspectors typically don't have this kind of equipment. An external cylindrical datum feature is usually placed in a three-jaw chuck or on a set of V-blocks.

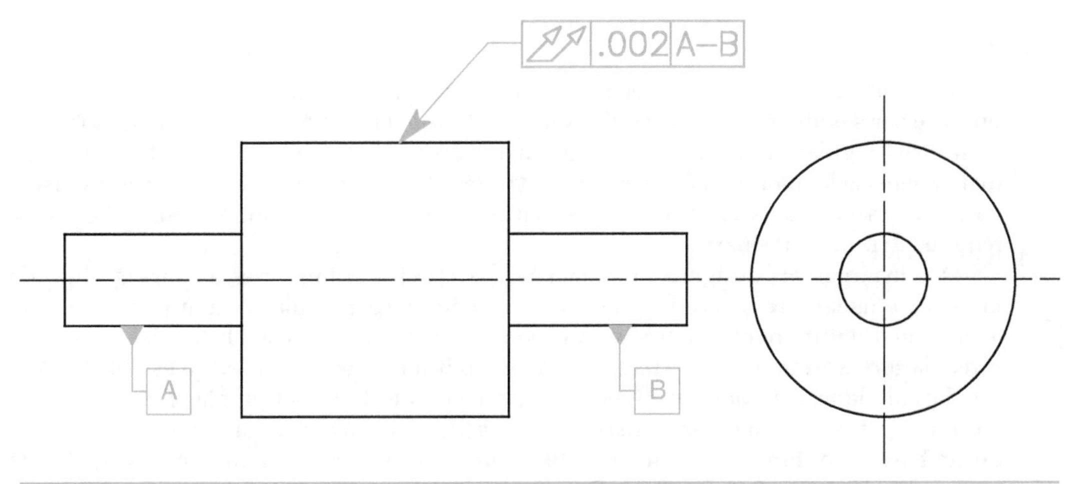

FIGURE 4-12 Common datum features A and B are of equal value.

An internal cylindrical datum feature is often placed on an adjustable mandrel. A part such as the one in Fig. 4-12 would probably be placed on a pair of V-blocks to inspect the total runout specified.

Partial Datum Features

Where a datum feature symbol is associated with a surface, the entire surface is considered to be the datum feature. If only a part of a surface is required to be the datum feature, such as datum features A and B in Fig. 4-13, a heavy chain line is drawn adjacent to the surface profile and dimensioned with basic dimensions. The tolerance for the basic dimensions are gage-makers' tolerance. The datum feature symbol is attached to the chain line.

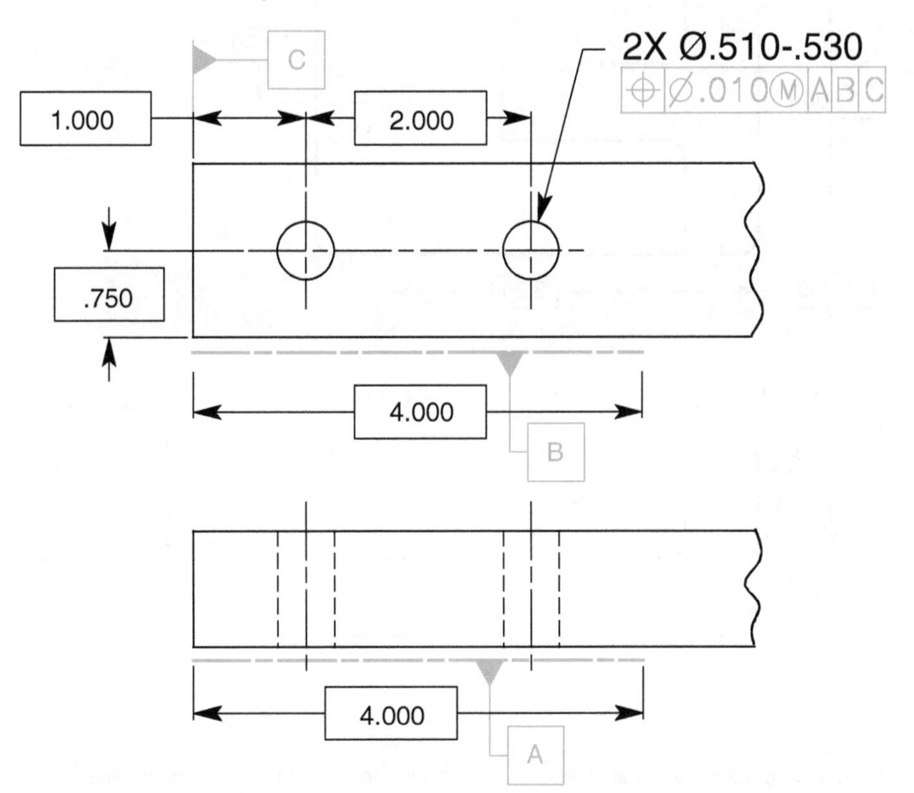

FIGURE 4-13 Partial surfaces as datum features.

Datum Targets

Some manufacturing processes, such as casting, forging, welding, and heat treating, are likely to produce parts with uneven or irregular surfaces. Datum targets may be used to immobilize parts with such irregular surfaces. Datum targets may also be used to support irregularly shaped parts that are not easily mounted in a simulated datum reference frame. Datum targets are used only when necessary because, once they are specified, costly manufacturing and inspection tooling is required to process the part.

Datum targets are designed to contact parts at specific points, lines, and areas. These datum targets are usually referenced from three mutually perpendicular datum planes to establish a simulated datum reference frame. A primary datum plane is established by a minimum of three datum targets not in a straight line. Two datum targets are used to establish a secondary datum plane. And one datum target establishes a tertiary datum plane. A combination of datum target points, lines, and areas may be used. Where a datum target area is required, the desired area is outlined by a phantom line and filled with section lines, as shown for datum targets A1, A2, and A3 in Fig. 4-14. Datum target points are represented on the drawing with datum target point symbols, as shown for datum targets B1 and B2. A datum target line is represented by a target point symbol on the edge of the part in the top view of the drawing and

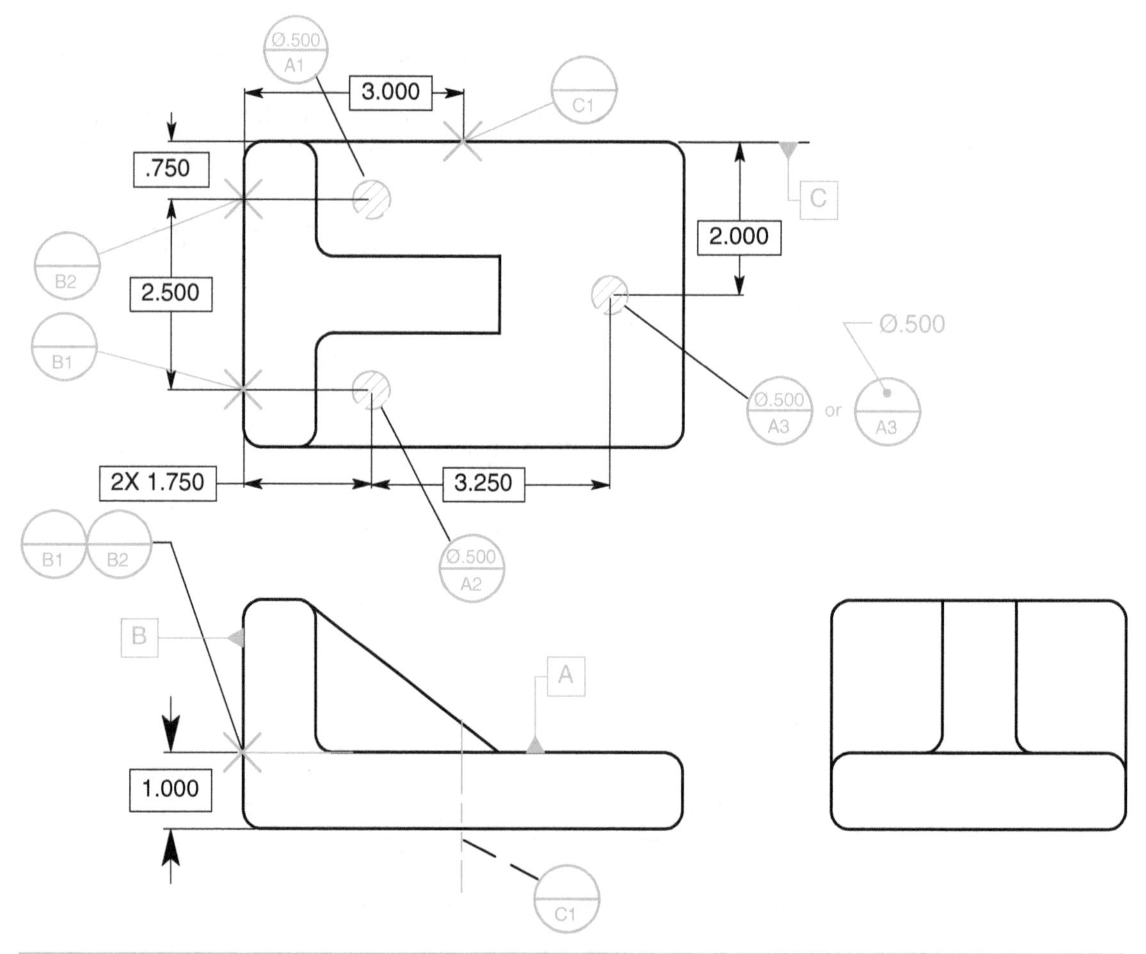

FIGURE 4-14 A drawing with datum target areas, datum target points, and a datum target line.

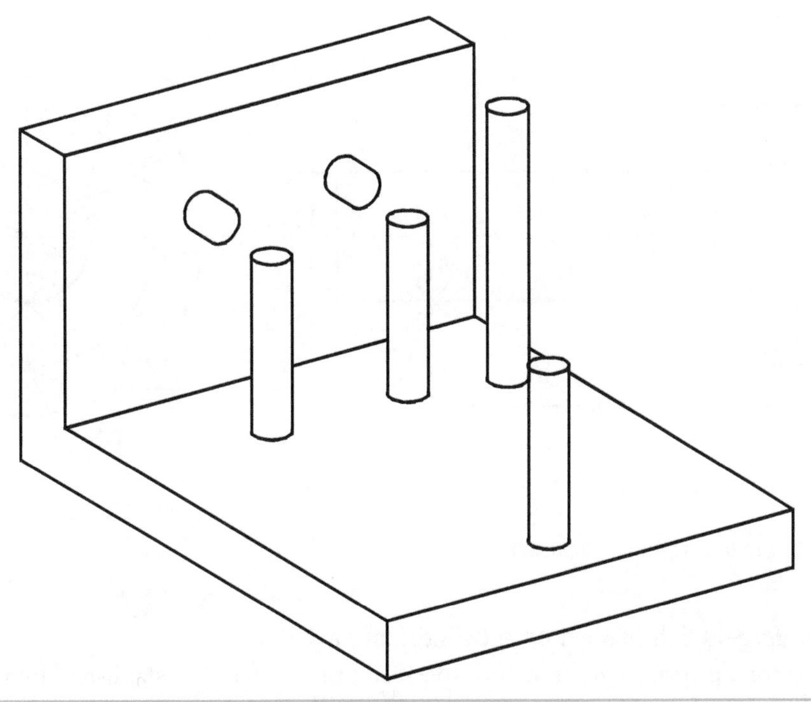

FIGURE 4-15 Fixture for the part shown in Fig. 4-14.

by a phantom line on the front view, datum target C1. Datum target points, lines, and areas are connected to datum target symbols with radial lines. A dashed radial line is used to connect the datum target line to the datum target symbol C1 since the datum target line is on the far side of the part. A fixture for this part is shown in Fig. 4-15.

Actual tooling areas on the fixture are pins the size and shape of the datum target areas. The pins that represent datum areas A1, A2, and A3 for the part in Fig. 4-14 are .500-diameter pins with flat surfaces cut across the top. The tooling points are not points at all but rather pins with hemispherical ends that contact the part at the highest point on the hemisphere, shown in Fig. 4-16. Datum target lines, also shown in Fig. 4-16, are generated with the edge of a cylindrical pin that contacts the part. All datum targets are dimensioned for location and size by either toleranced dimensions or basic dimensions. Basic dimensions are toleranced with gage-makers' tolerances.

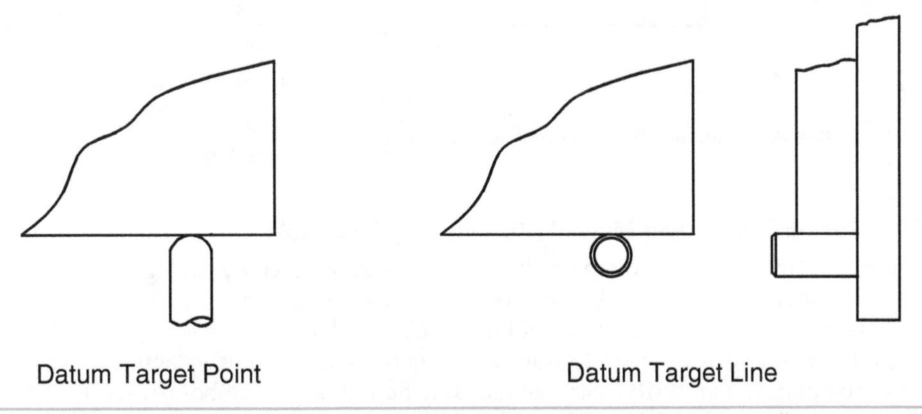

FIGURE 4-16 The actual tooling used to produce datum target points and datum target lines.

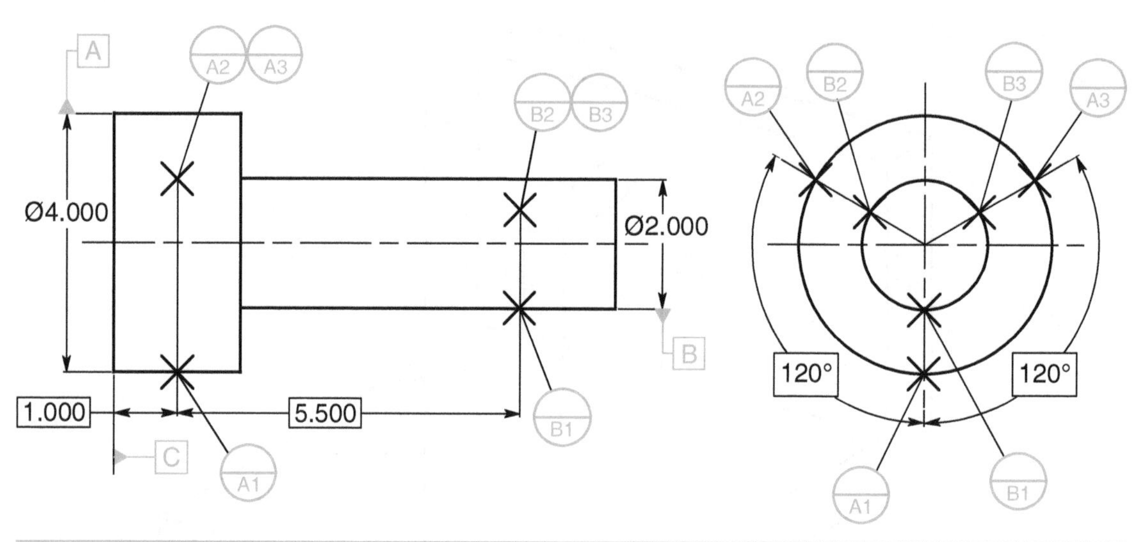

FIGURE 4-17 Datum targets on a cylindrical part.

Datum Targets Established on a Cylindrical Part

The axis of a primary datum feature specified at RMB may be established by two sets of three equally spaced datum targets, as shown in Fig. 4-17.

A datum target line is represented by a phantom line drawn across the cylinder; see datum target line A1 in Fig. 4-18. Where a datum target area is required, the desired area is bounded by phantom lines and filled with section lines, as shown for datum target area B1.

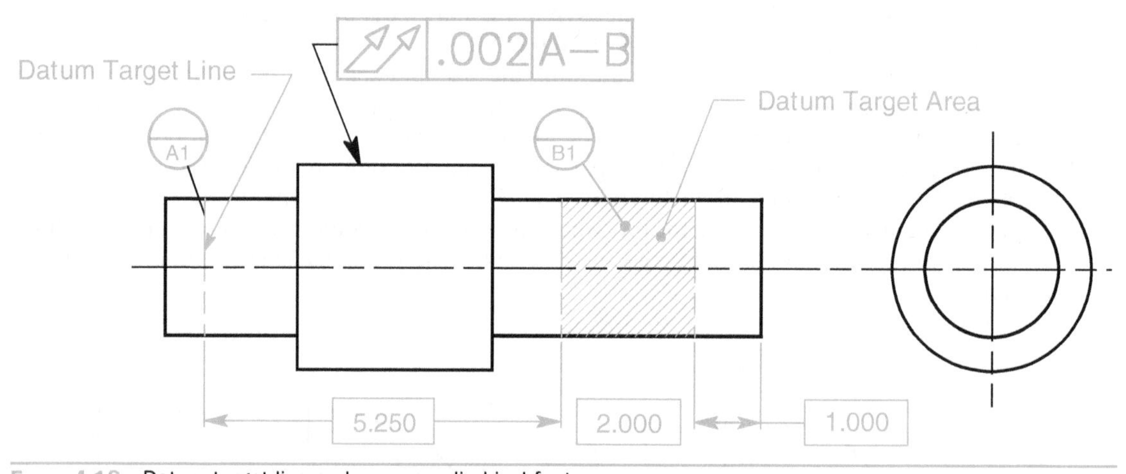

FIGURE 4-18 Datum target line and area on cylindrical features.

Step Datum Targets and Movable Datum Target Symbols

A datum plane may have a step or offset, such as datum feature A in Fig. 4-19. The step between datum points A1, A2, and A3 is specified with a basic dimension of 1.000 inch.

The movable datum target symbol is similar to a datum target symbol but is not fixed to its basic location and is allowed to translate. Where datum targets establish a center point, axis, or center plane on an RMB basis, the movable datum target symbol is used to indicate that the datum feature simulator moves normal to the true profile, as shown in Fig. 4-19.

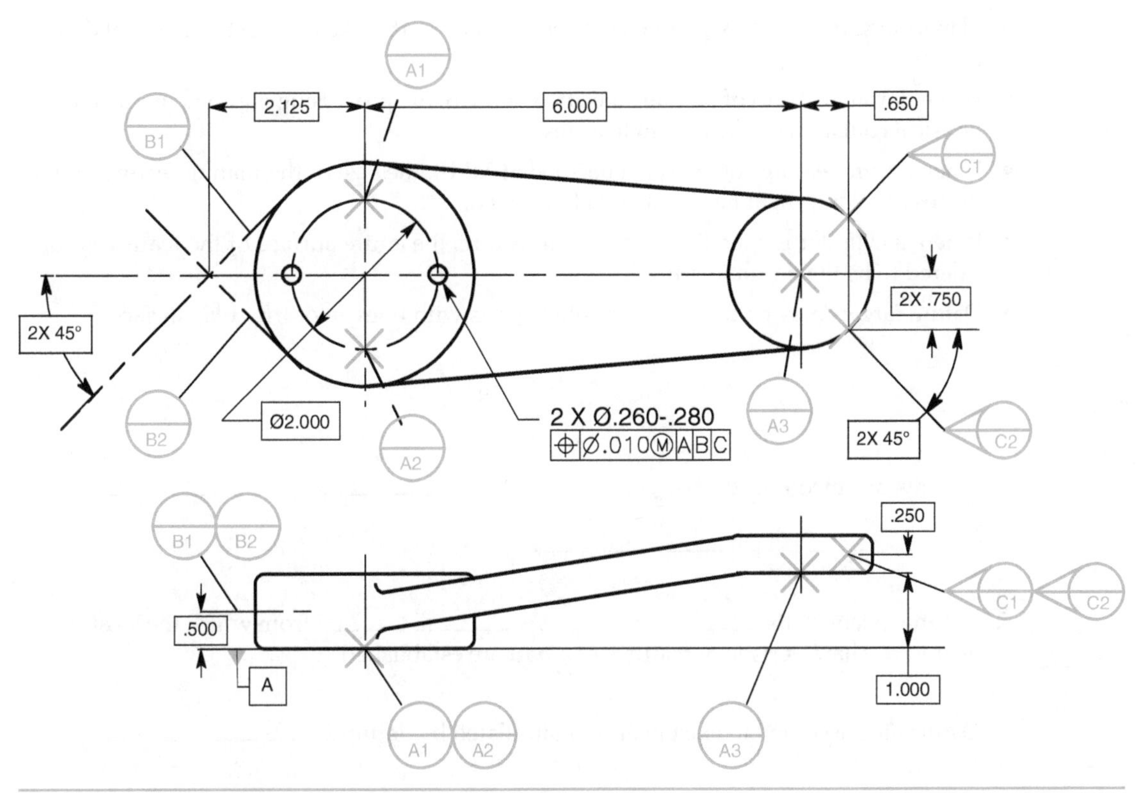

FIGURE 4-19 A stepped datum surface and movable datum targets.

Summary

- A datum is a theoretically perfect point, axis, line, plane, or combination thereof that establishes the origin of measurements. Datums exist within a structure of three mutually perpendicular intersecting datum planes known as a datum reference frame.
- Since measurements cannot be made from theoretical surfaces, datums are assumed to exist in and be simulated by the processing equipment.
- A part is oriented and immobilized relative to the three mutually perpendicular intersecting datum planes of a datum reference frame in a selected order of precedence.
- Datum features are specified in order of precedence as they appear from left to right in the feature control frame.
- Datum features are selected to meet design requirements. Functional surfaces, mating surfaces, readily accessible surfaces, and surfaces of sufficient size to allow repeatable measurements make good datum features.
- The datum feature symbol is used to identify physical features of a part as datum features. Datum feature symbols shall not be applied to centerlines, center planes, or axes.
- A cylindrical datum feature is always intersected by two theoretical planes meeting at right angles at its datum axis. Another datum feature may be established to control rotational orientation or clocking.

- Plane, flat surface features on a rectangular shaped part make the most convenient datum features.
- When datum features of size are specified at RMB, the processing equipment must make physical contact with the datum features.
- When datum features of size are specified at MMB, the size of the mating feature on the processing equipment has a constant boundary.
- Where a cylinder is specified as a datum feature, the entire surface of the feature is considered to be the datum feature.
- Datum targets may be used to immobilize parts with uneven or irregular surfaces.

Chapter Review

1. Datums are theoretically perfect ______________________________

   ______________________________.

2. Datums establish the ______________________________ from which the location or geometric characteristic of features of a part are established.

3. Datums are assumed to exist in and be simulated by datum ______________________________.

4. A datum reference frame consists of three mutually perpendicular ______________________________

   and three mutually perpendicular ______________________________ at the intersection of those planes.

5. A part is oriented and immobilized relative to the three mutually perpendicular intersecting datum planes of the datum reference frame in a selected order of ______________________________

   ______________________________.

6. The primary datum feature contacts the datum reference frame with a minimum of

   ______________________________ points of contact that are not in a straight line.

7. Datum features are specified in order of precedence as they appear from left to right in the ______________________________.

8. Datum feature letters need not be in ______________________________ order.

9. The primary datum feature controls ______________________________.

10. When selecting datum features, the designer should consider features that are:

____

____

____

____

11. Datum features must be identified with ____________________ or ____________________ and specified in a feature control frame.

12. Datum feature symbols must *not* be applied to ____________________ ____________________.

13. One method of tolerancing datum features at an angle to the datum reference frame is to place a datum feature symbol on the ____________________ and control that surface with an angularity tolerance and a basic angle.

14. A cylindrical datum feature is always associated with two ____________________ ____________________ meeting at right angles at its datum axis.

15. The two kinds of features specified as datum features are:

____

16. Datum features of sizes may apply at ____________________ ____________________.

17. Where datum features of sizes are specified at RMB, the processing equipment must make ____________________ with the datum features.

18. Where features of sizes are specified at MMB, the size of the processing equipment is equal to its ____________________.

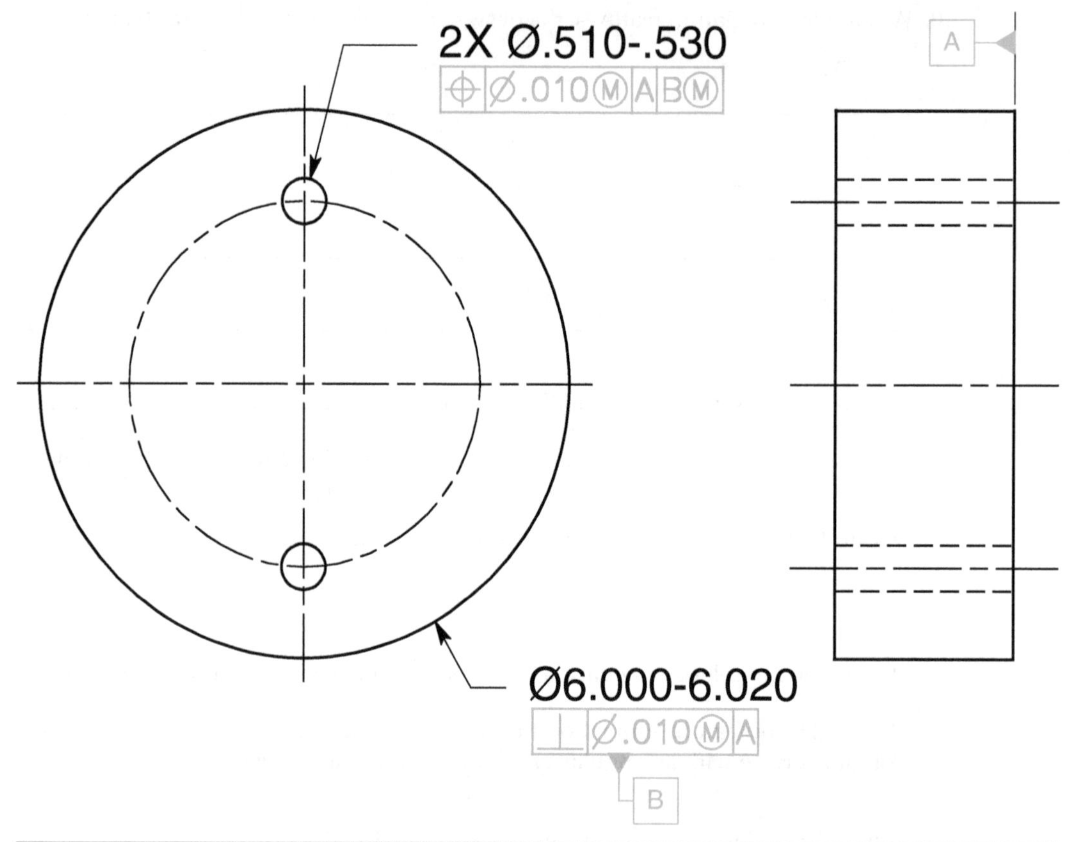

FIGURE 4-20 Datum feature of size: Questions 19 through 24.

Refer to Fig. 4-20 to answer Questions 19 through 24.

19. The two-hole pattern is perpendicular to what datum feature? ____________

20. The two -hole pattern is located to what datum feature? ____________

21. If inspected with a gage, what is the diameter of datum feature B on the gage? ____________

22. If inspected with a gage, what is the diameter of the two pins on the gage? ____________

23. If datum feature B had been specified at RFS, explain how the gage would be different.

____

____

____

24. If datum feature B had been specified as the primary datum feature at RFS, explain how the gage would be different.

____

____

____

25. If a datum feature symbol is in line with a dimension line, the datum feature is the ____________________ measured by the dimension.

26. Where more than one datum feature is used to establish a single datum, the __________ ____________________ and appropriate ____________________ are separated by a dash and specified in one compartment of the feature control frame.

27. Where cylinders are specified as datum features, the entire surface of the feature is considered to be the ____________________.

28. If only a part of a feature is required to be the datum feature, a ____________________ ____________________ is drawn adjacent to the surface profile and dimensioned with basic dimensions.

29. Datum targets may be used to immobilize parts with ____________________ ____________________.

30. Costly manufacturing and inspection ____________________ is required to process the part.

Problems

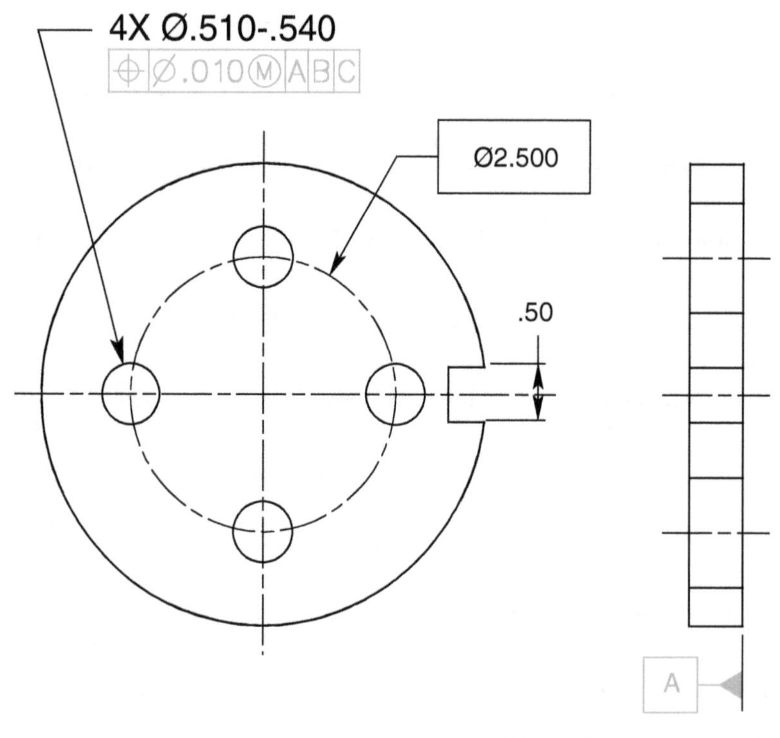

FIGURE 4-21 Placement of datum feature symbols: Prob. 1.

1. Attach the appropriate datum feature symbols on the drawing in Fig. 4-21.

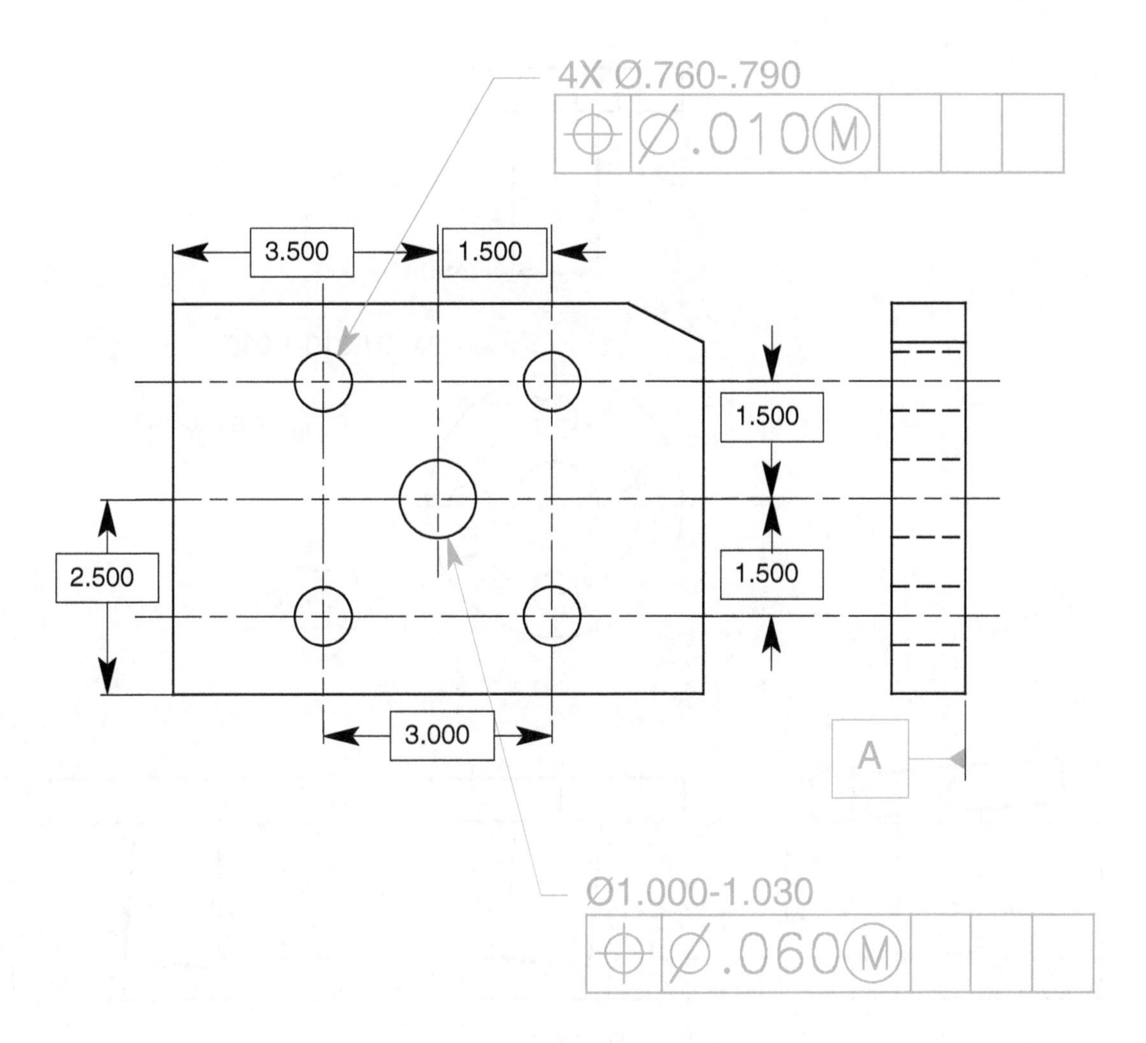

FIGURE 4-22 Placement of datum references in feature control frames: Prob. 2.

2. Provide the appropriate datum feature symbols and complete the feature control frames on the drawing in Fig. 4-22.

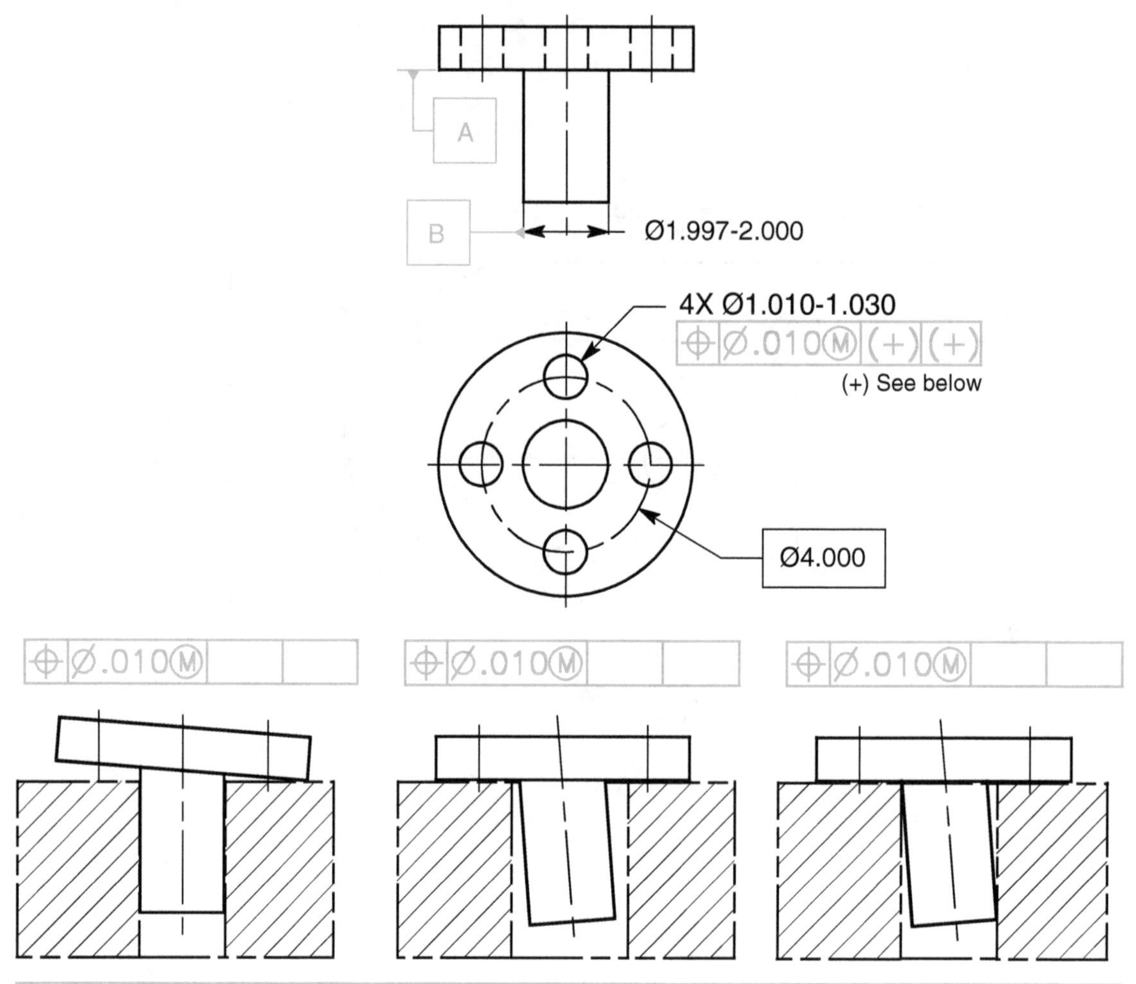

FIGURE 4-23 Datum features of size at MMB and RMB: Prob. 3.

3. Complete the feature control frames with datum references and material condition modifiers to reflect the drawings in Fig. 4-23.

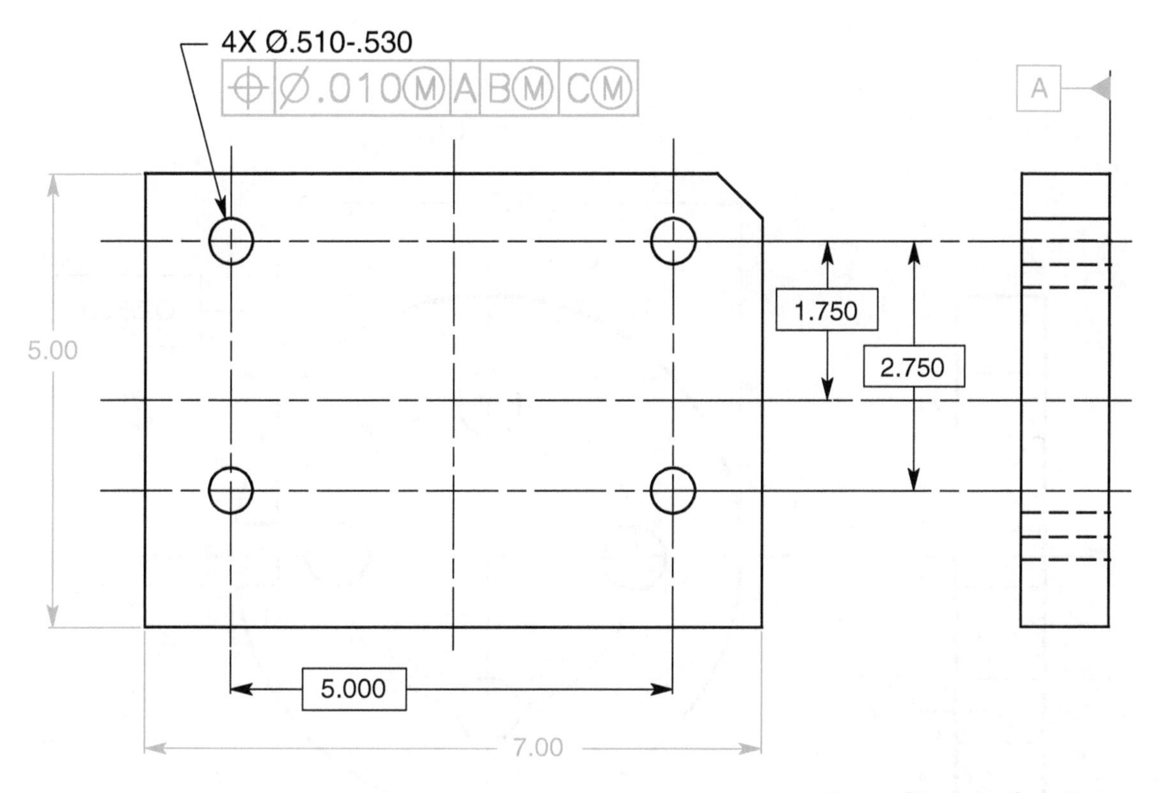

FIGURE 4-24 Datum features located to the center planes of the drawing: Prob. 4.

4. Specify the appropriate datum feature symbols to locate the four-hole pattern to the center planes of the drawing in Fig. 4-24.

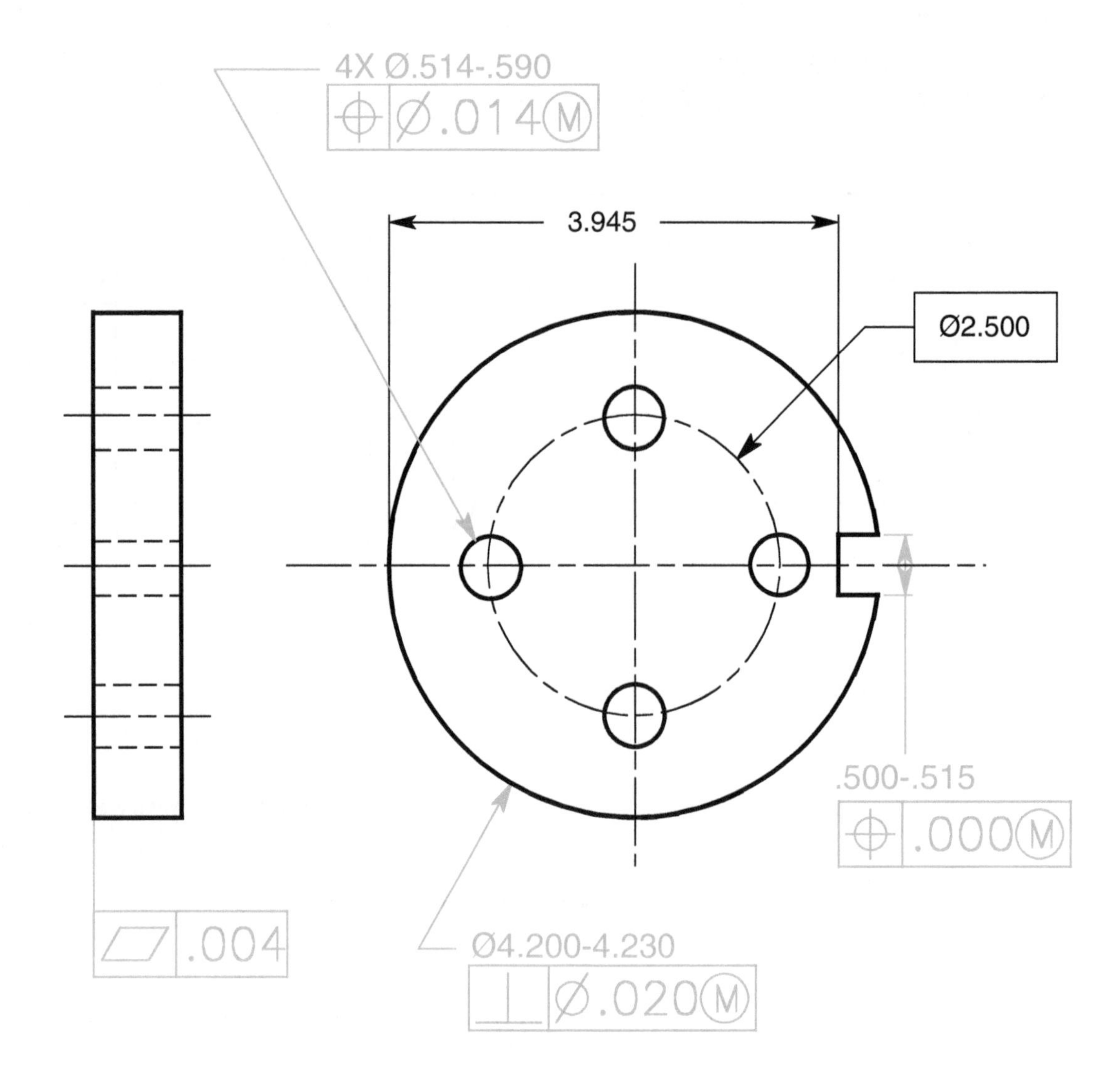

FIGURE 4-25 Specifying datum references and datum feature symbols: Probs. 5 and 6.

5. Specify the appropriate datum feature symbols and complete the feature control frames in the datum exercise in Fig. 4-25.
6. On Fig. 4-25, draw and dimension the gage used to inspect the part in Prob. 5.

CHAPTER 5

Form

All form tolerances apply to single, that is, individual, features; consequently, form tolerances are independent of all other features. No datums apply to form tolerances. The form of individual features is automatically controlled by the size tolerance, Rule #1. Where the size tolerance does not sufficiently control the form of a feature, a form tolerance may be specified as a refinement. Except for flatness of a median plane and straightness of a median line, all form tolerances are surface controls and are attached to feature surfaces with either a leader or an extension line. The feature control frame contains a form control symbol and a numerical tolerance; no cylindrical tolerance zones or material condition modifiers are appropriate for surface controls. Normally, no other symbols appear in a feature control frame for form controls except possibly the free state symbol (circle F), unit flatness, or unit straightness, as shown below.

Chapter Objectives

After completing this chapter, the learner will be able to:

- *Specify* and interpret flatness
- *Explain* the difference between flatness of a surface and flatness of a derived median plane
- *Specify* and interpret straightness
- *Explain* the difference between straightness of a cylindrical surface and straightness of a derived median line
- *Specify* and interpret circularity
- *Specify* and interpret cylindricity
- *Specify* and interpret free state variation

Flatness

Definition

Flatness is the condition of a surface or derived median plane having all elements in one plane.

Specifying Flatness Tolerance

In a view where the surface to be controlled appears as a line, a feature control frame is attached to the surface with a leader or extension line, as shown in Fig. 5-1. The feature control frame contains a flatness symbol and a numerical tolerance. Normally, nothing else appears in a feature control frame controlling flatness of a surface, except possibly the free state symbol or unit flatness, as shown in Fig. 5-5. Flatness tolerance is a refinement of the size tolerance, Rule #1, and must be less than the size tolerance. The thickness at each local size must fall within the limits of size; the feature of size may not exceed the boundary of perfect form at maximum material condition (MMC).

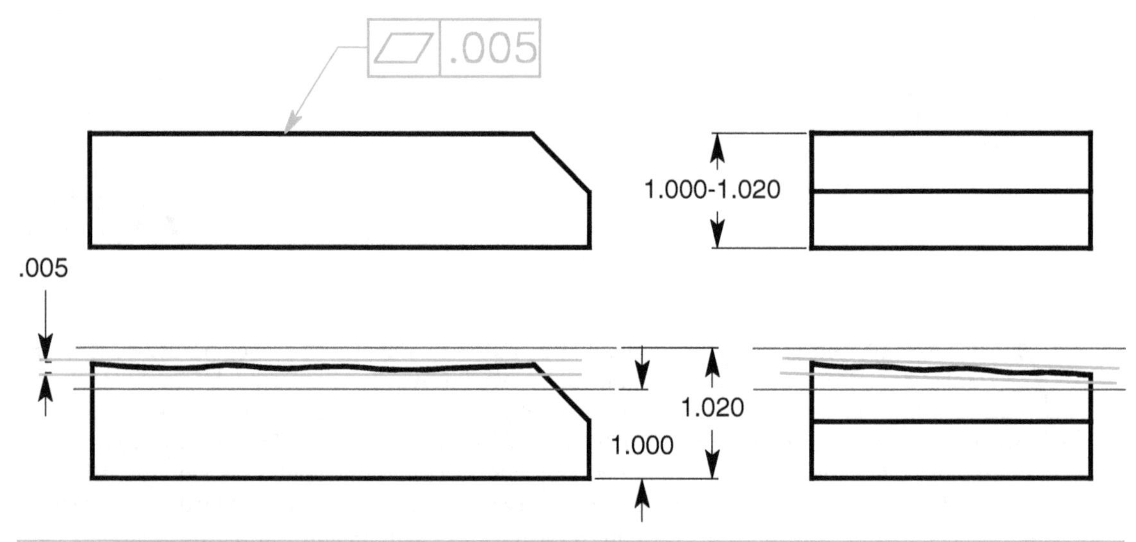

FIGURE 5-1 A flatness tolerance applied to a plane surface.

Interpretation

The surface being controlled for flatness in Fig. 5-1 must lie between two parallel planes separated by the flatness tolerance of .005 specified in the feature control frame. In addition, the surface must fall within the size tolerance, the two parallel planes .020 apart. The flatness tolerance zone does not need to be parallel to any other surface, as indicated in the right-side view. The standard states that the flatness tolerance must be less than the size tolerance, but the size tolerance applies to both top and bottom surfaces of the part. The manufacturer will probably make the part in Fig. 5-1 about 1.010 thick using only half of the size tolerance. Since the MMC of 1.020 minus the actual size of 1.010 is an automatic Rule #1 form tolerance of .010, a flatness tolerance refinement of .005, as specified in the feature control frame, seems appropriate. The entire part in Fig. 5-1 must fit between two parallel planes 1.020 apart. If the thickness of the part is produced anywhere between 1.015 and 1.020, the flatness of the part is controlled by Rule #1. If the thickness of the part is between 1.000 and 1.014, the geometric tolerance ensures that the top surface of the part does not exceed a flatness of .005, as shown in Table 5-1. The above discussion applies to a surface that has been bowed or warped. If the part in Fig. 5-1 is not bowed or warped and is produced at its MMC thickness of 1.020, the top surface may be gouged or eroded as much as .005.

Actual Part Size	Flatness Tolerance	Controlled By
1.020	.000	Rule #1
1.018	.002	
1.016	.004	
1.014	.005	Flatness tolerance
1.010	.005	
1.005	.005	
1.000	.005	

TABLE 5-1 Flatness Tolerances for the Part in Fig. 5-1

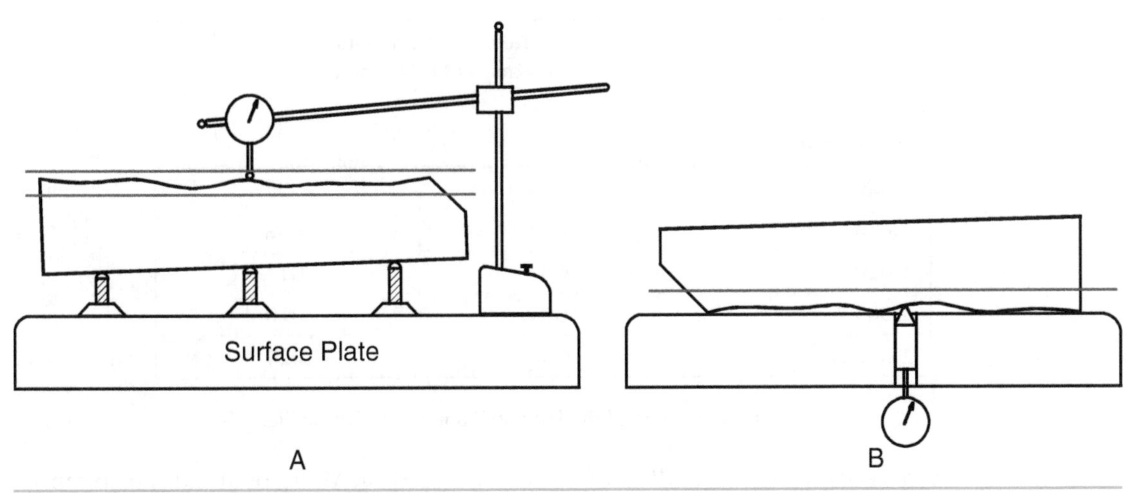

FIGURE 5-2 Two flatness verification techniques.

Inspection

When verifying flatness, the feature of size is first measured to verify that it falls within the limits of size. The measurement of surface variation in Fig. 5-2*A* is performed after the surface in question has been adjusted with jackscrews to remove any parallelism error. Then flatness verification is achieved by measuring the surface with a dial indicator, in all directions, to determine that the variation does not exceed the tolerance in the feature control frame. In Fig. 5-2*B*, flatness is measured by moving the part over a probe in the surface plate and reading the indicator to determine any flatness error. This is a relatively simple method of measuring flatness; no adjustment for parallelism error is needed. However, specialized equipment is required.

Specifying Flatness of a Derived Median Plane

When a feature control frame with a flatness tolerance is associated with a size dimension, the flatness tolerance applies to the derived median plane for a noncylindrical feature, such as the part shown in Fig. 5-3. The median plane derived from the surfaces of the noncylindrical feature may bend, warp, or twist in any direction away from a perfectly flat plane but must not exceed the flatness tolerance zone boundaries.

Interpretation

While each actual local size of the feature must fall within the size tolerance, the feature in Fig. 5-3 may exceed the boundary of perfect form at MMC due to bending or warping. A flatness control of a median plane will allow the feature to violate Rule #1. Flatness associated with a size

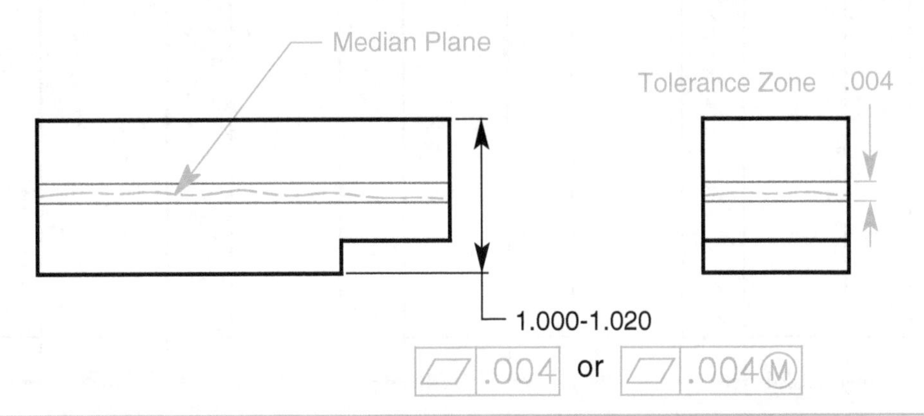

FIGURE 5-3 The flatness tolerance of a median plane is associated with the size dimension.

Feature Size	Noncylindrical Feature (Flatness of a Median Plane)	
	⏥ .004	⏥ .004 Ⓜ
1.020 MMC	.004	.004
1.015	.004	.009
1.010	.004	.014
1.005	.004	.019
1.000 LMC	.004	.024

TABLE 5-2 Flatness Tolerances of the Median Plane of the Part in Fig. 5-3

dimension may be specified at regardless of feature size (RFS) , at MMC, or at LMC. If specified at RFS, the flatness tolerance applies at any increment of size within the size limits. If specified at MMC, the total flatness tolerance equals the tolerance in the feature control frame plus any bonus tolerance. The bonus is equal to the departure from the MMC size of the feature toward the LMC size. A feature with a flatness control of a derived median plane has a virtual condition. The part in Fig. 5-3 has a virtual condition of 1.024, the sum of the MMC and the geometric tolerance, as shown in Table 5-2.

Inspection

First, a feature of size is measured to verify that it falls within its limits of size. Then verification of flatness of the median plane of a feature of size specified at MMC can be achieved by placing the part in a full form functional gage, as shown in Fig. 5-4. If a part goes all the way in the gage and satisfies the size requirements, the part is in tolerance. Flatness verification of a feature of size specified at RFS can be achieved by taking differential measurements on opposite sides of the part with a dial indicator to determine how much the median plane varies from a perfectly flat center plane. If the bow or warp of the part exceeds the tolerance in the feature control frame, at any point within the size tolerance, the part is not acceptable.

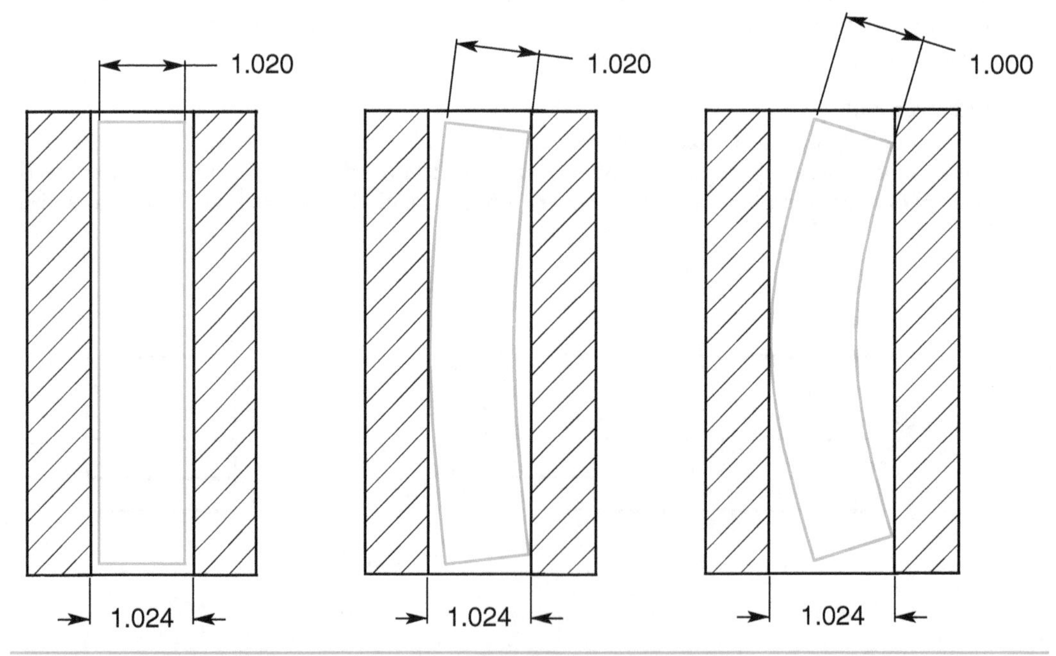

FIGURE 5-4 Verification of flatness of a feature of size at MMC with a gage.

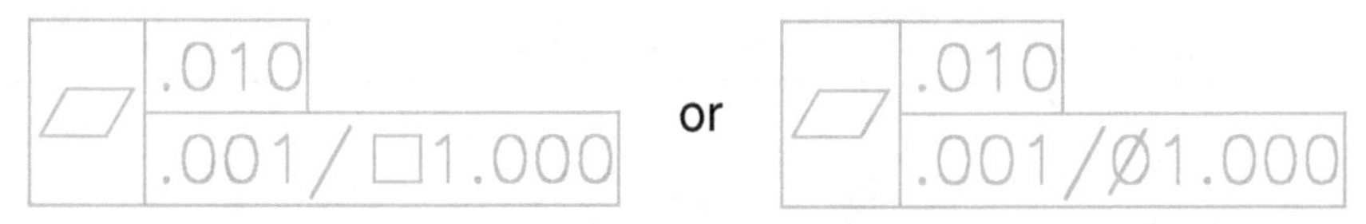

FIGURE 5-5 An overall flatness of .010 with unit flatness as a refinement.

Unit Flatness

Flatness may be applied on a unit basis to prevent abrupt surface variation in a relatively small area of the feature. The overall flatness of .010 in the feature control frame in Fig. 5-5 applies to the entire surface. The refinement of .001 per square inch applies to each and every square inch on the surface as an additional requirement to the overall flatness. The size of the unit area may be expressed as a square area, such as every 1.000-inch square, or a circular area 1.000 inch in diameter.

Straightness

Definition

Straightness is a condition where an element of a surface, or a derived median line, is a straight line.

Specifying Straightness of a Surface Tolerance

In a view where the line elements to be controlled appear as a line, a feature control frame is attached to the surface with a leader or extension line, as shown in Fig. 5-6. The feature control frame contains a straightness symbol and a numerical tolerance. Normally, nothing else appears in a feature control frame controlling straightness of a surface except possibly the free state symbol or unit straightness, as shown below. Straightness tolerance is a refinement of the size tolerance, Rule #1, and must be less than the size tolerance. The feature of size may not exceed the boundary of perfect form at MMC.

Interpretation

The line elements being controlled in Fig. 5-6 must lie within a tolerance zone that consists of two parallel lines separated by the straightness tolerance of .004 specified in the feature control frame and parallel to the view in which they are specified, the front view shown in Fig. 5-6. In addition, the line elements must fall within the size tolerance of .020. The straightness tolerance zone is not required to be parallel to the bottom surface or axis of the respective part. Each line element is independent of all other line elements. Straightness tolerance must be less than the size tolerance. The parts in Fig. 5-6 are both likely to be produced at a thickness/diameter of 1.010.

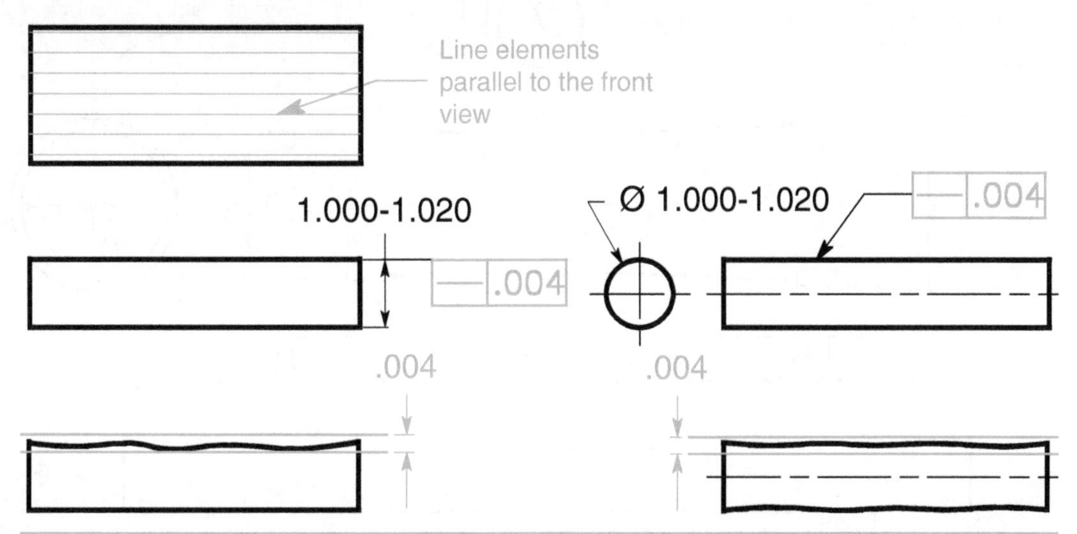

FIGURE 5-6 A straightness tolerance applied to a plane surface and a cylindrical surface.

Actual Part Size	Straightness Tolerance	Controlled By
1.020 MMC	.000	Rule #1
1.018	.002	
1.016	.004	
1.014	.004	Straightness tolerance
1.010	.004	
1.005	.004	
1.000 LMC	.004	

TABLE 5-3 Straightness Tolerances for the Parts in Fig. 5-6

Since the MMC of 1.020 minus the actual size of 1.010 is the automatic Rule #1 form tolerance of .010, the straightness tolerance refinement of .004 as specified in the feature control frame seems appropriate. The entire rectangular part in Fig. 5-6 must fit between two parallel planes 1.020 apart, and the entire cylindrical part must fit inside an internal cylinder 1.020 in diameter. Just as for flatness, if the thickness/diameter of the parts is produced anywhere between 1.016 and 1.020, the straightness of each part is controlled by Rule #1, as shown in Table 5-3.

Inspection

When inspecting straightness of a surface, the feature of size is first measured to verify that it falls within the limits of size. Then each line element in the surface, parallel to the view in which it is specified, is measured to determine that straightness variation does not exceed the tolerance indicated in the feature control frame. Straightness is verified with a dial indicator after the surface in question has been adjusted with jackscrews to remove any parallelism error. The measurement of surface variation for straightness is performed similar to the measurement for flatness. Straightness of a cylindrical surface is inspected by moving the dial indicator across the surface plate, against the edge of a precision parallel. Line elements on cylindrical surfaces are measured where the tolerance zone boundaries and the axis of the cylinder share a common plane, as indicated in Fig. 5-7. Each line element is independent of every other line element, and the surface may be readjusted to

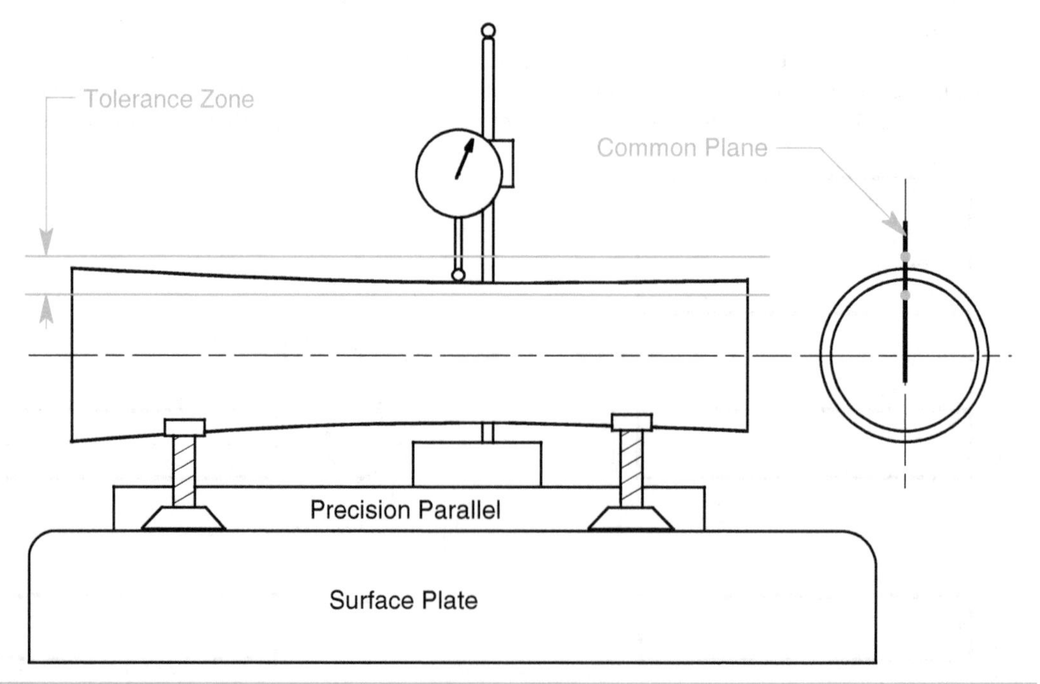

FIGURE 5-7 Inspection of straightness of a surface.

remove any parallelism error for the measurement of each subsequent line element. There are an infinite number of line elements on any surface. The inspector must measure a sufficient number of line elements to be convinced that all line elements fall within the tolerance specified.

Specifying Straightness of a Derived Median Line

Where a feature control frame with a straightness tolerance is associated with the size dimension of a cylinder, the straightness tolerance applies to the median line of the cylindrical feature, as shown in Fig. 5-8. A diameter symbol precedes the tolerance value indicating a cylindrical tolerance zone. The median line of a cylinder may bow in any direction away from a perfectly straight line but may not exceed the tolerance zone boundaries.

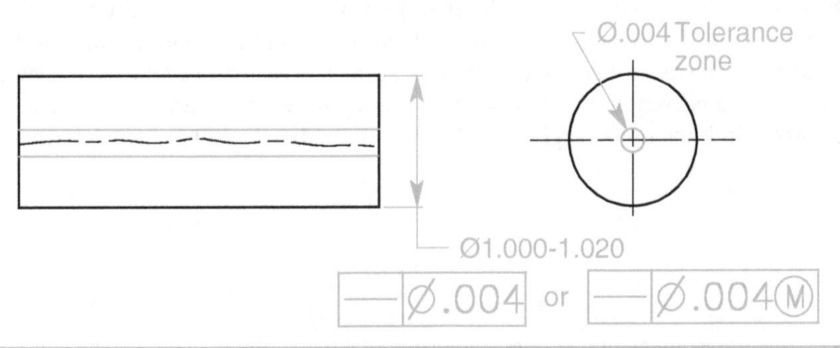

FIGURE 5-8 Straightness of a derived median line is associated with the size dimension.

Interpretation

While the actual local size of the feature must fall within the size tolerance, the feature in Fig 5-8 may exceed the boundary of perfect form at MMC due to bending. A straightness control of a median line will allow the feature to violate Rule #1. The tolerance value may be greater than the size tolerance; the boundary of perfect form at MMC does not apply. Straightness associated with a size dimension may be specified at RFS, MMC, or LMC. If specified at RFS, the tolerance applies at any increment of size within the limits of size. If specified at MMC, the total straightness tolerance equals the tolerance in the feature control frame plus any bonus tolerance, equal to the amount of departure from MMC toward LMC, as shown in Table 5-4. Consequently, a feature with a straightness control of a median line has a virtual condition. The cylinder in Fig 5-8 has a virtual condition of 1.024. Straightness of a derived median line may exceed the size tolerance if not used in conjunction with an orientation or position control. Where the straightness control is used in conjunction with an orientation or position tolerance, the straightness tolerance may not exceed the orientation or position tolerance.

	Cylindrical Feature (Straightness of a Median Line)	
Feature Size	— Ø.004	— Ø.004 Ⓜ
1.020 MMC	Ø .004	Ø .004
1.015	Ø .004	Ø .009
1.010	Ø .004	Ø .014
1.005	Ø .004	Ø .019
1.000 LMC	Ø .004	Ø .024

TABLE 5-4 Straightness Tolerances for the Cylinder in Fig. 5-8

Inspection

When inspecting straightness of a derived median line, the cylinder is first measured to verify that it falls within the limits of size. Then straightness verification specified at MMC can be achieved by placing the cylinder in a full form functional gage similar to the gage used to inspect flatness of a median plane shown in Fig. 5-4 except that the gage would be a full-length virtual condition internal cylinder. If a part fits all the way inside the gage and satisfies the size requirements, it is an acceptable part. Straightness verification of a feature of size specified at RFS can be achieved by taking differential measurements on opposite sides of the part with a dial indicator to determine how much the median line varies from a perfectly straight axis. If the bow or warp of the part exceeds the tolerance in the feature control frame, at any size within the size tolerance, the part is not acceptable.

Unit Straightness

Straightness may be applied on a unit basis to prevent abrupt surface variation in a relatively short length of the feature. The overall straightness of .010 in the feature control frames in Fig. 5-9 applies to the entire feature. The refinement of .001 per inch applies to each and every inch along the length of the feature as an additional requirement to the overall straightness. The feature control frame with the diameter symbols controls the unit straightness of a derived median line since the tolerance zone is a cylinder.

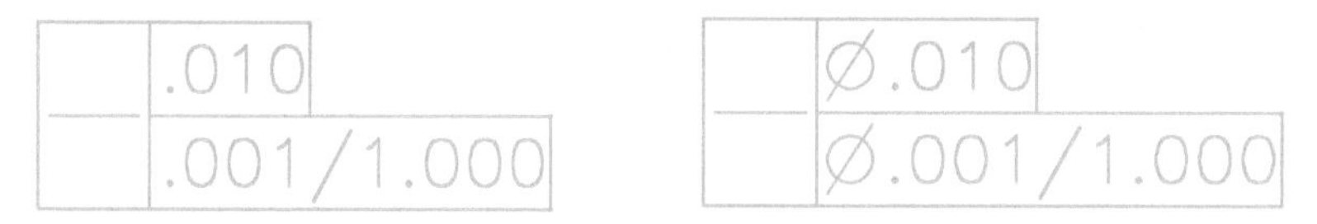

FIGURE 5-9 Unit straightness of a surface and unit straightness of a derived median line.

Circularity

Definition

Circularity has two definitions; one is for a surface of revolution, and the other is for a sphere. Circularity is:

- For a surface of revolution, all points of the surface intersected by a plane perpendicular to an axis or spine (curved line) are equidistant from that axis or spine.
- For a sphere, all points of the surface intersected by any plane passing through a common center are equidistant from that center.

Specifying Circularity Tolerance

Where specifying a circularity tolerance, a feature control frame is attached to the surface of the feature with a leader. The leader may be attached to the surface of a cylinder in the circular view or the longitudinal view, as shown in Fig. 5-10. For other surfaces of revolution, the leader is attached to the longitudinal view. The feature control frame contains a circularity symbol and a numerical tolerance. Normally, nothing else appears in a feature control frame controlling circularity except possibly the free state symbol. Circularity tolerance is a refinement of the size tolerance (Rule #1) and must be less than the size tolerance, except for parts subject to free state variation or the independency principle. If more information about circularity tolerance is required, a more complete discussion on the subject is available in the ANSI B89.3.1 *Measurement of Out-of-Roundness* and ASME Y14.5.1M *Mathematical Definition of Dimensioning and Tolerancing Principles.*

Interpretation

Circular elements in a plane perpendicular to the axis of the part on the surface being controlled must lie between two concentric circles, in which the radial distance between them is equal to the tolerance specified in the feature control frame. Each circular element is independent of every other circular element. That means that the part can look like a stack of pennies that is misaligned and yet can still satisfy a circularity inspection. Rule #1 requirements limit the misalignment of the stack of pennies.

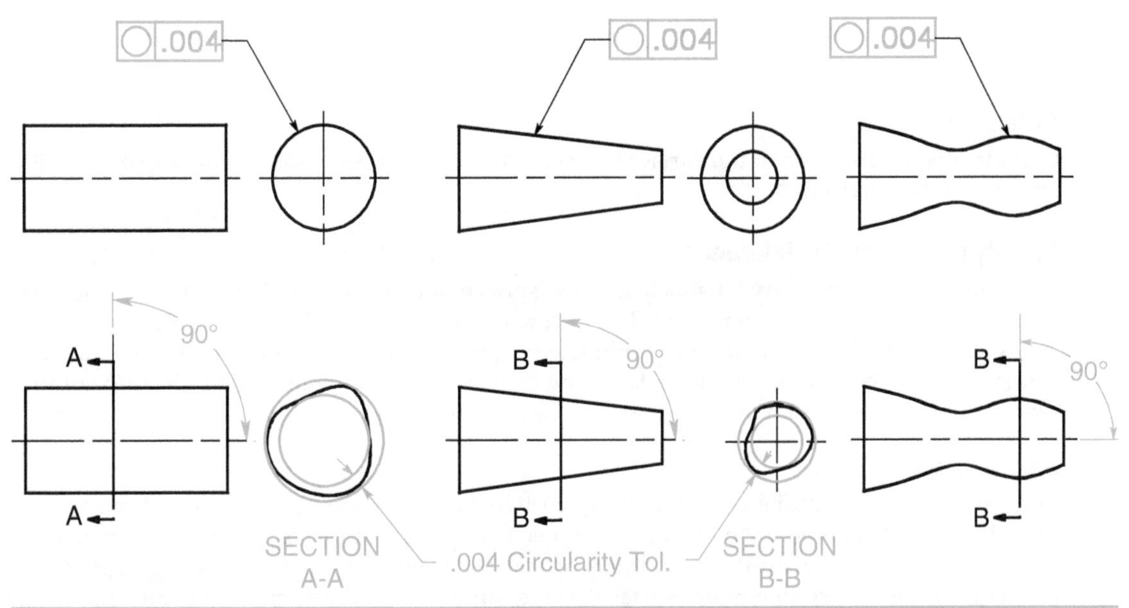

FIGURE 5-10 A circularity tolerance applied to a cylinder, a taper, and a profile of a surface of revolution.

Inspection

When verifying circularity, the feature of size is first measured to verify that it falls within the limits of size and Rule #1. Then the part is placed on the precision turntable of the circularity inspection machine and centered with the centering screws. The probe contacts the part while it is being rotated on the turntable. The path of the probe is magnified and plotted simultaneously on the polar graph as the part rotates. The circular path plotted on the polar graph in Fig. 5-11 falls within two circular elements on the graph. The measurement of this particular part is circular within a radial distance of .002.

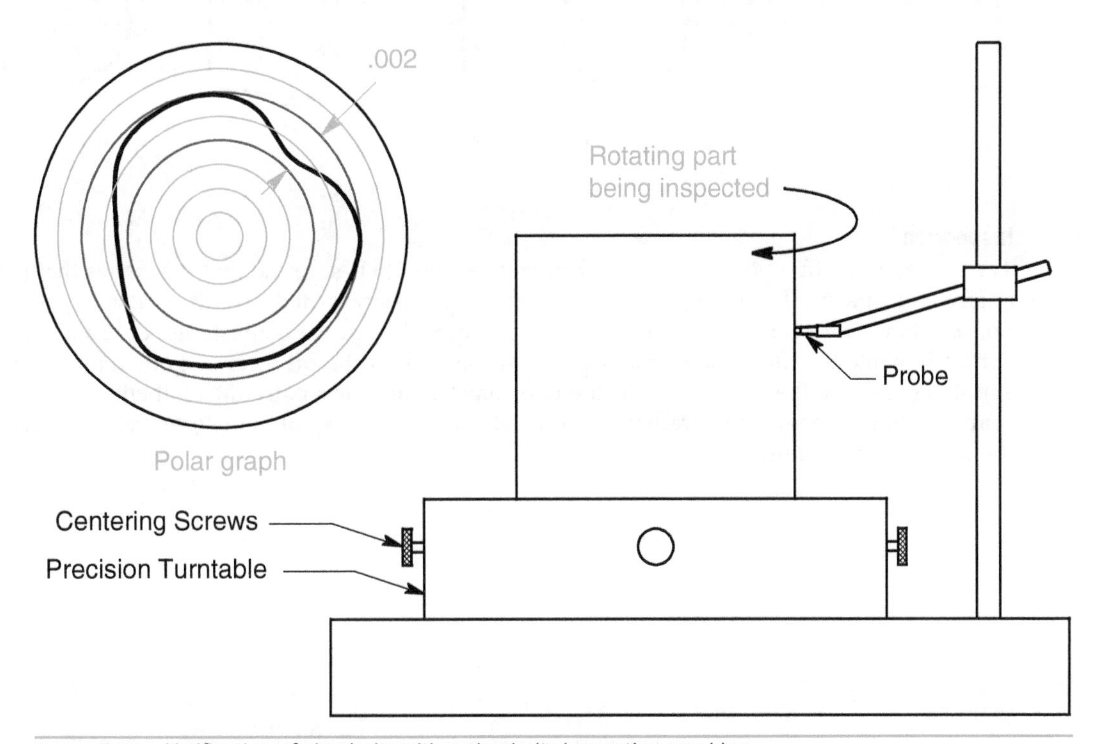

FIGURE 5-11 Verification of circularity with a circularity inspection machine.

Cylindricity

Definition

Cylindricity is a condition of the surface of a cylinder where all points of the surface are equidistant from a common axis.

Specifying Cylindricity Tolerance

A feature control frame may be attached to the surface of a cylinder with a leader in either the circular view or the rectangular view. The feature control frame contains a cylindricity symbol and a numerical tolerance. Normally, nothing else appears in a feature control frame controlling cylindricity, except possibly the free state symbol. Cylindricity tolerance is a refinement of the size tolerance (Rule #1) and must be less than the size tolerance.

Interpretation

The surface being controlled must lie between two coaxial cylinders in which the radial distance between them is equal to the tolerance specified in the feature control frame, as shown in Fig. 5-12. Unlike circularity, the cylindricity tolerance applies to both circular and longitudinal elements of the surface (the entire surface) at the same time. Cylindricity is a composite form tolerance that simultaneously controls circularity, straightness of a surface, and taper of cylindrical features.

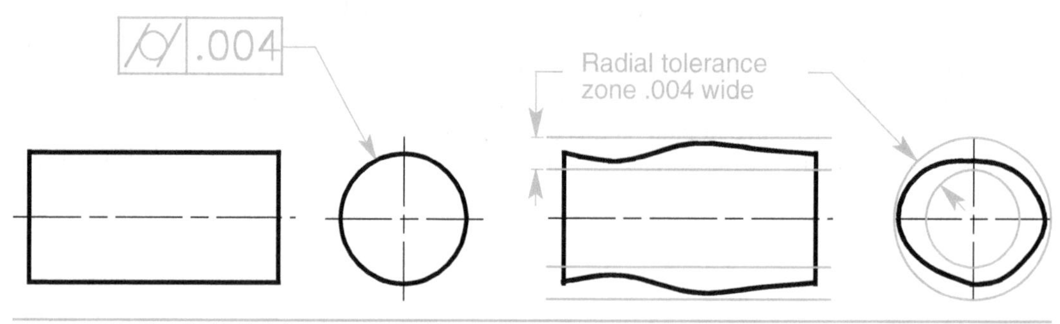

FIGURE 5-12 A cylindricity tolerance applied to a cylinder.

Inspection

The feature must first be measured at each cross section to determine that it satisfies the limits of size and Rule #1. Then the part is placed on the precision turntable of the circularity inspection machine and centered with the centering screws. The probe contacts the part and moves vertically while the turntable is rotating. The spiral path of the probe is magnified and plotted simultaneously on the polar graph as the part rotates. The spiral path must fall within two concentric cylinders in which the radial distance between them is equal to the tolerance specified in the feature control frame.

Average Diameter

An average diameter is the mathematical average of several diametric measurements across a circular or cylindrical feature. The size dimension and tolerance are followed by the abbreviation AVG indicating that the measurements are to be averaged. The individual measurements may violate the limits of size, but the average value must be within the limits of size. Enough measurements (at least four) should be taken to ensure the establishment of an average diameter. The circularity tolerance zone boundaries for the maximum and minimum averages are the averages plus and minus the circularity tolerance, .03, shown in Fig. 5-13 and in the table below.

	Maximum Average Dimension		Minimum Average Dimension	
Max. and min. average	20.02	20.02	20.00	20.00
Circularity tolerance	+ .03	− .03	+ .03	− .03
Upper and lower boundaries	20.05	19.99	20.03	19.97

If the average of measurements of a circular feature falls inside the average size tolerance range, the feature is in tolerance. If a particular part made from the drawing in Fig. 5-13 actually measures 20.03 in one direction and 19.99 in the other direction, the average diameter is 20.01 which falls within the average size tolerance range. Therefore, the average diameter is within the size tolerance. For clarity, only two measurements are shown here, but a minimum of four measurements must be taken to ensure accuracy. The upper and lower circularity boundaries of this feature are the average dimension, 20.01, plus and minus the circularity tolerance, of .03 which equals 20.04 and 19.98. Therefore, the circularity tolerance for this part is also satisfied.

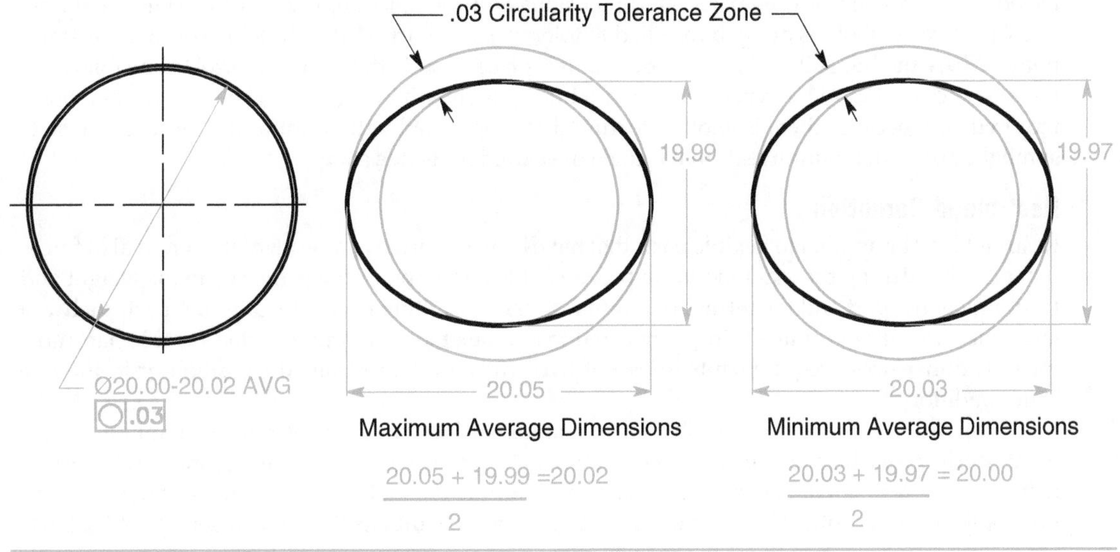

FIGURE 5-13 The circularity of this flexible part is to be measured in its free state. The free state symbol is not required here.

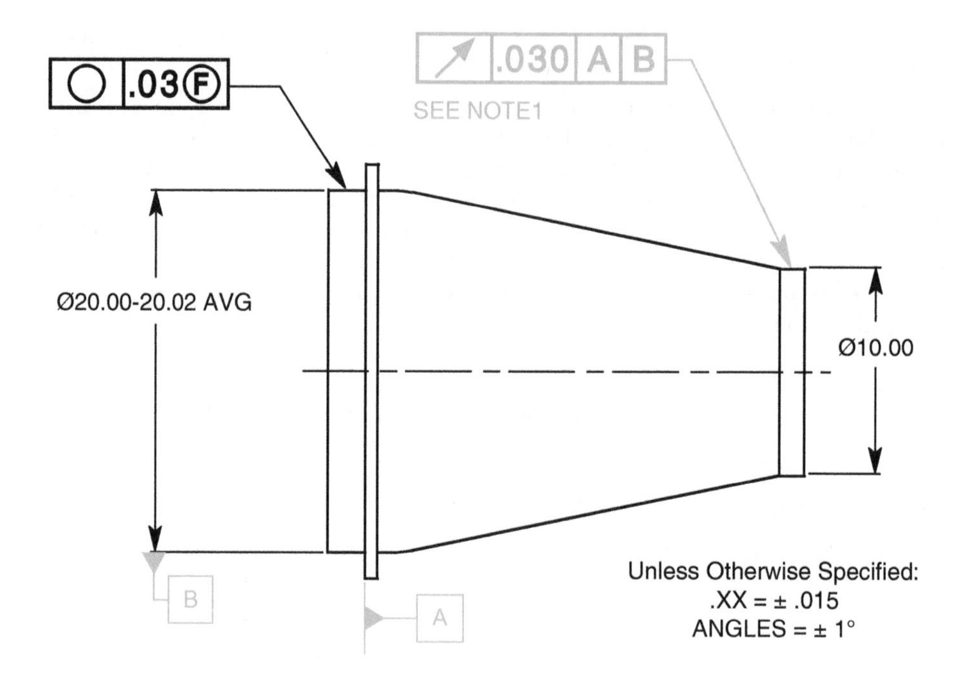

FIGURE 5-14 If specified in a note, a flexible part must be restrained before it is inspected.

Free State

Except for restrained flexible parts, all dimensions and tolerances apply in a free state condition. The free state symbol may only be applied to tolerances on parts that include one or more restraint notes. Where the free state symbol is required, it is placed inside the feature control frame following the tolerance or datum reference and any modifiers, as shown in Fig. 5-14. The free state symbol is required here because there is another feature that is specified with a restraint note. The free state symbol clarifies that datum feature B is to be measured in its free state.

Restrained Condition

Restraint may be applied to flexible parts that have been distorted as a result of the removal of forces that occurred during the manufacturing process. This distortion is due primarily to the weight and flexibility of the part and the release of internal stresses resulting from fabrication. A part, such as a large sheet metal tube or an O-ring, is referred to as a flexible or nonrigid part. A flexible part must meet its dimensional requirements in one of two ways: in the restrained condition or in the free state condition.

Where features are to be controlled for orientation or location in the restrained condition, a note must clearly state which features are to be restrained, how they are to be restrained, and to what extent they are to be restrained. Figure 5-14 contains an example of a note specifying the restrained condition for the runout control. The restrained condition should simulate the actual assembly of the part.

Summary

The surface controls of flatness, straightness, circularity, and cylindricity all share the same general requirements. Flatness of a surface and flatness of a median plane are two quite different controls. Table 5-5 compares some of these similarities and differences.

	⏥	⏤	Feature of Size ⏥	Feature of Size ⏤	○	⌭
1. For these controls, datums do not apply	X	X	X	X	X	X
2. For these controls, Rule #1 applies	X	X			X	X
3. These are surface controls	X	X			X	X
4. These controls may be specified with a leader	X	X			X	X
5. These are refinements of the size tolerance	X	X			X	X
6. These tolerances violate Rule #1			X	X		
7. These controls apply to features of size			X	X		
8. These controls are associated with the dimension			X	X		
9. These controls may exceed the size tolerance			X	X		
10. The Ø symbol is used				X		
11. The MMC (circle M) symbol may be used			X	X		

TABLE 5-5 Summary of the Application of Form Controls

Chapter Review

1. All form tolerances are independent of all other ______________________.

2. No ______________________ apply to form tolerances.

3. The form of individual features of size is automatically controlled by the ______________ ______________________.

4. Where the size tolerance does not sufficiently control the form of a feature, a form tolerance may be specified as a ______________________.

5. All form tolerances are surface controls except for ______________________ ______________________.

6. No ______________________ or ______________________ are appropriate for surface controls.

7. Flatness of a surface or derived median plane is a condition where all line elements of that surface are in ______________________.

8. For flatness, in a view where the surface to be controlled appears as a ________________ ________________________, a feature control frame is attached to the surface with a ____________________________________.

9. The feature control frame controlling flatness contains a ____________________________ and a ____________________________________.

10. The surface being controlled for flatness must lie between ______________________ separated by the flatness tolerance. In addition, the feature must fall within the ____________________________________.

11. The flatness tolerance zone does not need to be ____________________________________ to any other surface.

12. The feature of size may not exceed the ____ ____.

13. Specify a flatness tolerance of .006 in a feature control frame to the top surface of the part in Fig. 5-15.

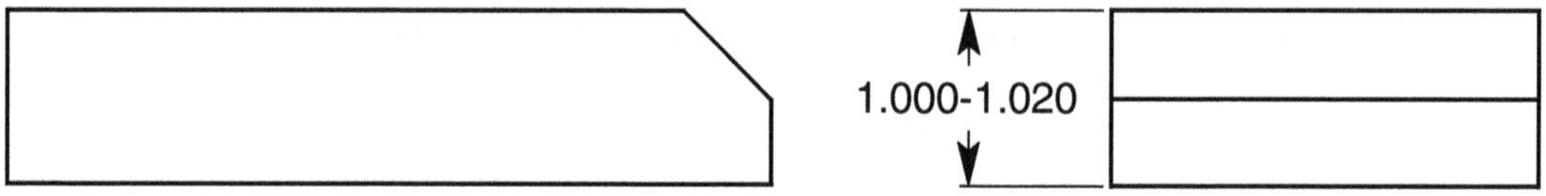

1.000-1.020

FIGURE 5-15 Specifying flatness: Question 13.

14. Draw a feature control frame below with an overall flatness of .015 and a unit flatness of .001 per square inch.

15. When verifying flatness, the feature of size is first measured to verify that it falls within the ____________________________________.

16. Then the surface is adjusted with jackscrews to remove any ________________ error.

17. Flatness verification is achieved by measuring the surface in all directions with a ________________.

18. Straightness is a condition where an element of a ________________ ________________ is a straight line.

19. In a view where the line elements to be controlled appear as a ________________, a feature control frame is attached to the surface with a ________________ ________________.

20. Straightness tolerance is a refinement of the ________________ and must be less than the ________________.

21. Complete Table 5-6 specifying the straightness tolerance and what controls it for the drawing in Fig. 5-6.

Actual Part Size	Straightness Tolerance	Controlled By
1.020		
1.018		
1.016		
1.014		
1.010		
1.005		
1.000		

TABLE 5-6 Control of Straightness: Question 21

22. The measurement of surface variation for straightness is performed similar to the measurement for ________________.

23. Each line element is ________________ of every other line element.

24. Where a feature control frame with a straightness tolerance is associated with a size dimension, the straightness tolerance applies to ________________ ________________.

25. While each actual local size of the feature must fall within the ______________________,

    the feature controlled with straightness of a median line may exceed the ______________

    ______________________________________ at MMC due to bending.

26. A straightness control of a median line will allow the feature to violate ______________

    ______________________________________.

27. If specified at MMC, the total straightness tolerance of a median line equals the tolerance

    in the feature control frame plus any ______________________________________.

28. Complete Table 5-7 specifying the appropriate tolerances for the sizes given.

Feature Size	Cylindrical Feature (Straightness of a Median Line)	
	— Ø.006	— Ø.006 Ⓜ
1.020 MMC		
1.015		
1.010		
1.005		
1.000 LMC		

TABLE 5-7 Straightness Tolerance: Question 28

29. Straightness verification of a feature of size specified at MMC can be achieved by

    ______________________________________.

30. Straightness verification of a feature of size specified at ______________________________ cannot be achieved by placing the part in a full form functional gage.

31. Circular elements must lie between two ______________________________ in which

    the ______________________________ between them is equal to the tolerance specified in the circularity feature control frame.

32. When verifying circularity, the feature of size is first measured to verify that it falls within

    the ______________________________ and ______________________________.

33. Circularity can be accurately inspected on a ______________________________.

34. Cylindricity is a condition of the surface of a cylinder where all points of the surface are

    ____.

35. The cylindricity tolerance consists of two ______________________ in which the ______________________ between them is equal to the tolerance specified in the ______________________.

36. Cylindricity is a ______________________ form tolerance that simultaneously controls ______________________ of cylindrical features.

37. In Table 5-8, place an X under the control that agrees with the statement.

	⏥	—	Feature of Size ⏥	Feature of Size —	○	⌭
1. For these controls, datums do not apply						
2. For these controls, Rule #1 applies						
3. These are surface controls						
4. These controls may be specified with a leader						
5. These are refinements of the size tolerance						
6. These tolerances violate Rule #1						
7. These controls apply to features of size						
8. These controls are associated with the dimension						
9. These controls may exceed the size tolerance						
10. The Ø symbol is used						
11. The MMC (circle M) symbol may be used						

TABLE 5-8 The Application of Form Controls: Question 37

38. Except for restrained flexible parts, all dimensions and tolerances apply in the ______________________.

39. A minimum of ______________________ must be taken to ensure the accuracy of an average diameter.

40. Restraint may be applied to flexible parts due to the distortion of a part after the removal of forces applied during the ______________________.

41. The restrained condition should simulate ______________________.

Problems

FIGURE 5-16 Flatness: Prob. 1.

1. Specify a flatness control of .005 for the top surface of the part in Fig. 5-16.

2. Below, specify a feature control frame with a unit flatness of .003 per square inch and an overall flatness of .015.

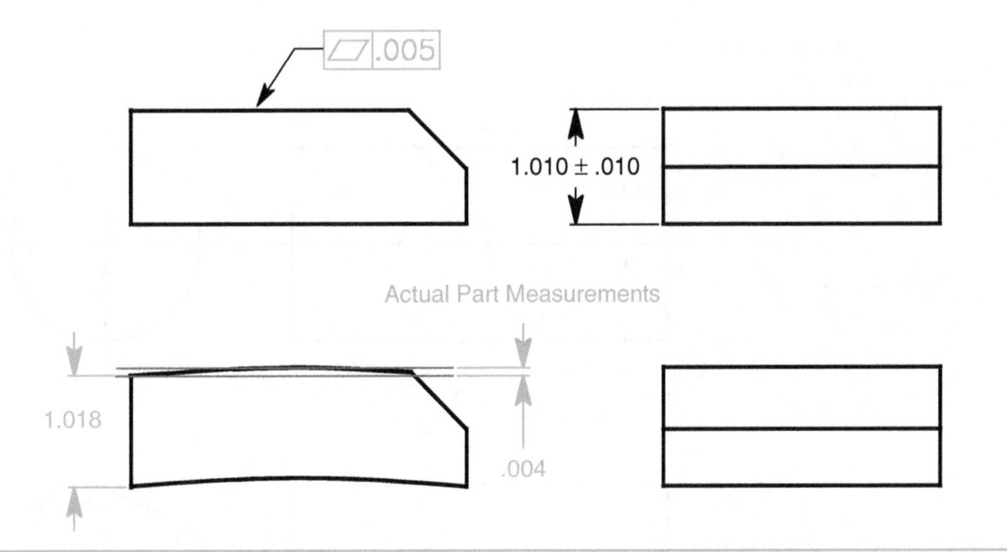

FIGURE 5-17 Flatness check: Prob. 3.

3. Is the part in Fig. 5-17 an acceptable part? Why or why not?

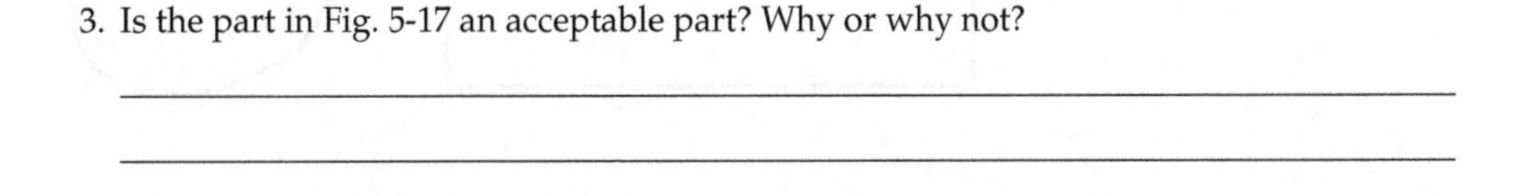

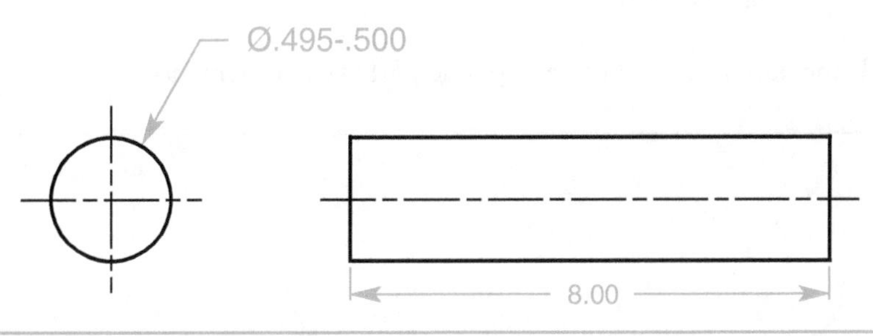

FIGURE 5-18 Straightness of a surface: Prob. 4.

4. Specify straightness of a surface of .002 on the cylinder in the drawing in Fig. 5-18. Draw and dimension the tolerance zone on the drawing.

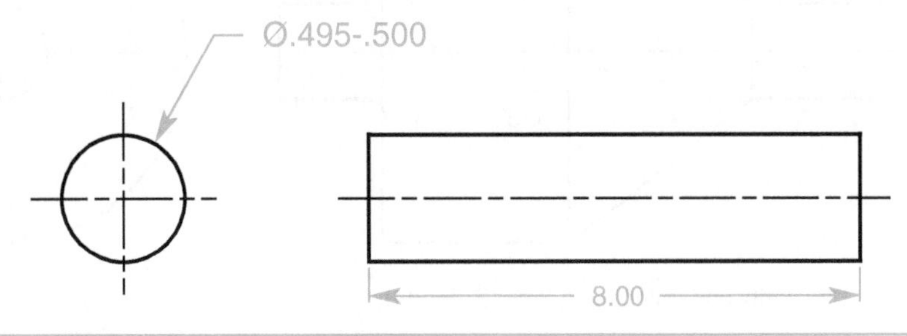

FIGURE 5-19 Straightness of a median line: Prob. 5.

5. On the cylinder in Fig. 5-19, specify straightness of a median line of .010 at MMC. Draw and dimension the tolerance zone on the drawing.

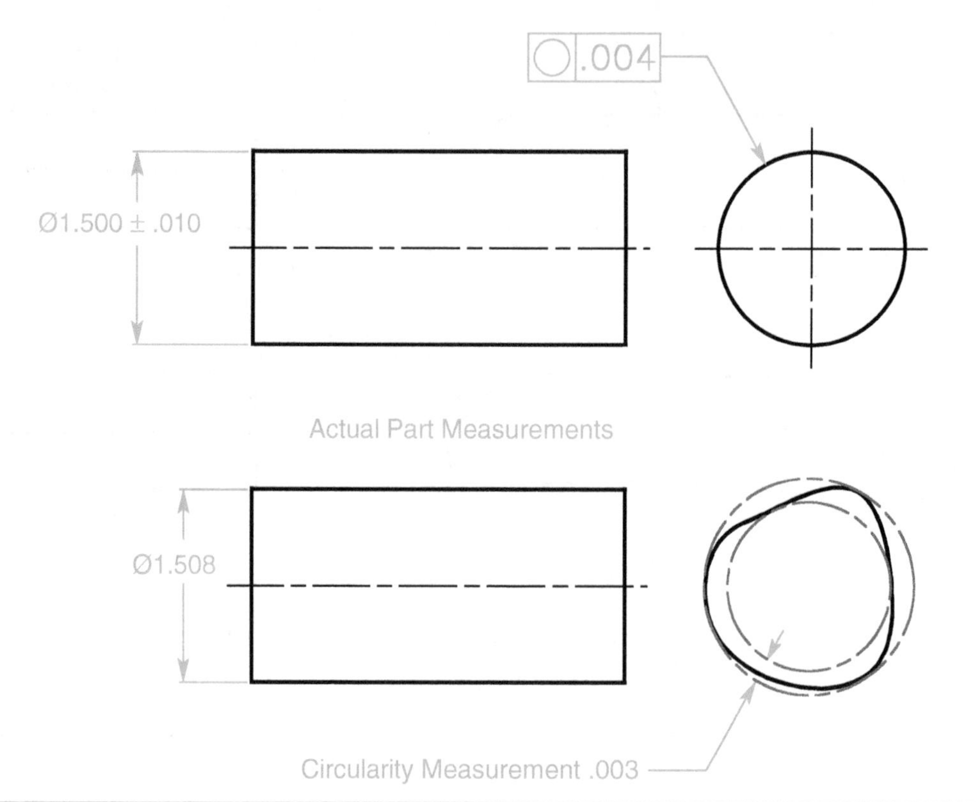

FIGURE 5-20 Circularity: Prob. 6.

6. Is the part in Fig. 5-20 an acceptable part? Why or why not?

____.

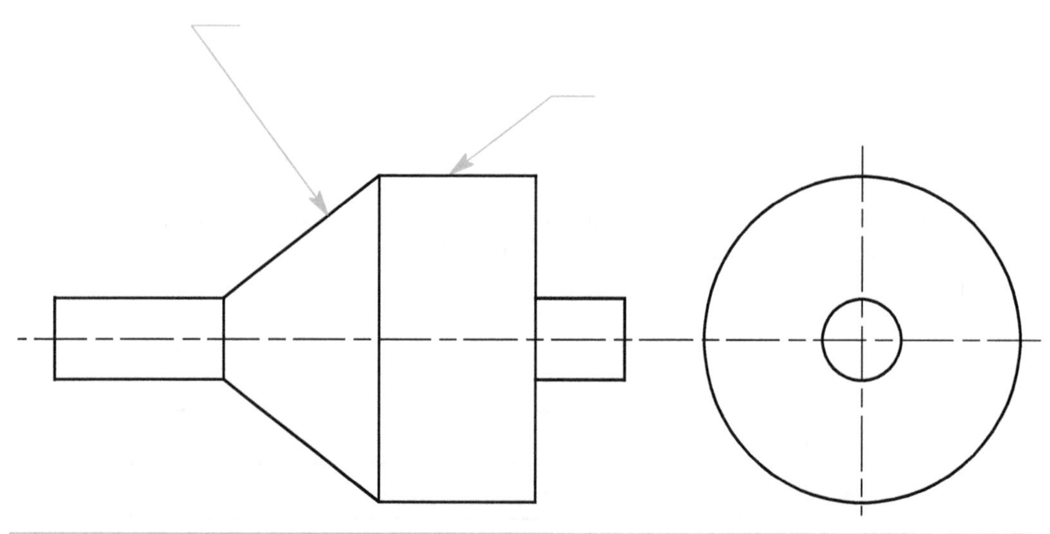

FIGURE 5-21 Circularity and cylindricity: Probs. 7 and 8.

7. Specify a circularity tolerance of .002 on the cone in the drawing in Fig. 5-21.
8. Specify a cylindricity tolerance of .0005 on the large cylinder on the drawing in Fig. 5-21.

CHAPTER 6

Orientation

Orientation is the general term used to describe the angular relationship between features. Orientation controls include perpendicularity, parallelism, angularity, and, in some cases, profile. All orientation controls must have datum features. It makes no sense to specify a pin, for instance, to be perpendicular. The pin must be perpendicular to some other feature. The other feature is the datum feature.

Chapter Objectives

After completing this chapter, the learner will be able to:

- *Specify* tolerances that will control flat surfaces perpendicular, parallel, and at some basic angle to a datum feature(s)
- *Specify* tolerances that will control axes perpendicular, parallel, and at some basic angle to a datum feature(s)

The orientation of a plane surface controlled by two parallel planes and an axis controlled by a cylindrical tolerance zone will be discussed in this chapter. When a plane surface is controlled with a tolerance zone consisting of two parallel planes, the entire surface must fall between the two planes. Since perpendicularity, parallelism, angularity, and profile control the orientation of a plane surfaces with a tolerance zone of two parallel planes, they also control flatness if a flatness tolerance is not specified. When it is desirable to control only the orientation of individual line elements of a surface, a note, such as EACH ELEMENT or EACH RADIAL ELEMENT, is placed beneath the feature control frame.

When a cylindrical tolerance zone controls an axis, the entire axis must fall inside the tolerance zone. Although axes and center planes of features of size may be oriented using two parallel planes, in most cases, they will be controlled by other controls, such as the position control, and will not be discussed in this chapter. All location controls are composite controls that control both location and orientation at the same time. Parallelism, perpendicularity, and angularity are often used to refine the orientation of other controls, such as the position control.

Perpendicularity

Definition

Perpendicularity is the condition of a surface, axis, or center plane that is at a 90° angle to a datum plane or datum axis.

Specifying Perpendicularity of a Flat Surface

In a view where the surface to be controlled appears as a line, a feature control frame is attached to the surface with a leader or extension line, as shown in Fig. 6-1. The feature control frame contains a perpendicularity symbol, a numerical tolerance, and at least one datum feature. The datum feature is identified with a datum feature symbol.

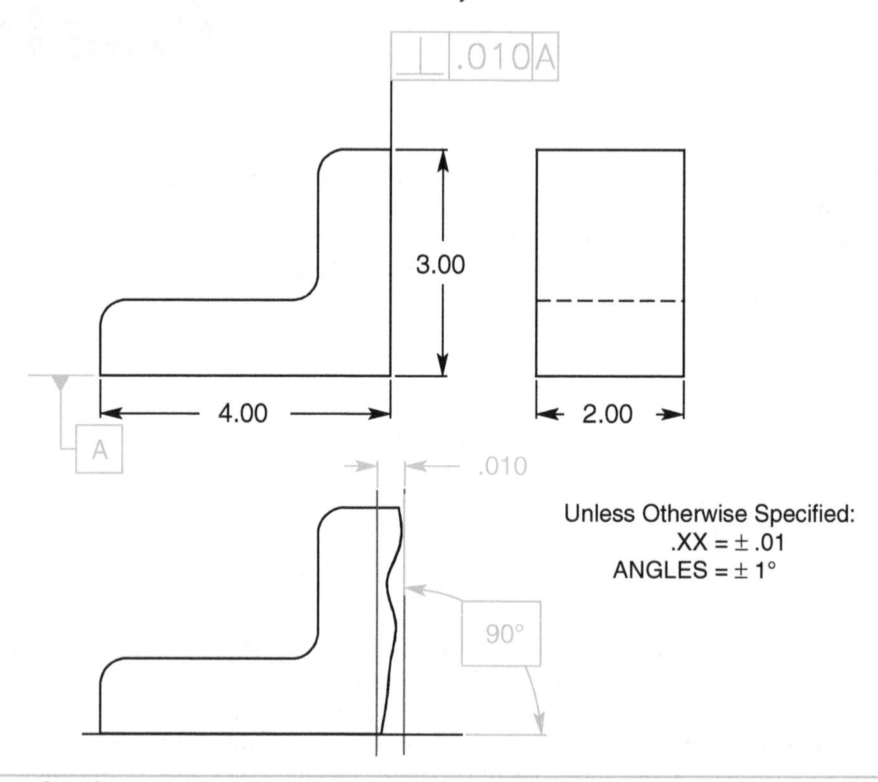

FIGURE 6-1 Specifying a plane surface perpendicular to a datum plane.

Interpretation

The surface being controlled must lie between two parallel planes, perpendicular to the datum plane, separated by the perpendicularity tolerance of .010 specified in the feature control frame. All features of size of the part must fall within the limits of size and may not exceed the boundary of perfect form at maximum material condition (MMC), Rule #1. There is no boundary of perfect orientation at MMC for perpendicularity. All of the other 90° angles on the part also have a tolerance. The title block angularity tolerance controls all angles, including 90° angles, that are not otherwise toleranced. Since this perpendicularity tolerance applies to a surface, no material condition symbol applies.

Inspection

When verifying the perpendicularity of a flat surface, the datum feature is clamped to an angle plate that sits on a surface plate. Then, as shown in Fig. 6-2, perpendicularity verification is achieved by using a dial indicator to measure the surface in all directions to determine that any variation does not exceed the tolerance specified in the feature control frame.

The Tangent Plane

The tangent plane symbol, circle T, in the feature control frame specifies that the tolerance applies to a precision plane contacting the high points of the surface. Even though the surface irregularities exceed the perpendicularity tolerance, if a precision plane contacting the high points of a surface falls inside the specified tolerance zone, the surface is in tolerance. The surface irregularities

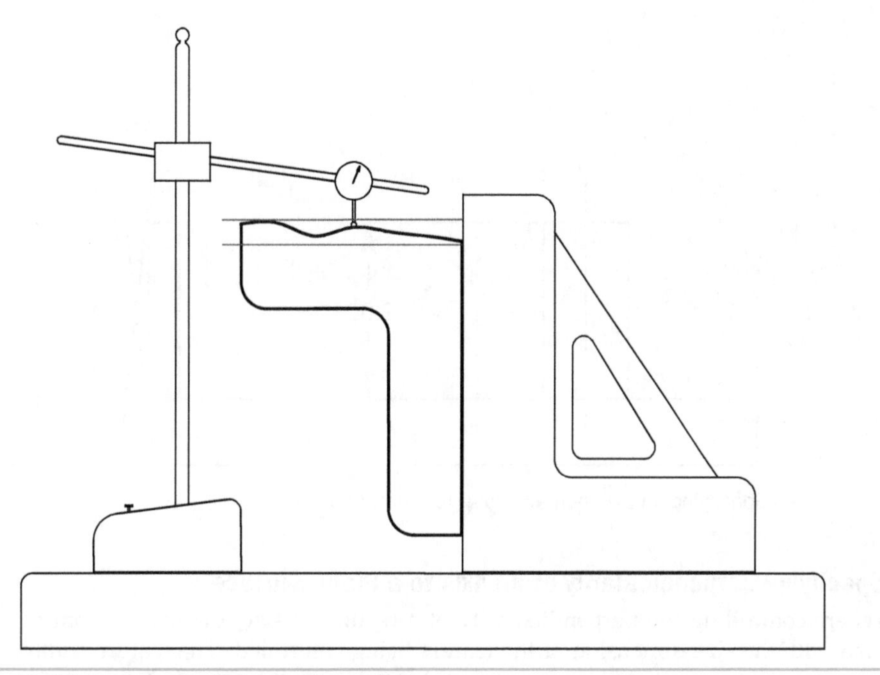

FIGURE 6-2 Verifying perpendicularity of a flat surface.

in Fig. 6-3 exceed the perpendicularity tolerance, but the tangent plane lies inside the tolerance zone. The circle T modifier maintains a small orientation tolerance but allows flatness to be a larger tolerance controlled by Rule #1. The tangent plane symbol may be applied to any orientation control of a surface. The tangent plane concept allows the acceptance of more parts.

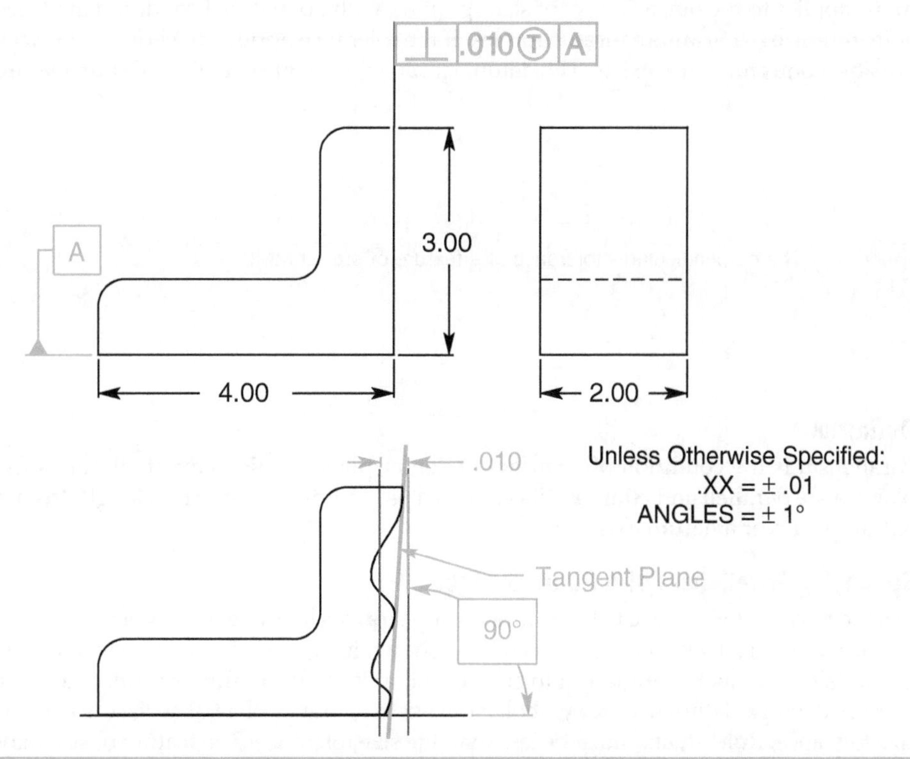

FIGURE 6-3 The tangent plane symbol specified in the feature control frame.

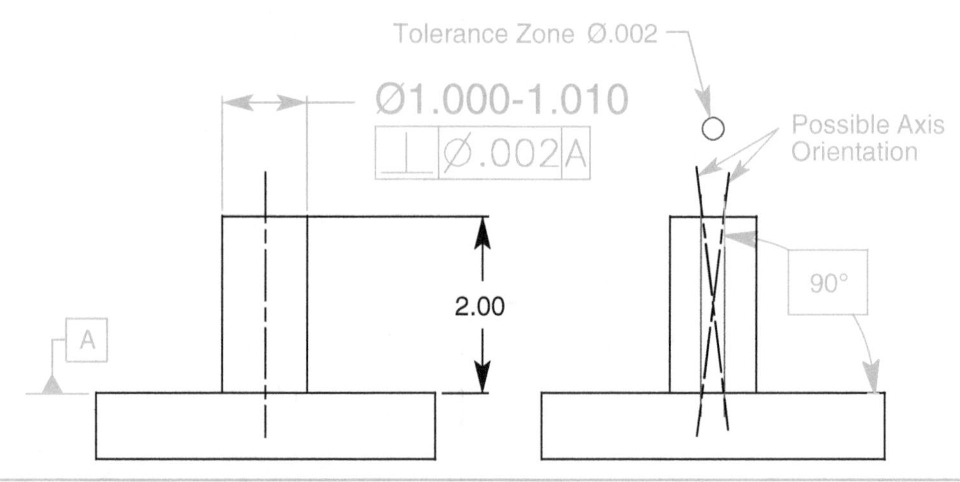

FIGURE 6-4 Specifying an axis perpendicular to a datum plane.

Specifying Perpendicularity of an Axis to a Plane Surface

Where controlling the perpendicularity of a feature of size, the feature control frame is associated with the size dimension of the feature being controlled. The feature control frame contains a perpendicularity symbol, a numerical tolerance, and at least one datum feature. If the element being controlled is an axis, the numerical tolerance is usually preceded by a diameter symbol, as shown in Fig. 6-4. A cylindrical tolerance zone that controls an axis perpendicular to a plane surface is perpendicular to that surface in all directions around the axis. There are some cases where an axis is controlled by two parallel planes, but these are very uncommon and would probably be toleranced with the position control employing a cylindrical tolerance zone. The perpendicularity tolerance may be larger or smaller than the size tolerance. Since the tolerance in the feature control frame applies to the pin, a feature of size specified with no material condition modifier, regardless of feature size (RFS) automatically applies. If the tolerance applies at MMC as shown in Fig. 6-5, a possible bonus tolerance exists. The datum feature(s) is identified with a datum feature symbol(s).

FIGURE 6-5 The perpendicularity tolerance of a feature of size at MMC.

Parallelism

Definition

Parallelism is the condition of a surface or center plane, equidistant at all points from a datum plane; also, parallelism is the condition of an axis, equidistant along its length from one or more datum planes or a datum axis.

Specifying Parallelism of a Plane Surface

In a view where the surface to be controlled appears as a line, a feature control frame is attached to the surface with a leader or extension line, as shown in Fig. 6-6. The feature control frame contains a parallelism symbol, a numerical tolerance, and at least one datum feature. The datum surface is identified with a datum feature symbol. Parallelism tolerance of a flat surface is a refinement of the size tolerance, Rule #1, and must be less than the size tolerance. The feature of size may not exceed the MMC boundary, and the thickness at each actual local size must fall within the limits of size.

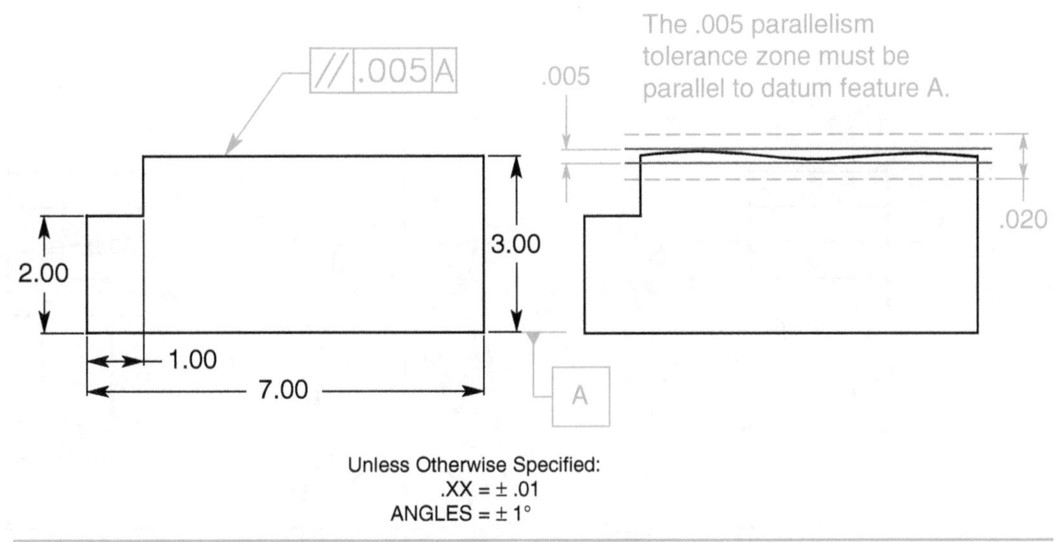

FIGURE 6-6 Specifying a plane surface parallel to a plane surface.

Interpretation

The surface being controlled in Fig. 6-6 must lie between two parallel planes, parallel to the datum plane, separated by the parallelism tolerance of .005 specified in the feature control frame. In addition, the surface must fall within the size tolerance, the two parallel planes .020 apart. The entire part in Fig. 6-6 must fit between two parallel planes 3.010 apart. The controlled surface may not exceed the boundary of perfect form at MMC, Rule #1. Where applied to a flat surface, parallelism is the only orientation control that requires a perfect angle (parallelism is a 0° angle) at MMC. No material condition symbol applies because both the parallelism control and the datum feature are plane surfaces.

Inspection

Verifying the parallelism of a plane surface is relatively easy. First, the feature of size is measured to determine that it falls within the limits of size. Next, the datum surface is placed on top of the surface plate. Then verification is achieved, as shown in Fig. 6-7, by using a dial indicator to measure the surface in all directions to determine that any variation does not exceed the tolerance specified in the feature control frame.

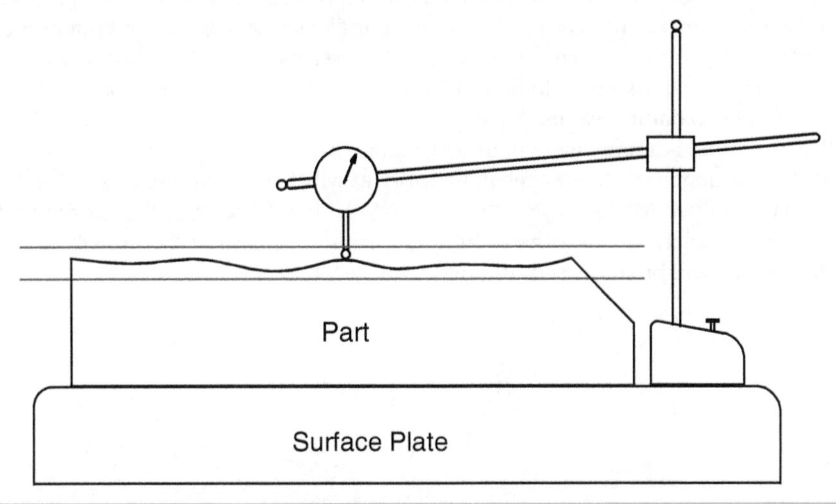

FIGURE 6-7 Verifying parallelism of a plane surface.

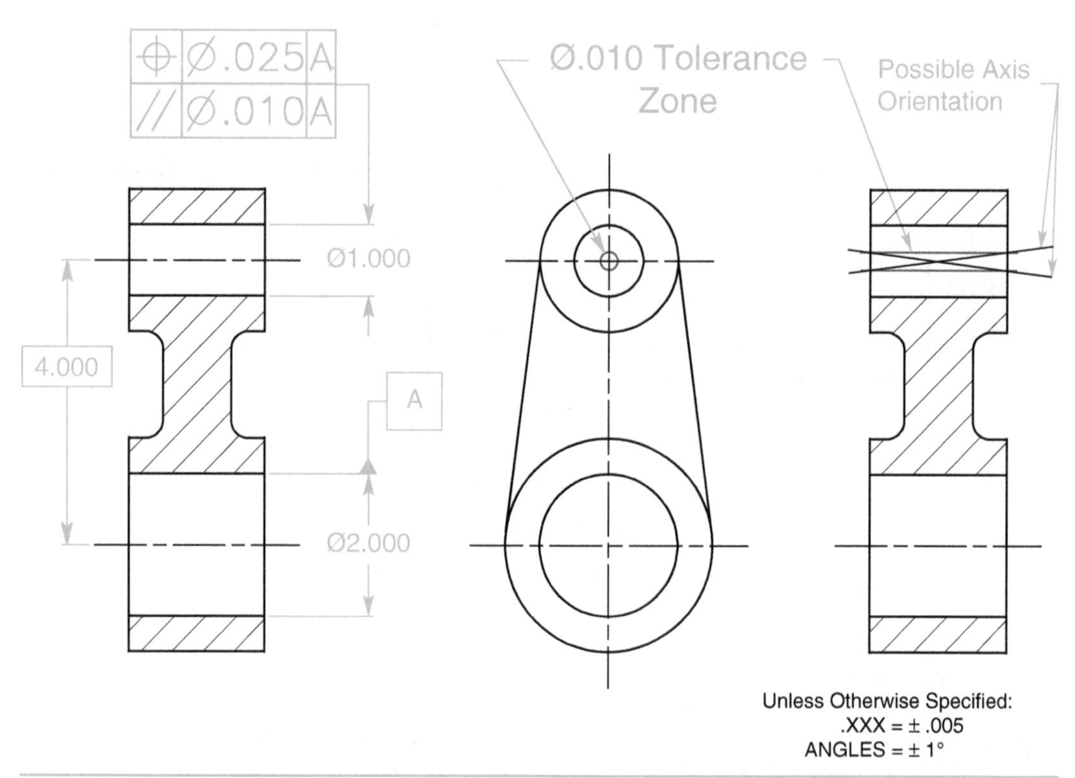

FIGURE 6-8 Controlling the axis of a feature of size parallel to a datum feature of size.

Specifying Parallelism of an Axis

Where controlling the parallelism of a feature of size, the feature control frame is associated with the size dimension of the feature being controlled. In Fig. 6-8, the parallelism feature control frame is attached to and is a refinement of the position control that is connected to an extension of the dimension line. The feature control frame contains a parallelism symbol, a numerical tolerance, and at least one datum feature. If the element being controlled is an axis, the numerical tolerance is usually preceded by a diameter symbol, as shown in Fig. 6-8. The controlled feature is a feature of size, and it applies at RFS since no material condition modifier is specified. The datum feature is identified with a datum feature symbol.

A parallelism tolerance may be used by itself to control the axis of a feature parallel to a datum feature. However, it makes no sense to control the parallelism of a feature but not its location. The position control in Fig. 6-8 first orients and then locates the axis of the controlled feature with a cylindrical tolerance zone .025 in diameter to datum feature A; the parallelism control refines the orientation of the axis parallel to datum feature A within a smaller cylindrical tolerance zone .010 in diameter. The parallelism control is a refinement of the orientation aspect of the position control.

All three orientation tolerances may apply at MMC for features of size, and datum features of size may apply at MMB, as shown in Fig. 6-9. If this is the case, the geometric tolerance has a possible bonus tolerance, and the datum feature has a possible shift tolerance. Both bonus and shift tolerances will be discussed in more detail in Chap. 7.

FIGURE 6-9 The parallelism tolerance at MMC and the datum feature at MMB.

Angularity

Definition

Angularity is the condition of a surface, axis, or center plane at any specified angle from a datum plane or datum axis.

Specifying Angularity of a Plane Surface

In a view where the surface to be controlled appears as a line, a feature control frame is attached to the surface with a leader or extension line. If an extension line is used, it only needs to contact the feature control frame at a corner, as shown in Fig. 6-10. The feature control frame contains an angularity symbol, a numerical tolerance, and at least one datum feature. The numerical tolerance for the surface being controlled is specified as a linear dimension because it generates a uniform shaped tolerance zone. A plus or minus angularity tolerance is not used because it generates a nonuniform, fan shaped tolerance zone. The datum feature is identified with a datum feature symbol.

Interpretation

The surface being controlled in Fig. 6-10 must lie between two parallel planes separated by the angularity tolerance of .010 specified in the feature control frame. The tolerance zone must be at the specified basic angle of 30° to the datum plane. All features of size of the part must fall

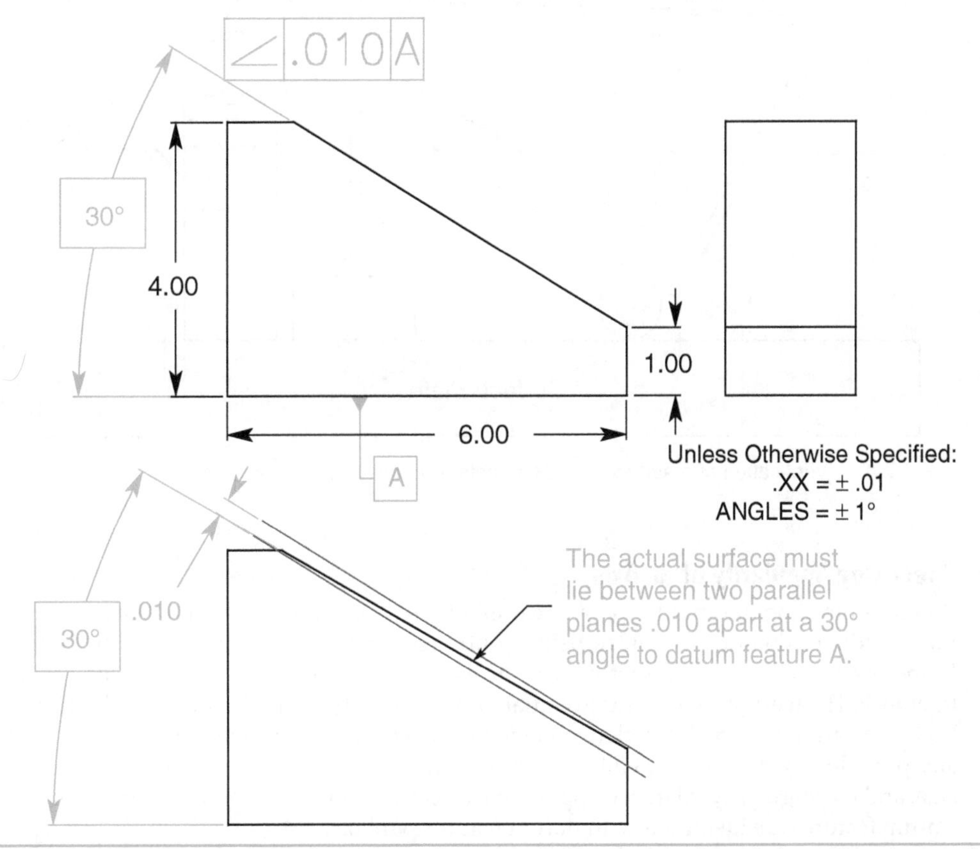

FIGURE 6-10 Specifying an angularity tolerance for a plane surface at a basic angle to another plane surface.

within the limits of size and may not exceed the boundary of perfect form at MMC, Rule #1. There is no boundary of perfect orientation at MMC for angularity. The 90° angles on the part also have a tolerance. The title block angularity tolerance controls all angles, including 90° angles, unless otherwise specified. Since the angularity control applies to a surface, no material condition symbol applies.

Inspection

The datum surface may be placed on a sine plate. The sine plate sits on a surface plate at a very accurate 30° angle produced by a stack of gage blocks. The basic angle between the tolerance zone and the sine plate is assumed to be perfect. Inspection equipment is not perfect, but inspection instrument error is very small compared to the geometric tolerance. As shown in Fig. 6-11, once the datum surface is positioned at the specified angle, angularity verification is achieved by using a dial indicator to measure the surface in all directions to determine that any variation does not exceed the tolerance specified in the feature control frame.

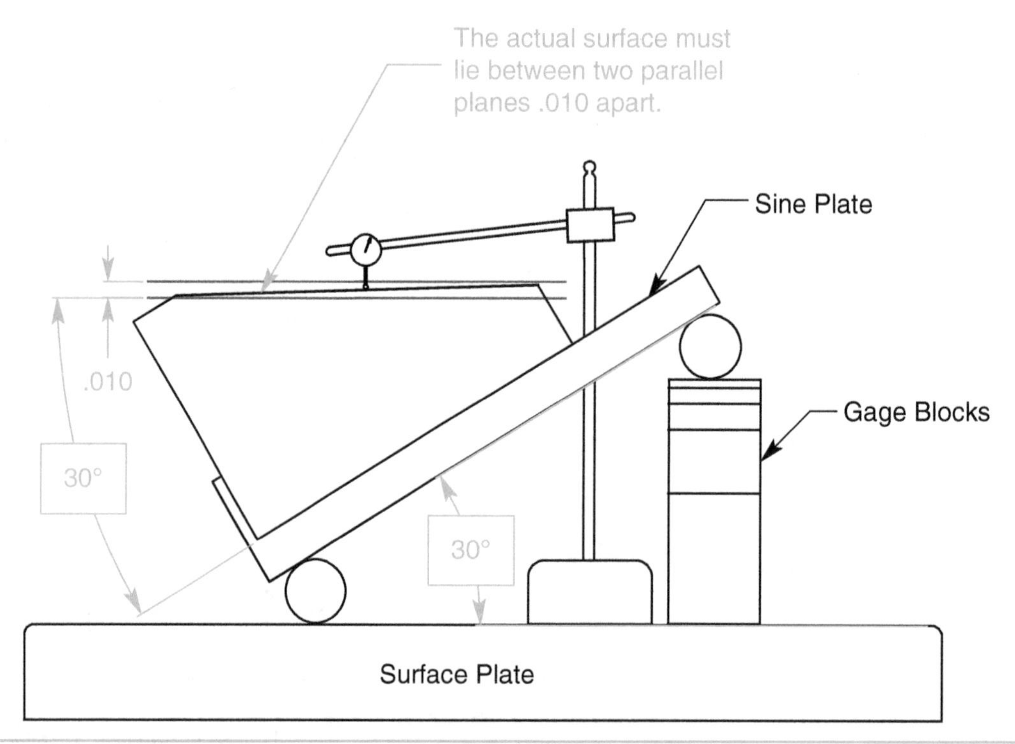

FIGURE 6-11 Verification of a surface at a 30° angle to a flat datum surface.

Specifying Angularity of an Axis

Where controlling the angularity of a feature of size, the feature control frame is associated with the size dimension of the feature being controlled. In Fig. 6-12, the angularity feature control frame is attached to and is a refinement of the position control that is associated with the size tolerance. The feature control frame contains an angularity symbol, a numerical tolerance, and at least one datum feature. If the element being controlled is an axis, the numerical tolerance is usually preceded by a diameter symbol, as shown in Fig. 6-12. The controlled feature is a feature of size, and the angularity tolerance applies at RFS since no material condition modifier is specified. Datum features are identified with datum feature symbols.

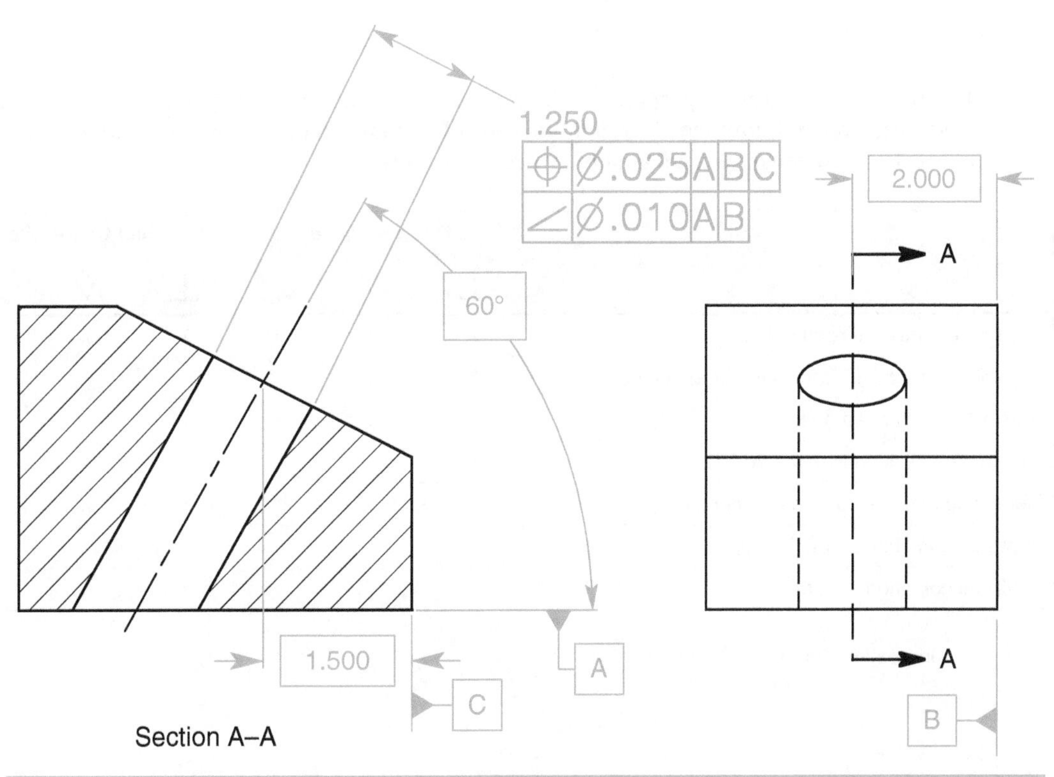

FIGURE 6-12 Specifying an axis at a basic angle to a datum plane.

An angularity tolerance may be used by itself to control the axis of a feature at some angle to a datum feature. However, it makes no sense to control the angularity of a feature and not control its location. The position control in Fig. 6-12 first orients the feature at a basic 60° angle to datum feature A and parallel to datum feature B, then it locates the feature to datum features B and C, all within a cylindrical tolerance zone .025 in diameter. The angularity control is a refinement of the orientation of the axis at a basic 60° angle to datum feature A and parallel to datum feature B within a smaller cylindrical tolerance zone .010 in diameter. The angularity control is a refinement of the orientation aspect of the position control.

If the design requires an angularity refinement of the position tolerance and will allow all of the tolerance available, the angularity tolerance may be specified at MMC. If the angularity tolerance applies at MMC, as shown in Fig. 6-13, it has a possible bonus tolerance.

Alternative Practice

As an alternative practice, the angularity symbol may be used to control parallel and perpendicular relationships.

FIGURE 6-13 The angularity tolerance specified at MMC is shown as a refinement to the position control.

Summary

The orientation controls of perpendicularity, parallelism, and angularity of flat surfaces all share the same general requirements. Orientation controls of features of size are quite different controls. Table 6-1 compares some of these similarities and differences.

	Plane Surfaces			Axes and Center Planes		
	⊥	//	∠	⊥	//	∠
Datum features are required	X	X	X	X	X	X
Controls flatness if flatness is not specified	X	X	X			
Circle T modifier can apply	X	X	X			
Tolerance specified with a leader or extension line	X	X	X			
Tolerance associated with a dimension				X	X	X
Material condition modifiers apply				X	X	X
A virtual condition applies				X	X	X

TABLE 6-1 Orientation Summary

Chapter Review

1. Orientation is the general term used to describe the ______________________ relationship between features.

2. Orientation controls include ____

____.

3. All orientation controls must have ______________________________________.

4. In a view where the surface to be controlled appears as a line, the perpendicularity feature control frame is attached to the surface with a ______________________________.

5. The datum feature is identified with a ______________________________________.

6. A surface being controlled with a perpendicularity tolerance must lie between ________ ______________________________ separated by the perpendicularity tolerance specified in the feature control frame. The tolerance zone must also be ______________________________ to the datum plane.

7. A tangent plane symbol, circle T, in the feature control frame specifies that the tolerance applies to a precision plane contacting the ______________________________ of the surface.

8. Where controlling the perpendicularity of a feature of size, the feature control frame is associated with the ________________________________ of the feature being controlled.

9. If the tolerance in the feature control frame applies to a feature of size and no material condition symbol is specified, ____________________________ automatically applies.

10. If the tolerance applies at MMC, then a possible ____________________________ tolerance exists.

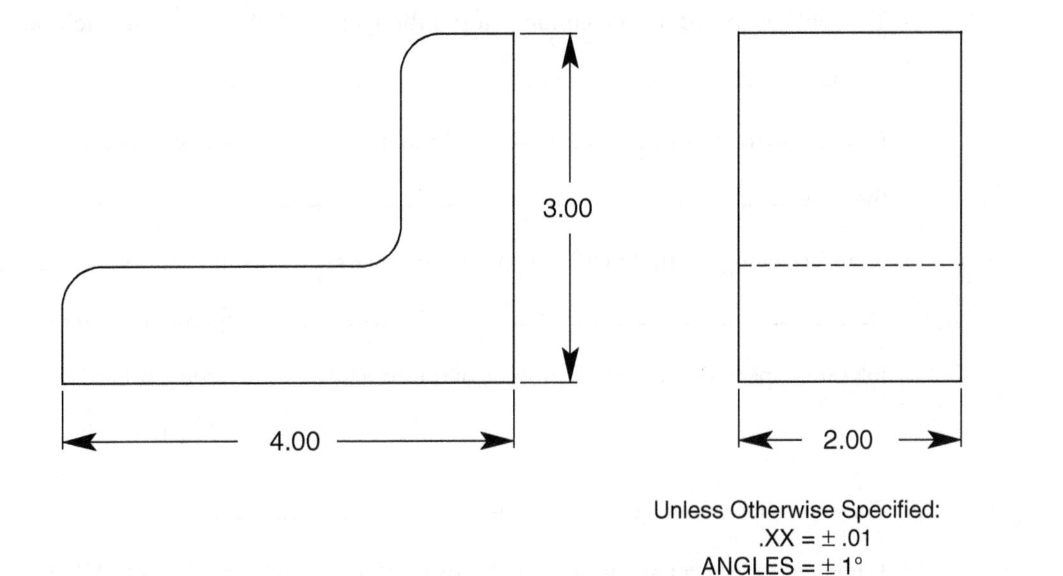

FIGURE 6-14 Specifying perpendicularity of a plane surface: Question 11.

11. Supply the appropriate geometric tolerance on the drawing in Fig. 6-14 to control the 3.00-inch vertical surface of the part perpendicular to the bottom surface within .005.

12. Supply the appropriate geometric tolerance on the drawing in Fig. 6-15 to control the 1.00-inch-diameter vertical pin perpendicular to the bottom surface of the plate within .005 at RFS.

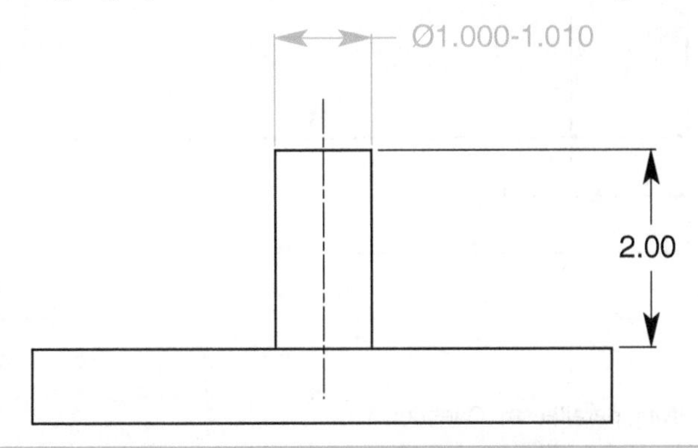

FIGURE 6-15 Specifying perpendicularity of a feature of size: Question 12.

FIGURE 6-16 Perpendicularity specified at MMC: Question 13.

13. If the pin in Fig. 6-15 were produced at a diameter of 1.004 and toleranced with the feature control frame in Fig. 6-16, what would the total perpendicularity tolerance be?

    ____

14. The feature control frame contains a parallelism symbol, a numerical tolerance, and at least one ______________________________.

15. Parallelism tolerance of a flat surface is a refinement of the size tolerance and must be less than the ______________________________.

16. A surface being controlled with a parallelism tolerance must lie between ______________ ______________________________ separated by the parallelism tolerance specified in the feature control frame. The tolerance zone must also be ______________ ______________________________ to the datum feature.

17. The controlled surface may not exceed the ______________________________.

18. Where applied to a flat surface, parallelism is the only orientation control that requires perfect orientation (parallelism is a 0° angle) at ______________________________.

19. Supply the appropriate geometric tolerance on the drawing to control the top surface of the part in Fig. 6-17 parallel to the bottom surface within .010.

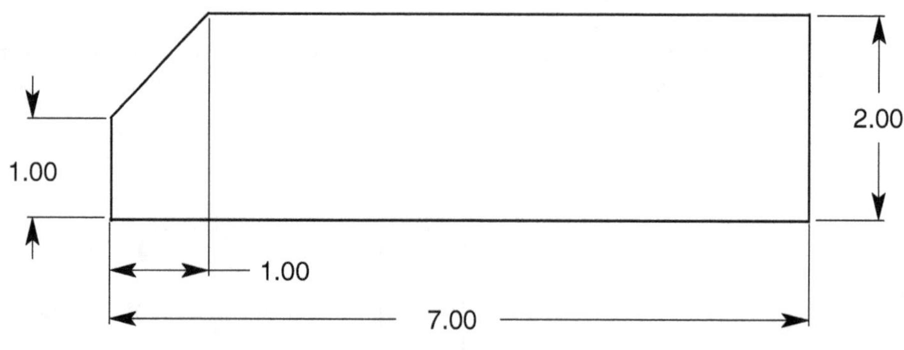

FIGURE 6-17 Specifying parallelism: Question 19.

20. Where controlling the parallelism of a feature of size, the feature control frame is associated with the ______________________________ of the feature being controlled.

21. If the element being controlled is an axis, the numerical tolerance is usually preceded by a ______________________________.

22. The numerical tolerance for angularity of a surface is specified as a linear dimension because it generates a ______________________________ shaped tolerance zone.

23. A plus or minus angularity tolerance is not used because it generates a ______________ shaped tolerance zone.

24. Where controlling the angularity of a feature of size, the feature control frame is associated with the ______________________________ of the feature being controlled.

25. If the element being controlled is an axis, the numerical tolerance is usually preceded by a ______________________________ symbol.

26. The angularity control is a refinement of the orientation of an axis at a basic angle to a ______________________________.

27. In Table 6-2, mark an **X** in the box that indicates the control that applies to the statement at the left.

	Plane Surfaces			Axes and Center Planes		
	⊥	//	∠	⊥	//	∠
Datum features are required						
Controls flatness if flatness is not specified						
Circle T modifier can apply						
Tolerance specified with a leader or extension line						
Tolerance associated with a dimension						
Material condition modifiers apply						
A virtual condition applies						

TABLE 6-2 The Application of Orientation Controls: Question 27

Problems

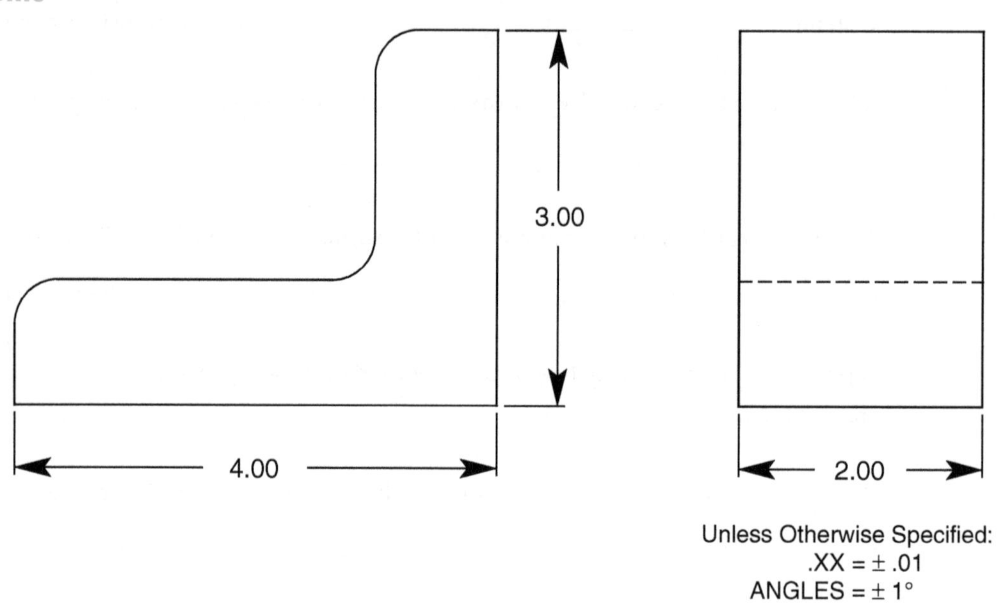

FIGURE 6-18 Perpendicularity of a plane surface: Probs. 1 and 2.

1. Specify the 3.00-inch surface of the part in Fig. 6-18 to be perpendicular to the bottom and back surfaces within a tolerance of .010. Draw and dimension the tolerance zone.

2. Specify a feature control frame that would require a precision plane surface placed against the 3.00-inch surface of the part in Fig. 6-18 to be perpendicular to the bottom and back surfaces within a tolerance of .010.

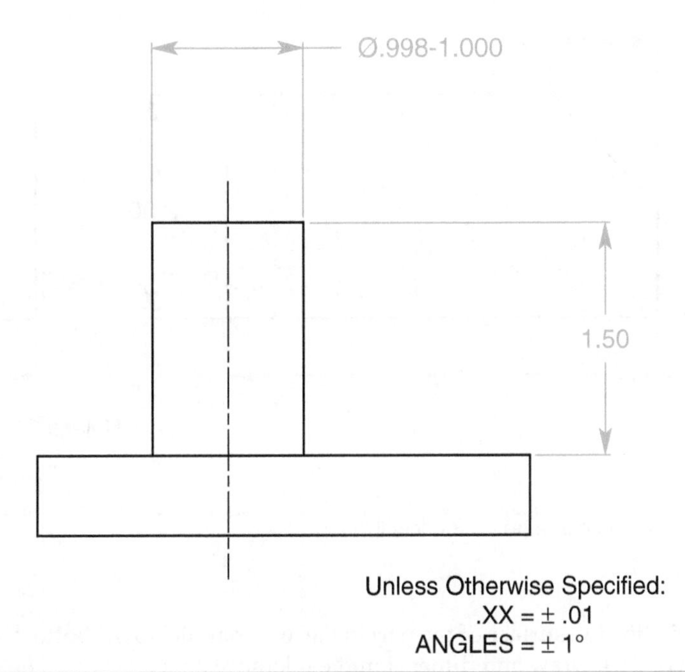

FIGURE 6-19 Perpendicularity of a pin to a plane surface: Prob. 3.

3. Specify the 1-inch-diameter pin perpendicular to the top surface of the horizontal plate in Fig. 6-19 within a tolerance of .015 at MMC. On the drawing, sketch and dimension a gage used to inspect this part.

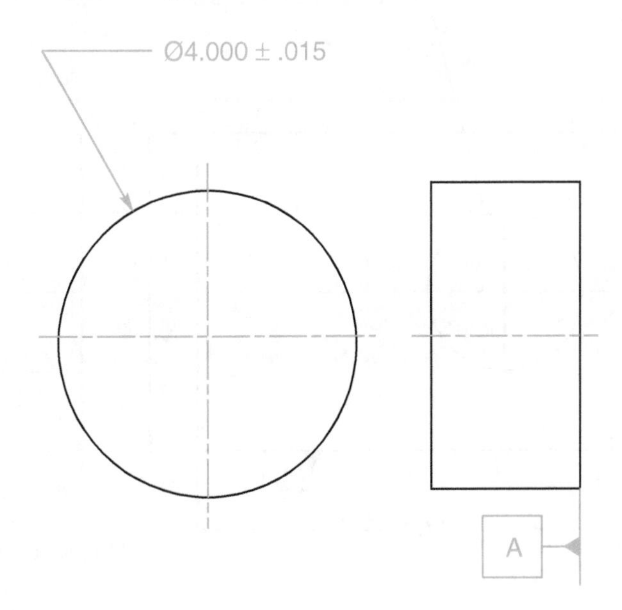

FIGURE 6-20 Perpendicularity of a cylinder to a plane surface: Prob. 4.

4. Specify the Ø4.000 cylinder perpendicular to its back surface, datum feature A, within a tolerance of .010 at MMC in Fig. 6-20.

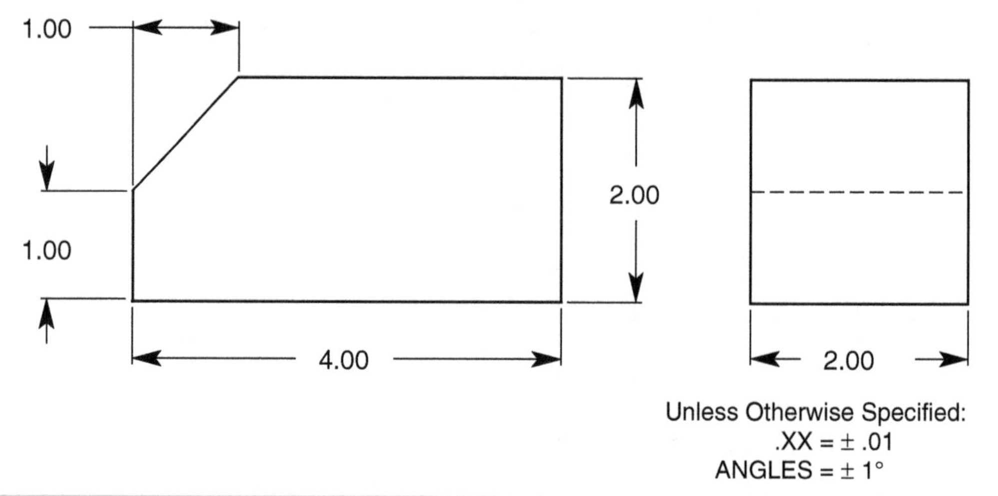

FIGURE 6-21 Parallelism of a plane surface: Prob. 5.

5. Specify the top surface of the part in Fig. 6-21 parallel to the bottom surface within a tolerance of .004. Draw and dimension the tolerance zone.

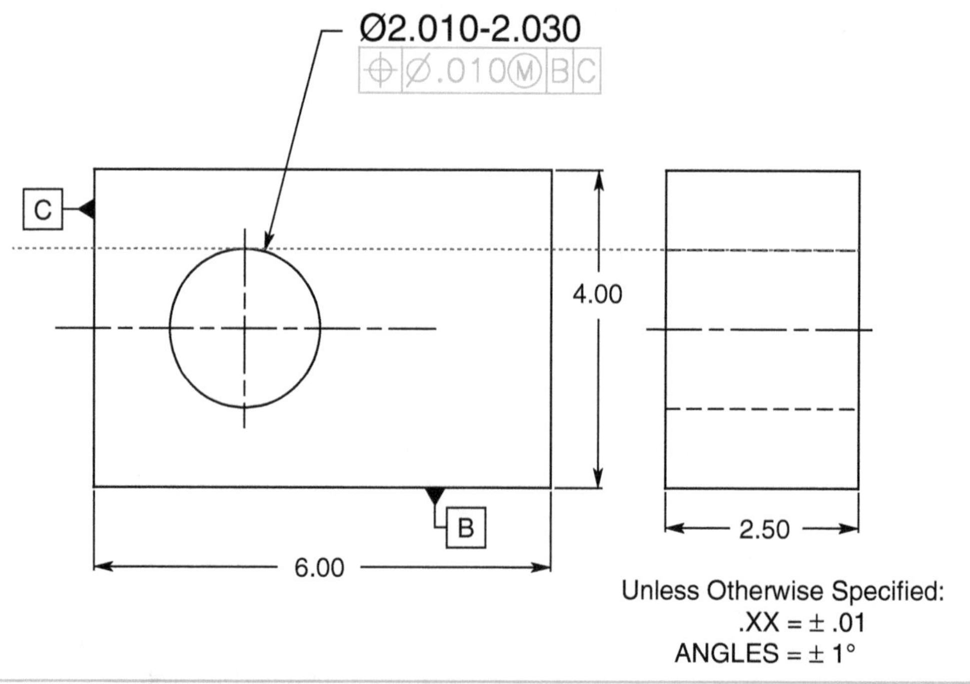

FIGURE 6-22 Parallelism of a cylindrical feature of size: Prob. 6.

6. Specify the hole in the part in Fig. 6-22 with a parallelism refinement to datum features B and C within a tolerance of .002 at MMC.

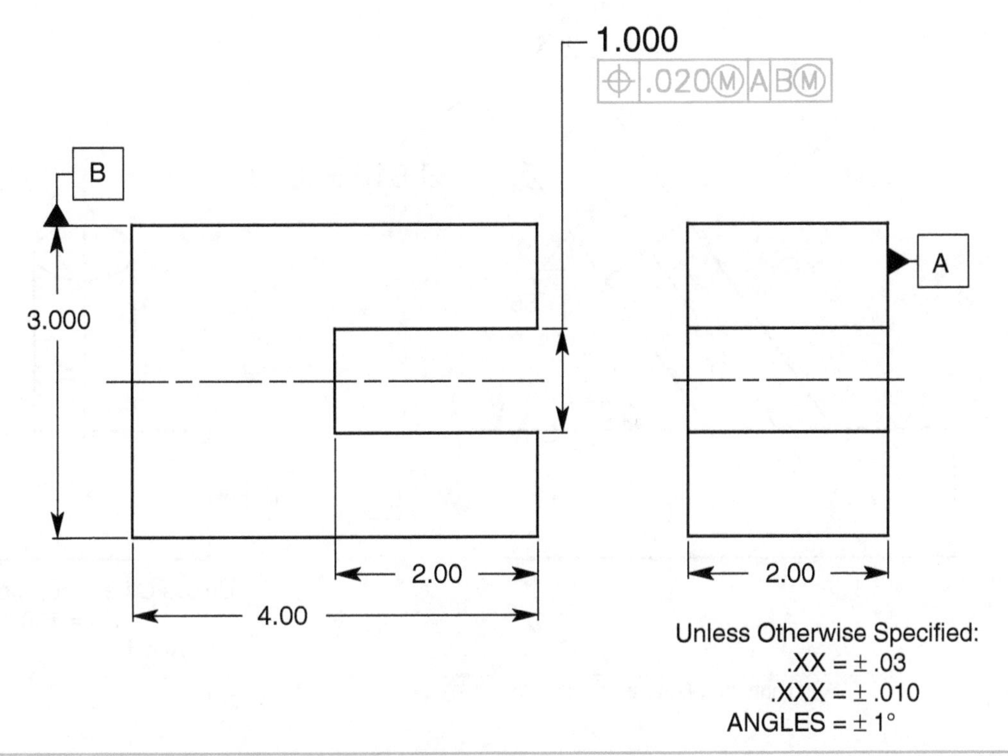

FIGURE 6-23 Perpendicularity of a noncylindrical feature of size: Prob. 7.

7. For the slot in Fig. 6-23, refine the perpendicularity to datum feature A within a tolerance of .002 at MMC.

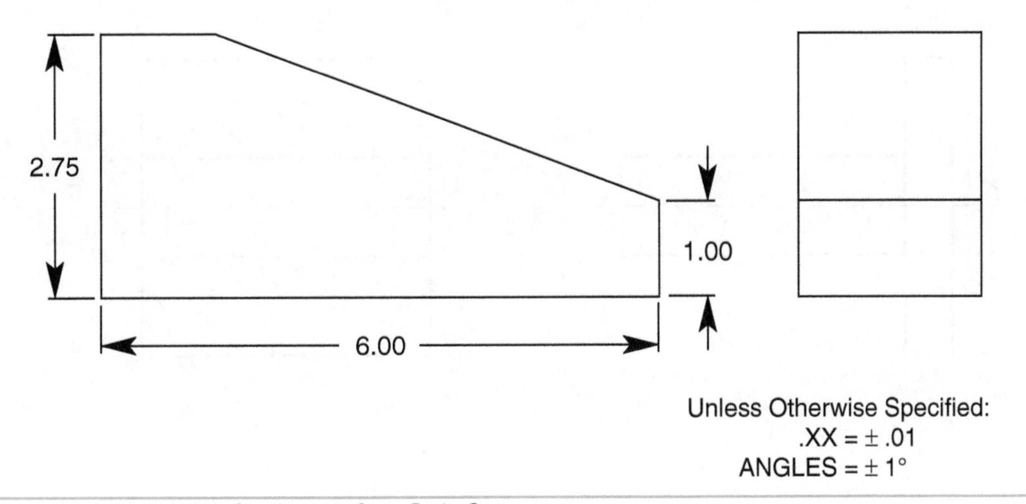

FIGURE 6-24 Angularity of a plane surface: Prob. 8.

8. Specify the top surface of the part in Fig. 6-24 to be at an angle of 20° to the bottom surface within a tolerance of .003. Draw and dimension the tolerance zone.

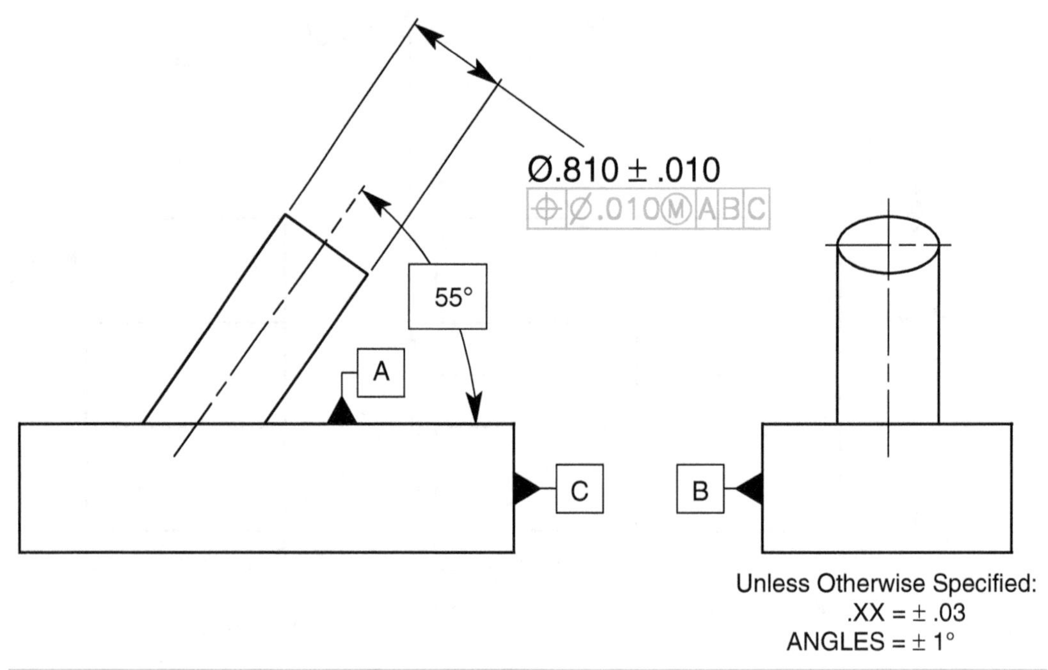

FIGURE 6-25 Angularity of a feature of size: Prob. 9.

9. For the pin in Fig. 6-25, refine the angularity to datum feature A and parallelism to datum feature B within a tolerance of .002 at MMC.

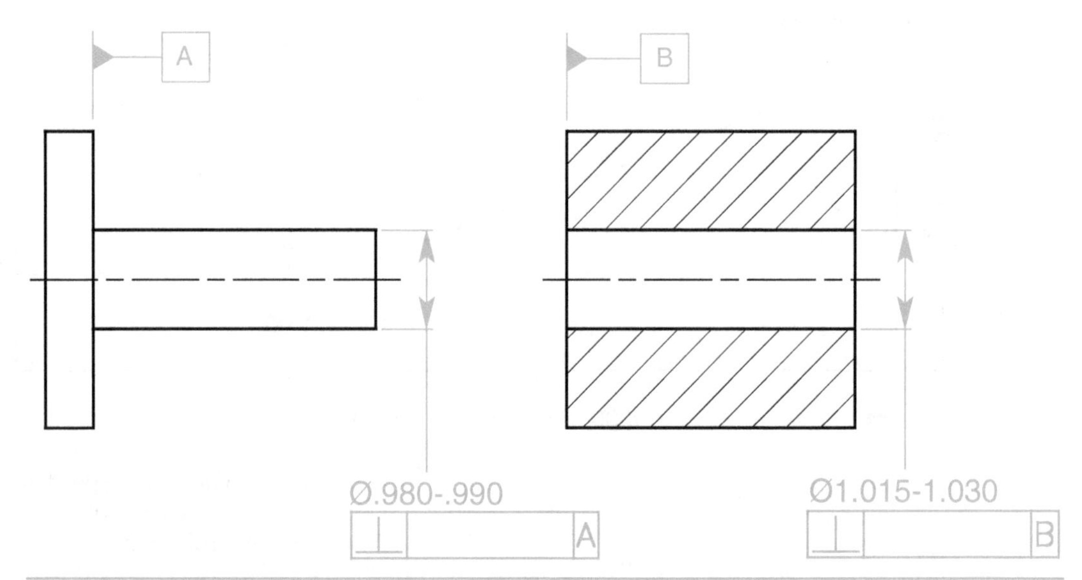

FIGURE 6-26 Orientation: Prob. 10.

10. Complete the feature control frames in Fig. 6-26 so that the two parts will always assemble, datum features A and B will meet, and the part can be produced using the most cost-effective design. The pin is machined in a lathe, and the hole is drilled.

CHAPTER 7

Position, General

Position is a composite tolerance that controls both the location and the orientation of features of sizes at the same time. It is the most frequently used of the twelve geometric controls. The position tolerance significantly contributes to part function, part interchangeability, the optimization of tolerance, and the communication of design requirements.

Chapter Objectives

After completing this chapter, the learner will be able to:

- *Explain* how the tolerance of position works
- *Specify* positional tolerance for the location and orientation of a feature of size
- *Interpret* tolerance specified at regardless of feature size (RFS)
- *Calculate* bonus tolerances for features of size specified at maximum material condition (MMC)
- *Specify* datum features of size at regardless of material boundary (RMB)
- *Specify* datum features of size at maximum material boundary (MMB)
- *Specify* a position tolerance at least material condition (LMC)
- *Calculate* minimum wall thickness at LMC
- *Calculate* boundary conditions
- *Calculate* tolerances specified with zero positional tolerance at MMC

Definition

The tolerance of position may be viewed in either of two ways:

- A *theoretical tolerance zone* of the toleranced feature located at true position within which the center point, axis, or center plane of the feature may vary from true position.
- A *virtual condition boundary* of the toleranced feature, when specified at maximum material condition (MMC) or least material condition (LMC) and located at true position, which may not be violated by the surface or surfaces of the considered feature of size.

The Tolerance of Position

A feature of size, such as the pin shown in Fig. 7-1, has four geometric characteristics that must be controlled. These characteristics are size, form, orientation, and location. Both size and form are controlled by the limits of size. Rule #1 states that an individual regular feature of size may not exceed the boundary of perfect form at MMC. If the actual local size of the pin is smaller than the MMC size, it may bow or be out of round by the amount that it departs from MMC. The sum of the actual local size and any form error equals the actual mating envelope.

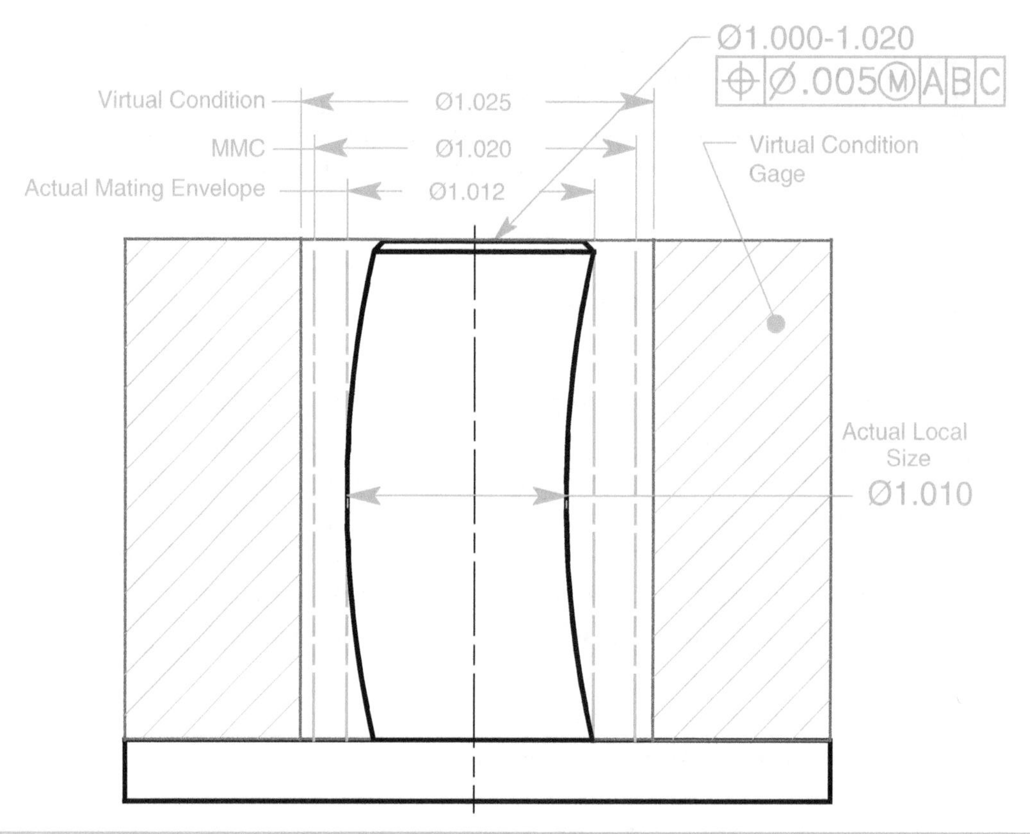

FIGURE 7-1 Both the limits of size and the tolerance of position control size, form, orientation, and location.

Both the orientation and the location of a feature are controlled by the tolerance of position. The total positional tolerance is equal to the tolerance specified in the feature control frame plus any bonus tolerance. The pin in Fig. 7-1 has a positional tolerance of .005 in diameter at MMC plus the bonus tolerance. Bonus tolerance is the difference between MMC and the actual mating envelope. Because MMC is a diameter of 1.020 and the actual mating envelope is a diameter of 1.012, the bonus tolerance is .008; the total positional tolerance is equal to the geometric tolerance, .005, plus the bonus tolerance, .008, or a total of .013 as shown in Table 7-1. The combination of the form error and the total positional tolerance is equal to the difference between the virtual condition and the actual local size of the pin.

- MMC – Actual Mating Envelope = Bonus
- Bonus + Geometric Tolerance = Total Positional Tolerance

MMC	–	Actual Mating Envelope	=	Bonus	+	Geometric Tolerance	=	Total Positional Tolerance
1.020		1.012		.008		.005		.013

TABLE 7-1 The Calculation of Bonus Tolerance for an External Feature

Specifying the Position Tolerance

Specifying the Position Tolerance at RFS

Since the position tolerance controls only features of sizes such as pins, holes, tabs, and slots, the feature control frame is usually associated with a size dimension. In Fig. 7-2, the hole is oriented and located with the position control. In this case, the feature control frame is placed beneath the local note describing the diameter and size tolerance of the hole. The true position of the hole, the theoretically perfect location of the axis, is specified with basic dimensions from the datum features indicated in the feature control frame. Once the feature control frame is assigned, an imaginary tolerance zone is defined and located about true position. The datum features are identified by datum feature symbols. Datum features A, B, and C specified in the feature control frame identify the datum reference frame in which the part is to be positioned for processing.

The feature control frame is the sentence in the language of geometric dimensioning and tolerancing (GD&T); it must be specified correctly in order to properly communicate design requirements. The feature control frame in Fig. 7-2 tells the location tolerancing story for the hole in this part. The diameter symbol and the tolerance specify a cylindrical tolerance zone, .010 in diameter, the full length of the feature, perfectly perpendicular to datum feature A, and located about true position. True position is located a basic 2.000 inches up from datum feature B and a basic 3.000 inches over from datum feature C. Tolerance zones are theoretical and do not appear on drawings. However, the tolerance zone has been shown here for illustration purposes only.

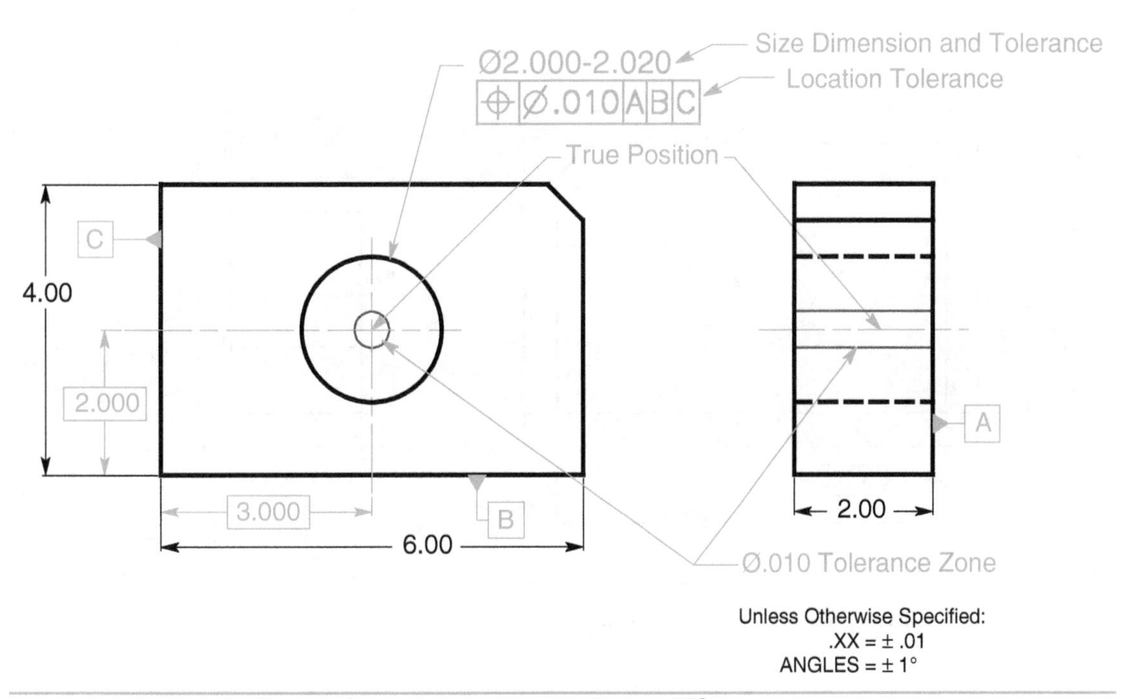

FIGURE 7-2 Location of a feature of size with a position tolerance at RFS.

The RFS modifier automatically applies for features of size where no material condition symbol is specified after the tolerance. Since no material condition symbol is specified in the feature control frame in Fig. 7-2, RFS automatically applies to the orientation and location tolerance of the hole. In other words, the position tolerance zone is .010 in diameter no matter what size the hole happens to be. The actual mating envelope may be anywhere between 2.000 and 2.020 in diameter, but the tolerance zone size remains .010 in diameter. No bonus tolerance is allowed.

Inspection

Inspection starts with measuring the hole diameter. Suppose the hole is 2.012 in diameter; if that is the case, it is within the size tolerance of 2.000 to 2.020 in diameter. The next step is to measure the hole orientation and location. The part is clamped in a datum reference frame by bringing a minimum of three points on the surface of the primary datum feature into contact with the primary datum plane, a minimum of two points on the surface of the secondary datum feature into contact with the secondary datum plane, and a minimum of one point on the surface of the tertiary datum feature into contact with the third datum plane as shown in Fig. 7-3. Next, the largest pin gage that will fit inside the hole is used to simulate the actual mating envelope. The actual mating envelope for an internal feature of size is the largest, similar, perfect, feature counterpart that can be inscribed within the feature so that it just contacts the surface of the hole at the highest points.

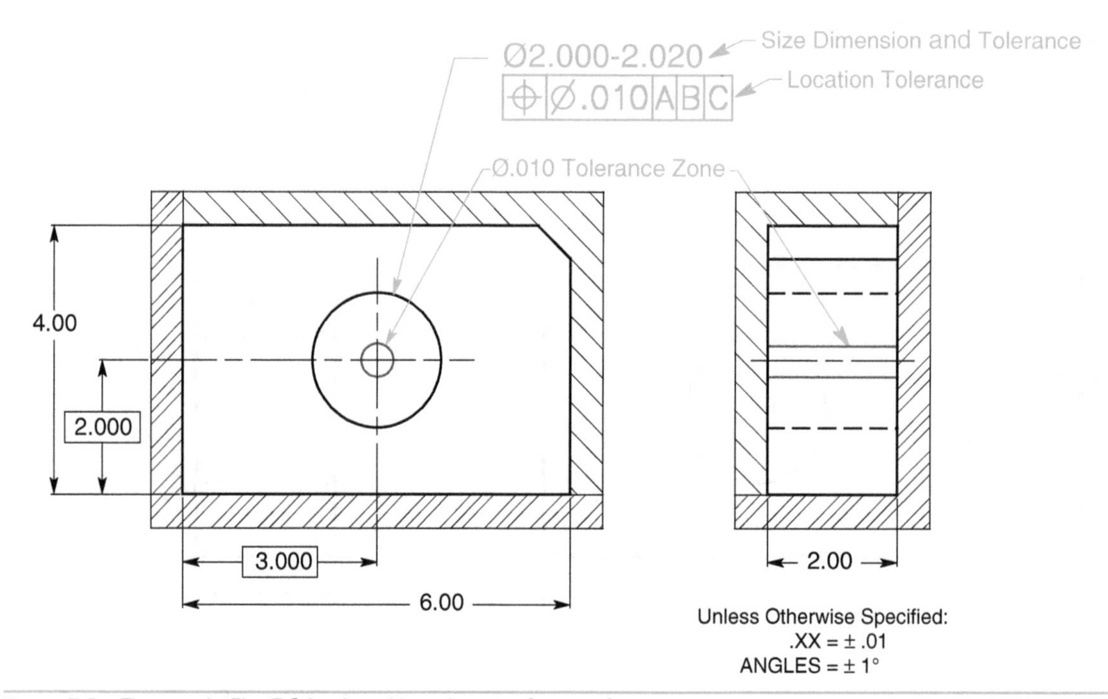

FIGURE 7-3 The part in Fig. 7-2 is placed in a datum reference frame.

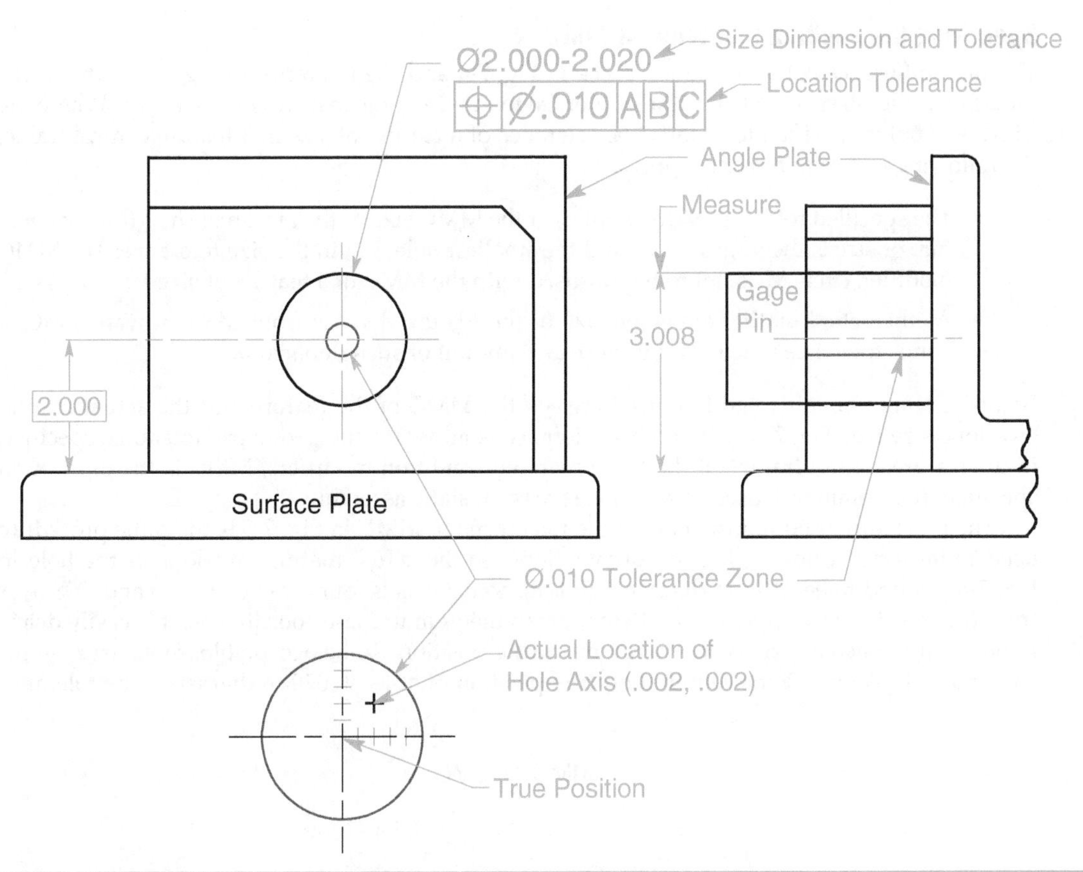

FIGURE 7-4 Inspection of the hole location using a theoretical tolerance zone at RFS.

As shown in Fig. 7-4, the distance from the surface plate, datum feature B, to the top edge of the pin gage is measured. Measurements are taken along the pin gage to determine that the hole is within the perpendicularity tolerance to the angle plate, datum feature A. Suppose the distance from the surface plate to the top edge of the pin gage is 3.008. This measurement minus half of the diameter of the pin gage equals the distance from datum feature B to the actual axis of the hole:

- Half of the diameter of the pin gage — 2.012/2 = 1.006
- The distance from the surface plate to the axis of the hole — 3.008 – 1.006 = 2.002
- The actual axis of the hole minus true position — 2.002 – 2.000 = .002

The distance, then, from true position to the actual axis of the hole in the vertical direction is .002. With the part still clamped to the angle plate, it is rotated 90° counterclockwise, and the distance from datum feature C to the actual axis of the hole is measured by repeating the previous measurement. If the distance from true position to the actual axis in the horizontal direction is also .002, the actual axis is .002 up and .002 over from true position, requiring a tolerance zone of less than .006 in diameter, well within the .010-diameter cylindrical tolerance zone shown in Fig. 7-4. The hole is within tolerance.

Specifying the Position Tolerance at MMC

The only difference between the tolerance in Fig. 7-4 and the tolerance in Fig. 7-5 is the MMC modifier, circle M, specified after the numerical tolerance in the feature control frame. Where the MMC symbol is specified to modify the tolerance of a feature of size in a feature control frame, the following two requirements apply:

- The specified tolerance value applies at the MMC size of the feature. (The MMC of a feature of size is the largest shaft and the smallest hole within the size tolerance. The MMC modifier, circle M, is not to be confused with the MMC of a feature of size.)
- As the actual mating envelope size of the feature departs from MMC toward LMC, a bonus tolerance is achieved in the exact amount of such departure.

Bonus tolerance equals the difference between the MMC of the feature and the actual mating envelope size (see Fig. 7-1). The bonus tolerance is added to the geometric tolerance specified in the feature control frame. Of the three material condition modifiers, MMC is the most often specified. It is commonly used to tolerance parts for static assembly.

The procedure used to inspect the hole toleranced at MMC in Fig. 7-5 is the same procedure used to inspect the hole in Fig. 7-4 above. Suppose the actual mating envelope of the hole in Fig. 7-5 is found to be 2.012 in diameter and the actual axis is found to be .006 up and .008 over from true position. By applying the Pythagorean theorem to these coordinates, it is easily determined that the actual axis is .010 away from true position. To be acceptable, this part requires a cylindrical tolerance zone centered on true position of at least .020 in diameter. The tolerance

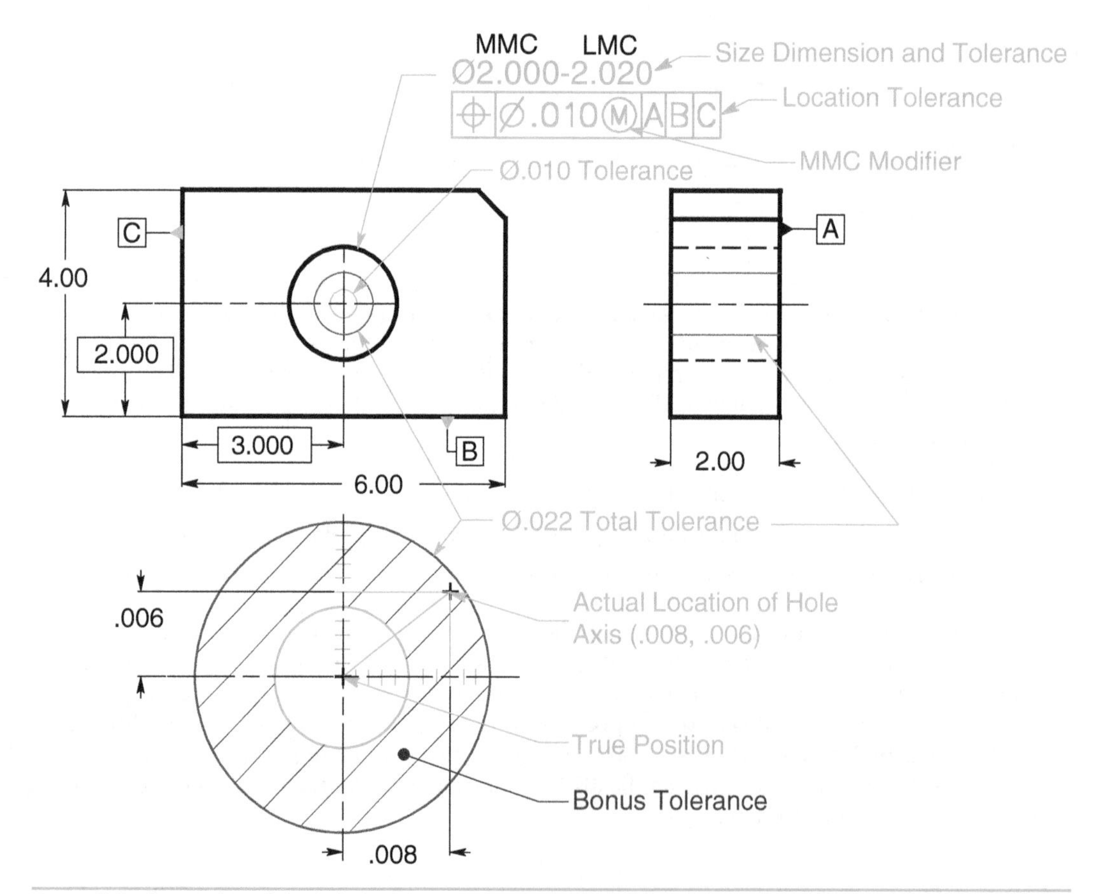

FIGURE 7-5 Inspection of the hole location using a theoretical tolerance zone at MMC.

specified is only .010 in diameter, but it includes an MMC modifier; consequently, bonus tolerance is available. The following formulas are used to calculate the bonus tolerance and total positional tolerance:

- Actual Mating Envelope – MMC = Bonus
- Bonus + Geometric Tolerance = Total Positional Tolerance

Actual Mating Envelope	–	MMC	=	Bonus	+	Geometric Tolerance	=	Total Positional Tolerance
2.012		2.000		.012		.010		.022

TABLE 7-2 The Calculation of Bonus Tolerance for an Internal Feature

Calculations in Table 7-2 show a total positional tolerance of .022 in diameter, sufficient tolerance to make the hole location in Fig. 7-5 acceptable. Bonus tolerance is the positive difference or the absolute value between the actual mating envelope and MMC.

Inspection with a Functional Gage

Another way of inspecting the hole specified at MMC is with a functional gage, shown in Fig. 7-6. A functional gage for this part is a datum reference frame with a virtual condition pin positioned perpendicular to datum feature A, located a basic 2.000 inches up from datum feature B and a basic 3.000 inches over from datum feature C. If the part can be set over the pin and placed against the datum reference frame in the proper order of precedence, the hole is in tolerance.

Although features toleranced with GD&T may be inspected using any appropriate inspection technique, functional gages are very convenient for checking large numbers of parts or when inexperienced operators are required to inspect parts. A functional gage represents the worst-case mating part. Dimensions on gage drawings may be either toleranced or dimensioned

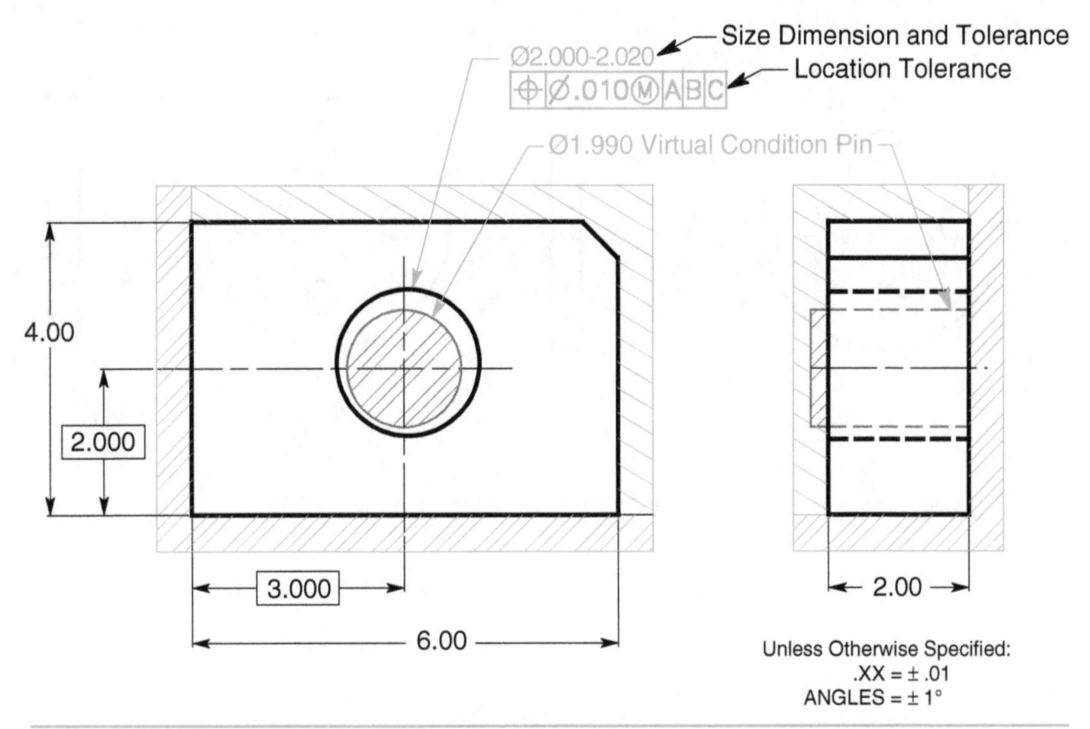

FIGURE 7-6 Inspection of a feature of size with a position tolerance at MMC using a functional gage.

with basic dimensions. Tolerances for basic dimensions on gage drawings are gage-makers' tolerances. Tolerances specified on gages are usually no more than 10% of the tolerance of the part. All of the tolerance for the gage comes from the tolerance for the part. In other words, a gage may not accept a bad part, but it can reject a marginally good part. If a part is inspected with a gage and also with some other inspection method and the two methods yield contradictory results, the gage is the more reliable inspection method because the gage more closely simulates the mating part.

Datum Features of Size Specified with an RMB Modifier

A datum feature of size applies at RMB if no material condition modifier follows the datum feature. Where datum features of size are specified at RMB, the datum is established by physical contact between the surface(s) of the datum feature and the surface(s) of the processing equipment. There is no shift tolerance for datum features specified at RMB. A holding device that can be adjusted to fit the size of the datum feature, such as a chuck, vise, or adjustable mandrel, is used to position the part. In Fig. 7-7, the outside diameter of the part, datum feature B, is specified at RMB. The pattern of features is inspected by placing the outside diameter in an adjustable gage, a three-jaw chucking device, and the hole pattern is placed over a set of virtual condition pins. If the part can set inside this gage and all the feature sizes are within size tolerance, the pattern is acceptable. The gage in Fig. 7-7 is shown here to more clearly illustrate the inspection process of a feature of size specified at RMB. A gage like this is complicated and expensive. In actual practice, this part would be inspected with an open setup or a coordinate measuring machine. If it is necessary to inspect this part with a gage, the drawing could be modified to specify datum feature B at MMB.

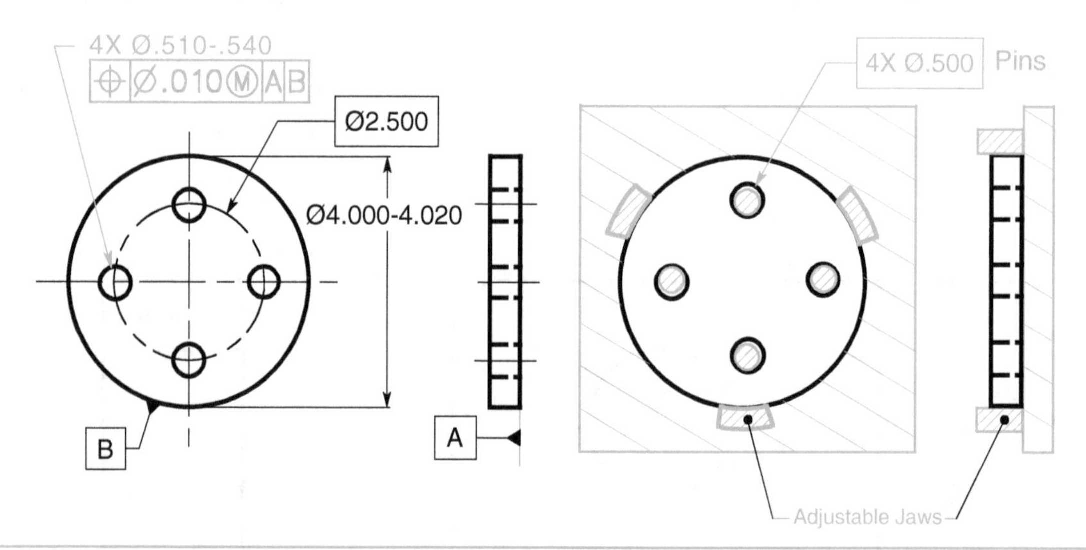

FIGURE 7-7 Inspection of a hole pattern controlled to a datum feature of size at RMB.

Datum Features of Size Specified with an MMB Modifier

Where a datum feature of size, such as datum feature B in Fig. 7-8, is toleranced with a geometric tolerance and is referenced in a feature control frame at MMB, circle M, the resulting MMB for datum feature B is equal to its virtual condition with respect to the preceding datum feature. In Fig. 7-8*A*, datum feature B applies at its virtual condition with respect to datum feature A, which is its MMC plus the perpendicularity tolerance, or 4.020 + .010 = 4.030. Because datum feature B at MMB on the part applies at a diameter of 4.030, datum feature simulator B on the gage is produced at a diameter of 4.030. If datum feature B, on a part, is actually produced at a diameter of 4.010, the four-hole pattern, as a group, can shift .020 in any direction inside the 4.030-diameter gage, as shown in Fig. 7-8*A*. If other inspection techniques are used, the axis of datum feature B and consequently the four-hole pattern can shift within a cylindrical tolerance zone .020 in diameter centered on true position of datum feature B. (See Chap. 13 for the inspection procedure of a pattern of features controlled to a datum feature of size.)

Datum feature B in Fig. 7-8*B* applies at its virtual condition with respect to the previous datum feature, but there is no previous datum feature. Consequently, datum feature B applies at its MMC, or 4.020 in diameter, and datum feature B simulator on the gage in Fig. 7-8*B* is produced at a diameter of 4.020. As the actual size of datum feature B departs from MMC toward LMC, a shift tolerance, of the pattern as a group, is allowed in the exact amount of such departure. The possible shift equals the difference between the actual size of the datum feature and the inside diameter of the gage as shown on the drawings in Fig. 7-8. If datum feature B, on a part, is actually produced at a diameter of 4.010, it can shift .010 in any direction inside the 4.020-diameter gage, as shown in Fig. 7-8*B*.

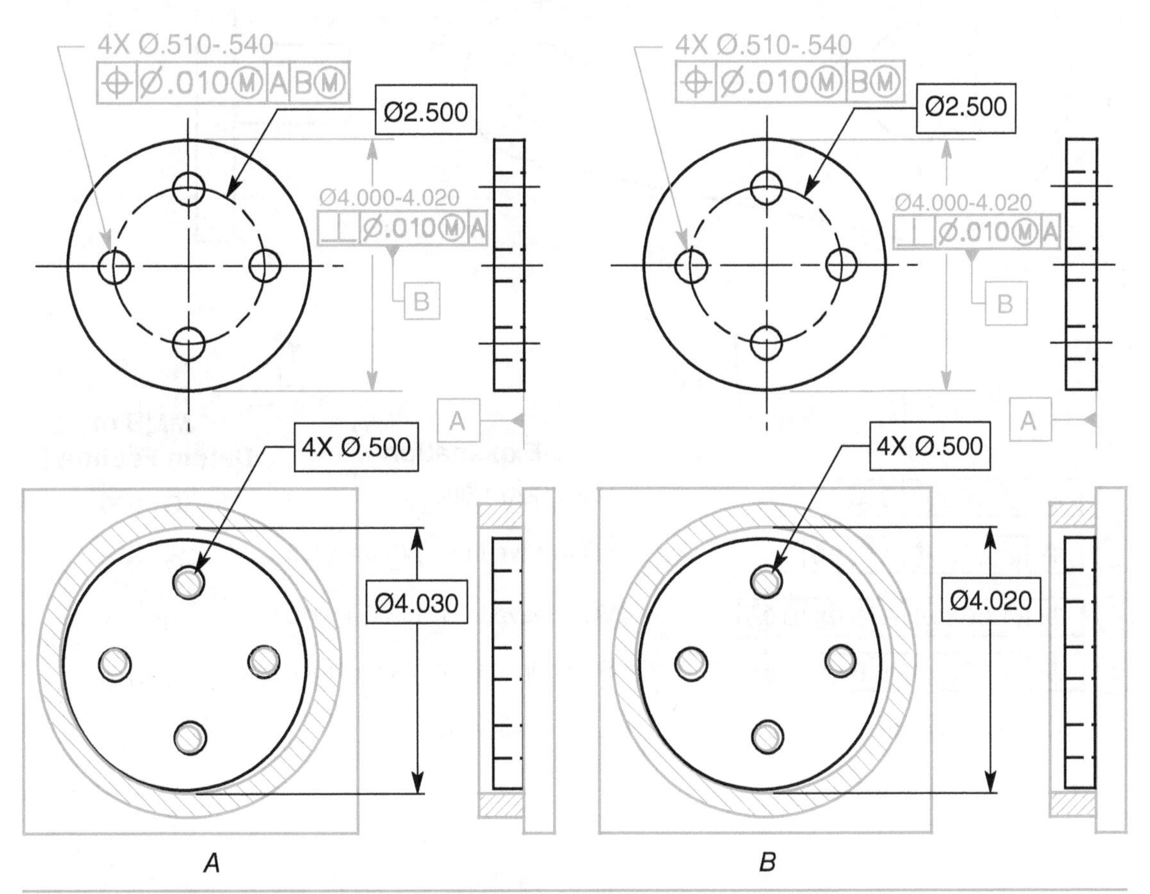

FIGURE 7-8 Inspection of a hole pattern controlled to a datum feature of size at MMB.

MMB Modifier Explained in More Detail

Where a datum feature(s) of size is controlled by an orientation or location tolerance and is referenced in a feature control frame at MMB, circle M, the resulting MMB for that feature of size is equal to its MMC or virtual condition with respect to the preceding datum feature. This concept can be easily understood by sketching the gage used to inspect the location of the feature in question to the appropriate datum feature(s).

The following discussion and accompanying gages explain the meanings of the four possible feature control frames shown in Fig. 7-9. For an external feature of size such as datum feature D, the MMB is the smallest value that will contain the datum feature while respecting datum feature precedence. Each feature control frame in question and the appropriate gage used to inspect

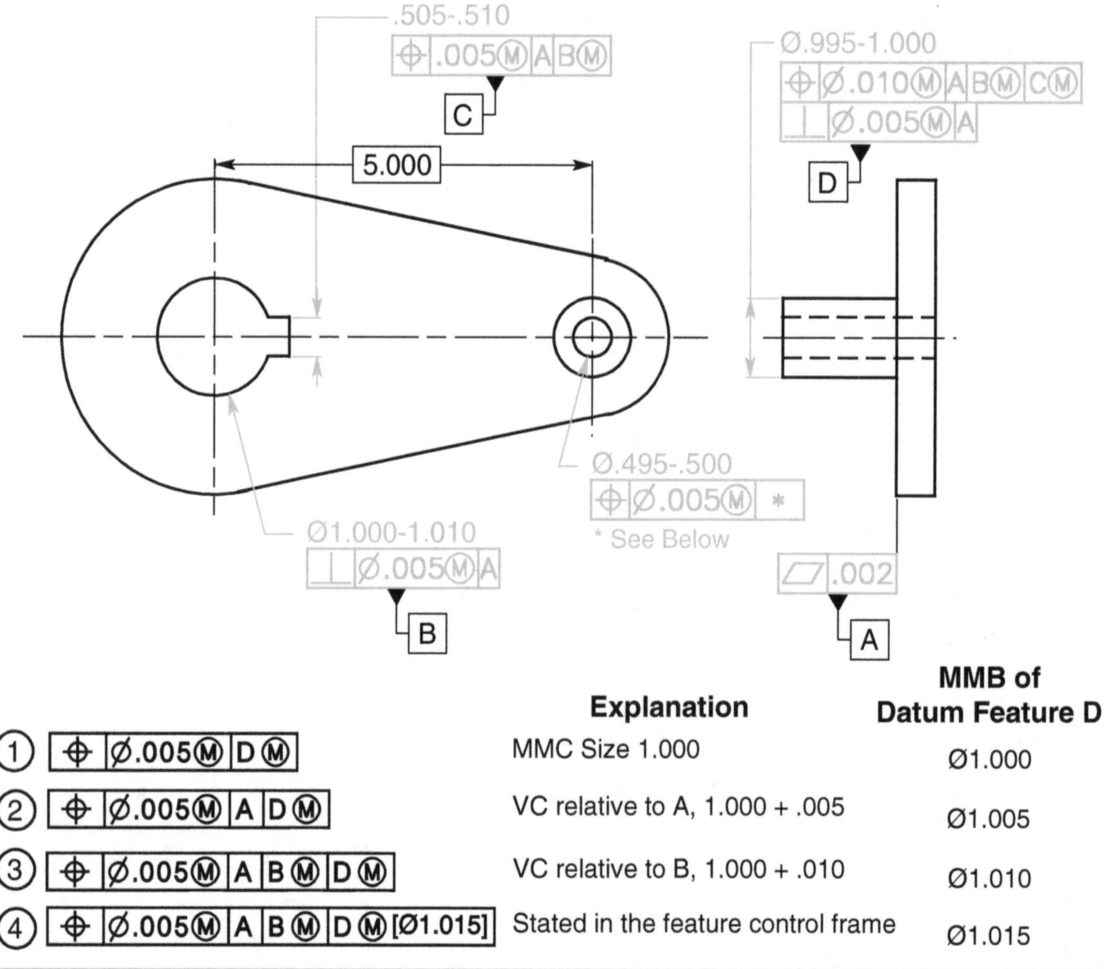

FIGURE 7-9 Features controlled to datum features of size specified at MMB.

the half-inch-diameter hole is shown below. (Gages are used here and throughout this book for illustration purposes only; parts toleranced with GD&T may be inspected with any appropriate inspection technique.)

Feature control frame #1, shown in Fig. 7-9, has only one datum feature: datum feature D. The 1-inch-diameter shaft applies at its virtual condition with respect to the previous datum feature, but there is no previous datum feature. In this case, datum feature D applies at its MMC diameter of 1.000. The gage, shown in Fig. 7-10, consists of datum feature simulator D at the 1.000-inch MMC diameter of datum feature D and a pin produced at the virtual condition of the half-inch-diameter hole, which is .495 – .005, or a diameter of .490.

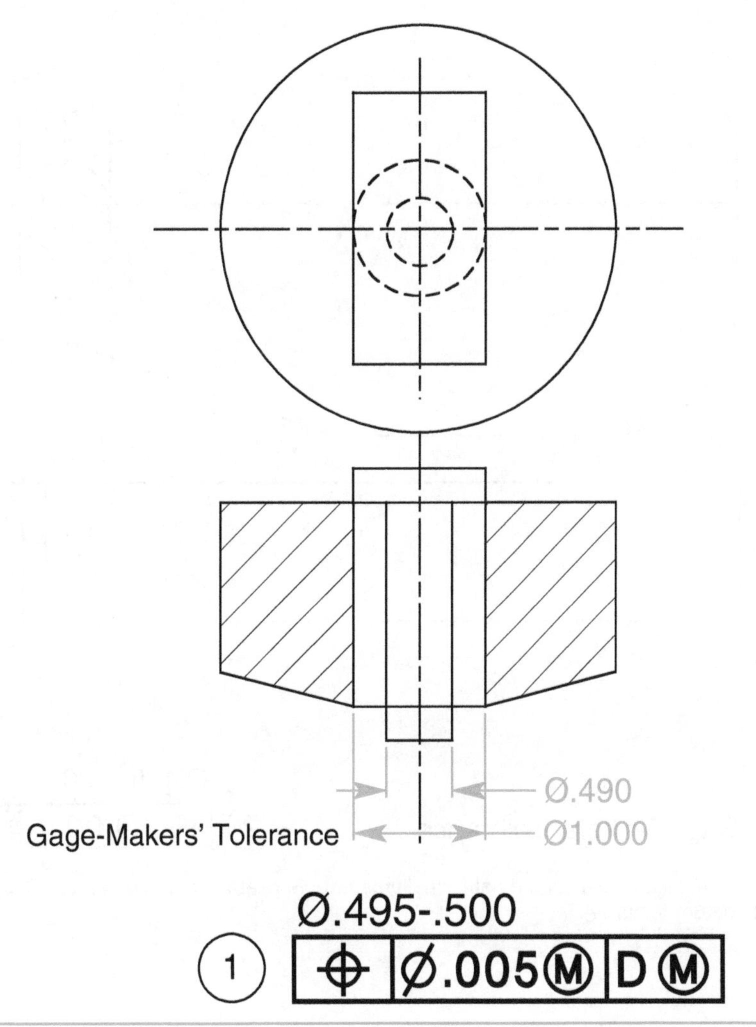

FIGURE 7-10 A gage to inspect the half-inch-diameter hole relative to datum feature D at its MMC size.

Feature control frame #2 in Fig. 7-9 has two datum features: datum features A and D. The 1-inch-diameter shaft applies at its virtual condition with respect to the previous datum feature: datum feature A. The geometric tolerance of datum feature D has a perpendicularity refinement of .005 to datum feature A. Consequently, the virtual condition of datum feature D with respect to datum feature A is 1.005. The gage, shown in Fig. 7-11, must first mate with datum feature A. Because datum feature D can be out of perpendicularity to datum feature A within a tolerance of .005, datum feature simulator D on the gage is produced at a diameter of 1.005, its virtual condition with respect to datum feature A.

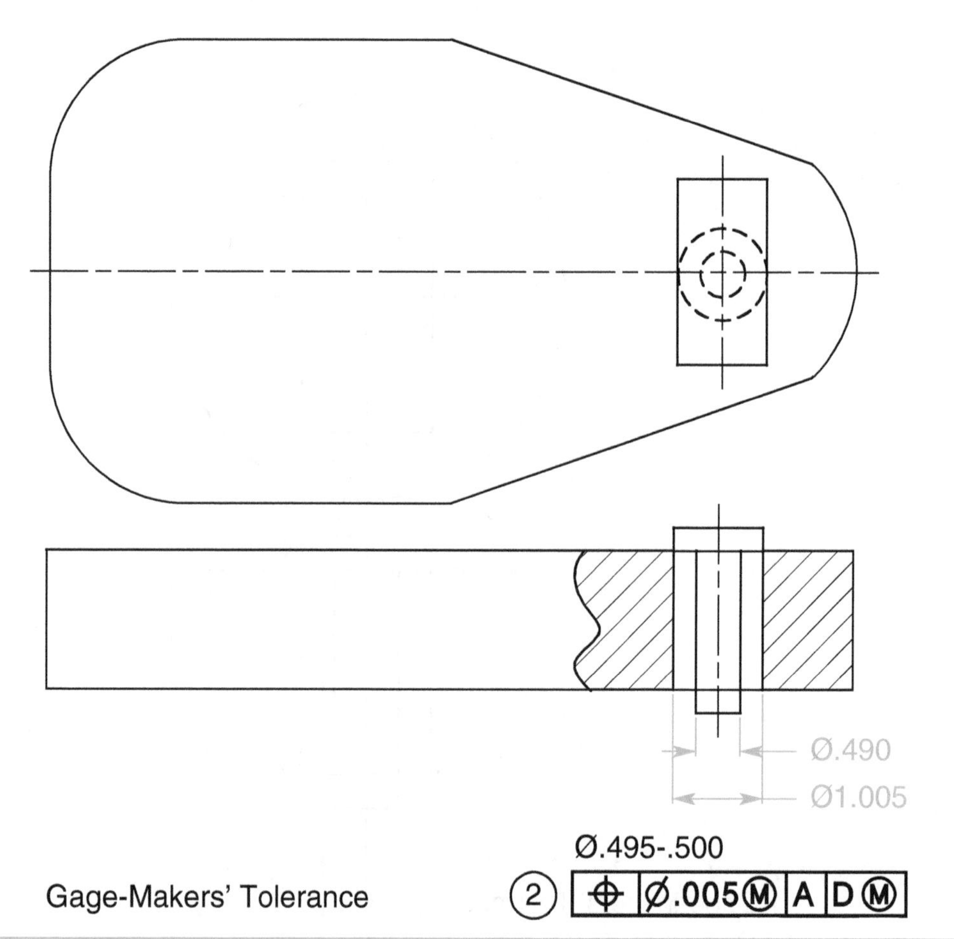

FIGURE 7-11 A gage to inspect the half-inch-diameter hole relative to datum feature D at virtual condition with respect to datum feature A.

Feature control frame #3 in Fig. 7-9 has three datum features: datum features A, B, and D. In this case, datum feature D applies at its virtual condition with respect to the previous datum feature: datum feature B. The geometric tolerance for datum feature D has a position tolerance of .010 to datum feature B. Consequently, the virtual condition of datum feature D with respect to datum feature B is a diameter of 1.010. The gage, shown in Fig. 7-12, must first mate with datum feature A and then with datum feature B. Because datum feature D is located to datum feature B with a basic dimension within a tolerance of .010, datum feature simulator D on the gage is produced at a diameter of 1.010, the virtual condition of datum feature D with respect to datum feature B.

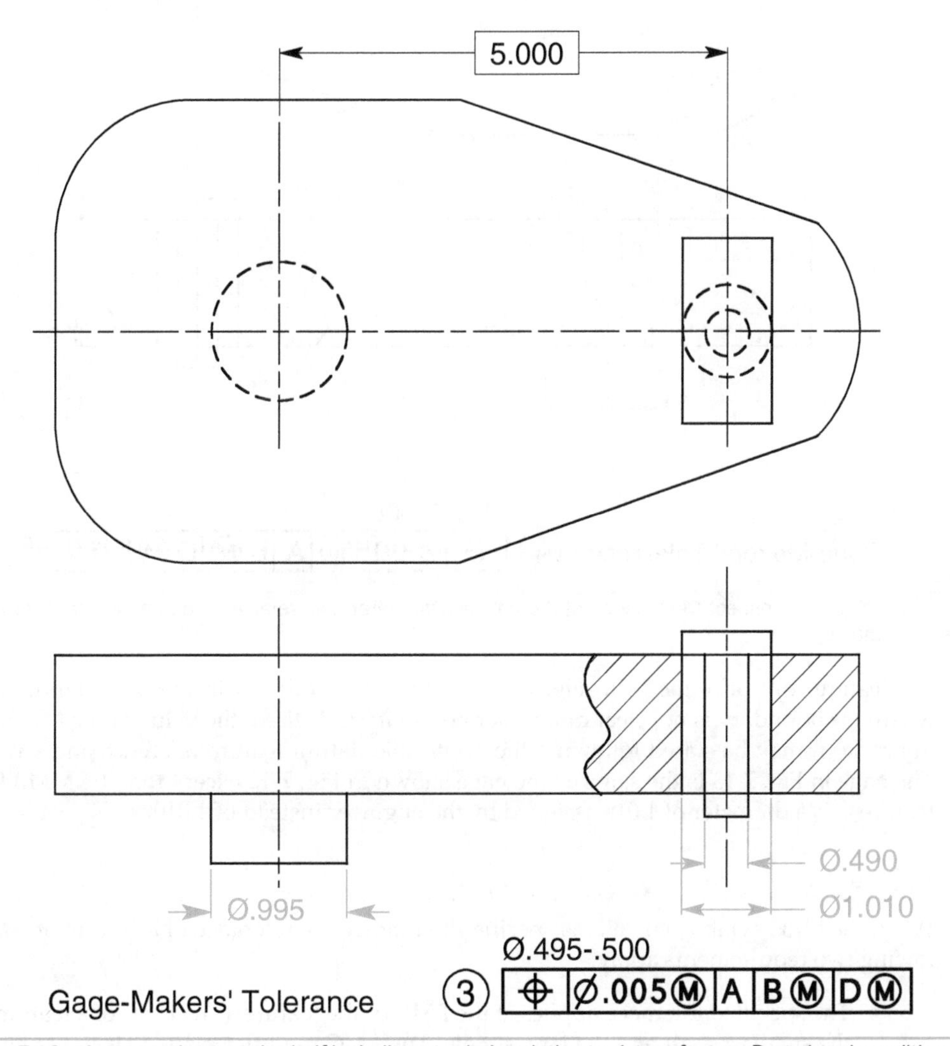

FIGURE 7-12 A gage to inspect the half-inch-diameter hole relative to datum feature D at virtual condition with respect to datum feature B.

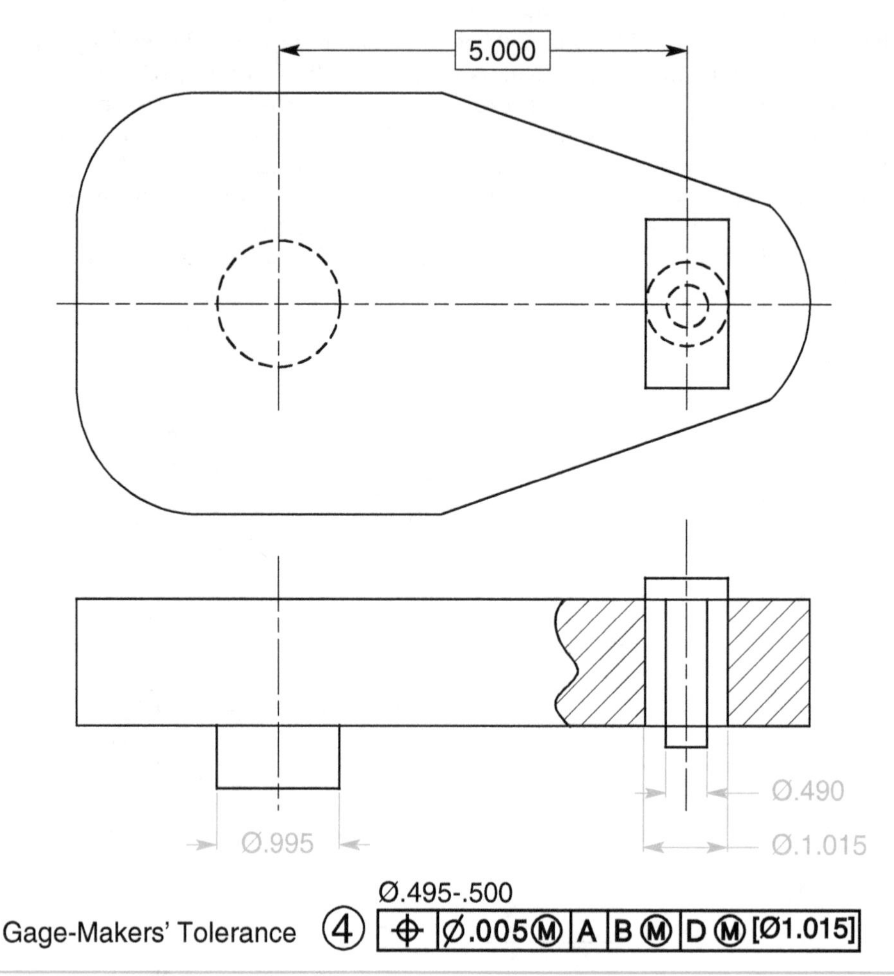

FIGURE 7-13 A gage to inspect the location of the half-inch-diameter hole relative to datum feature D at an MMB assigned by the engineer.

Feature control frame #4 in Fig. 7-9 is a concept introduced in the 2009 standard. In cases where the boundary is not clear or another boundary is desired, the value of the MMB, enclosed in brackets, may be stated following the applicable datum feature reference and any modifier. The gage in Fig. 7-13 is the same as the gage shown in Fig. 7-12 except that the MMB for datum feature D is a diameter of 1.015 assigned by the engineer instead of 1.010.

Locating Features of Size with an LMC Modifier

When the LMC symbol, circle L, is specified to modify the tolerance of a feature of size, the following two requirements apply:

- The specified tolerance applies at the LMC of the feature. (The LMC of a feature of size is the smallest shaft and the largest hole. The LMC modifier, circle L, is not to be confused with the LMC size of a feature of size.)
- As the actual mating envelope size of the feature departs from LMC toward MMC, a bonus tolerance is achieved in the exact amount of such departure.

Bonus tolerance specified at LMC equals the difference between the actual mating envelope size and the LMC size of the feature. The bonus tolerance is added to the geometric tolerance specified in the feature control frame. Where specified, it is used to maintain a minimum wall thickness or maintain a minimum distance between features. The LMC modifier is just the opposite in its

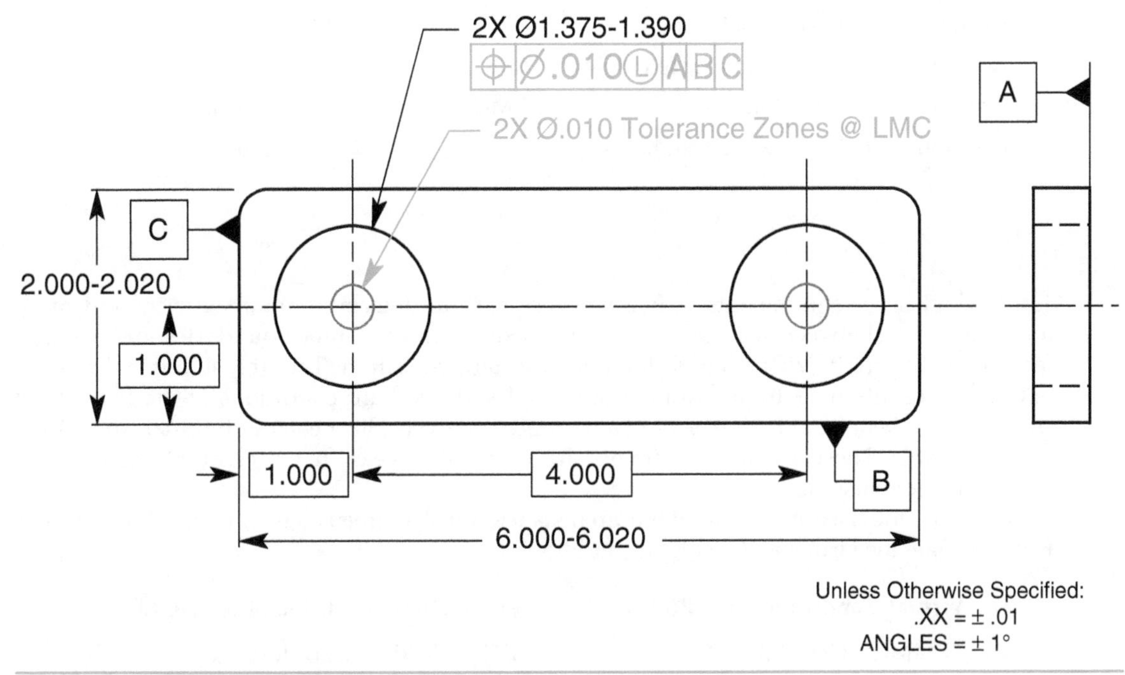

FIGURE 7-14 Location of a feature of size with a position tolerance at LMC.

effects of the MMC modifier. Even the form requirement of a feature of size at LMC is opposite the form requirement at MMC. When a tolerance is specified with an LMC modifier, the feature may not exceed the boundary of perfect form at LMC. Finally, features toleranced at LMC cannot be inspected with functional gages. Virtual condition for an internal feature at LMC is equal to the LMC plus the geometric tolerance.

The calculation for the virtual condition of the holes in Fig. 7-14 follows:

LMC	1.390
Geometric tolerance	+ .010
Virtual condition at LMC	1.400

It is not possible to put a 1.400 virtual condition pin through a 1.390 hole. Inspection of features specified at LMC may be done with a coordinate measuring machine, an open setup, or graphic analysis. LMC is the least used of the three material condition modifiers.

Minimum Wall Thickness at LMC

What is the minimum distance between the holes and the ends of the part in Fig. 7-14?

The distance from datum feature C to the axis of the first hole	1.000
Half of the diameter of the hole at LMC	– .695
Half of the tolerance of the hole at LMC	– .005
The minimum wall thickness	.300

The length of the part at LMC	6.000
The distance from datum feature C to the axis of the second hole	– 5.000
Half of the diameter of the hole at LMC	– .695
Half of the tolerance of the hole at LMC	– .005
The minimum wall thickness	.300

The distance between holes	4.000
Two times half of the diameter of the hole at LMC	– 1.390
Two times half of the tolerance of the hole at LMC	– .010
The minimum distance between holes	2.600

Boundary Conditions

To satisfy design requirements, it is often necessary to determine the maximum and minimum distances between features. The worst-case inner and outer boundaries, or loci, of a feature of size are its virtual condition and resultant condition. The first step in calculating boundary conditions is to determine the virtual and resultant conditions of the features of size; they are beneficial in performing a tolerance analysis. Start each individual calculation with a formula. These calculations are not difficult, but it is easy to omit a number and arrive at a wrong conclusion.

Calculate the maximum and minimum distances for the dimensions X, Y, and Z in Fig. 7-15. First calculate the virtual and resultant conditions:

The **Virtual Condition** of the **PIN**

$V.C._P = MMC + Geo.\ Tol. =$

$V.C._P = 1.000 + .010 =$

$\mathbf{V.C._P = 1.010}$ $\quad \mathbf{V.C._P/2 = .505}$

The **Virtual Condition** of the **HOLE**

$V.C._H = MMC - Geo.\ Tol. =$

$V.C._H = .520 - .020 =$

$\mathbf{V.C._H = .500}$ $\quad \mathbf{V.C._H/2 = .250}$

Resultant Condition of the PIN

$R.C._P = LMC - Geo.\ Tol. - Bonus =$

$R.C._P = .996 - .010 - .004 =$

$\mathbf{R.C._P = .982}$ $\quad \mathbf{R.C._P/2 = .491}$

Resultant Condition of the HOLE

$R.C._H = LMC + Geo.\ Tol. + Bonus =$

$R.C._H = .560 + .020 + .040 =$

$\mathbf{R.C._H = .620}$ $\quad \mathbf{R.C._H/2 = .310}$

The maximum and minimum distances for dimension **X**:

$X_{Max} = Location - R.C._P/2 =$

$X_{Max} = 2.000 - .491 =$

$\mathbf{X_{Max} = 1.509}$

$X_{Min} = Location - V.C._P/2 =$

$X_{Min} = 2.000 - .505 =$

$\mathbf{X_{Min} = 1.495}$

The maximum and minimum distances for dimension **Y**:

$Y_{Max} = Location - R.C._P/2 - V.C._H/2 =$

$Y_{Max} = 2.000 - .491 - .250 =$

$\mathbf{Y_{Max} = 1.259}$

$Y_{Min} = Location - V.C._P/2 - R.C._H/2 =$

$Y_{Min} = 2.000 - .505 - .310 =$

$\mathbf{Y_{Min} = 1.185}$

The maximum and minimum distances for dimension **Z**:

$Z_{Max} = Length_{MMC} - Loc. - V.C._H/2 =$

$Z_{Max} = 6.010 - 4.000 - .250 =$

$\mathbf{Z_{Max} = 1.760}$

$Z_{Min} = Length_{LMC} - Loc. - R.C._H/2 =$

$Z_{Min} = 5.990 - 4.000 - .310$

$\mathbf{Z_{Min} = 1.680}$

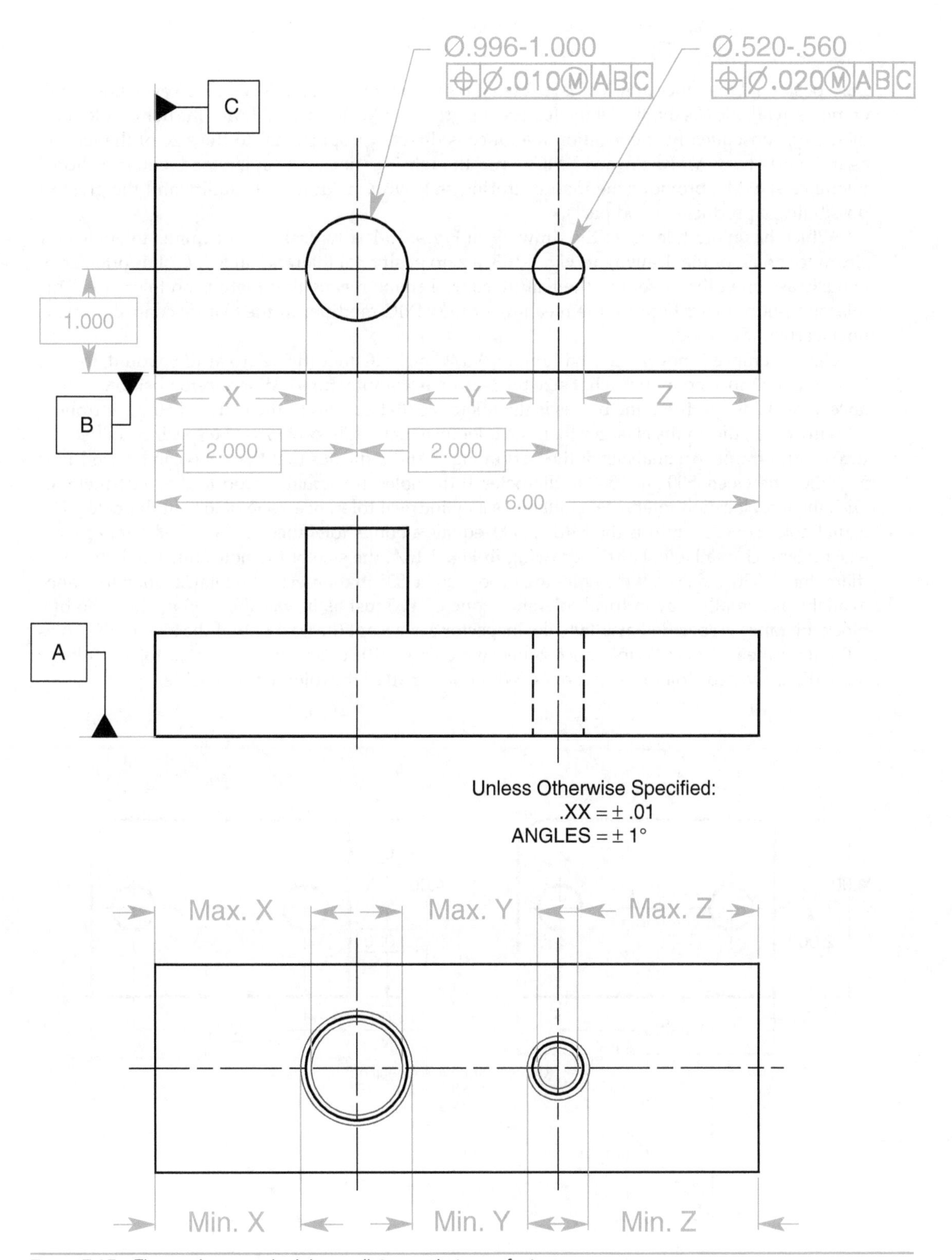

FIGURE 7-15 The maximum and minimum distances between features.

Zero Positional Tolerance at MMC

Zero positional tolerance at MMC is just what it says: no tolerance at MMC. However, bonus tolerance is available. As the size of the feature departs from MMC toward LMC, the bonus tolerance increases; consequently, the location tolerance is directly proportional to the size of the feature as it departs from MMC toward LMC. Even though it is not well understood, zero positional tolerance at MMC provides the manufacturing staff with maximum flexibility and the greatest possibility of producing good parts.

Which has more tolerance: the drawing in Fig. 7-16*A*, a typical plus or minus tolerance for clearance holes, or the drawing in Fig. 7-16*B*, a zero positional tolerance at MMC? It is often erroneously assumed that a zero in the feature control frame means that there is no tolerance. This misconception occurs because the meaning of the MMC modifier in the feature control frame is not clearly understood.

Zero tolerance is never used without an MMC or LMC modifier. Zero at RFS would, in fact, be a zero tolerance no matter what size the feature is manufactured. Where zero positional tolerance at MMC is specified, the bonus is the tolerance that applies to the feature being controlled. In many cases, the bonus is larger than the tolerance that might otherwise be specified in the feature control frame. An analysis of the part in Fig. 7-16*B* indicates that the holes can be produced anywhere between .500 and .540 in diameter. If the holes are actually produced at a diameter of .535, the total location tolerance available is a cylindrical tolerance zone of .035 in diameter. The actual hole size, .535, minus the MMC, .500, equals a bonus tolerance of .035. GD&T reflects the exact tolerance available. For the drawing in Fig. 7-16*A*, the size of the holes must be between a diameter of .530 and .540. If the holes are produced at .535 in diameter, the total location tolerance available is actually a cylindrical tolerance zone of .035 just as it was above. But, since the title block tolerance is specified at ± .005, the inspector can accept the part only if the axes of the holes fall within their respective tolerance zones, which are .010 square. In this case, a tolerance of at least .025 is wasted. Tolerance is money. Why not use all of the tolerance available?

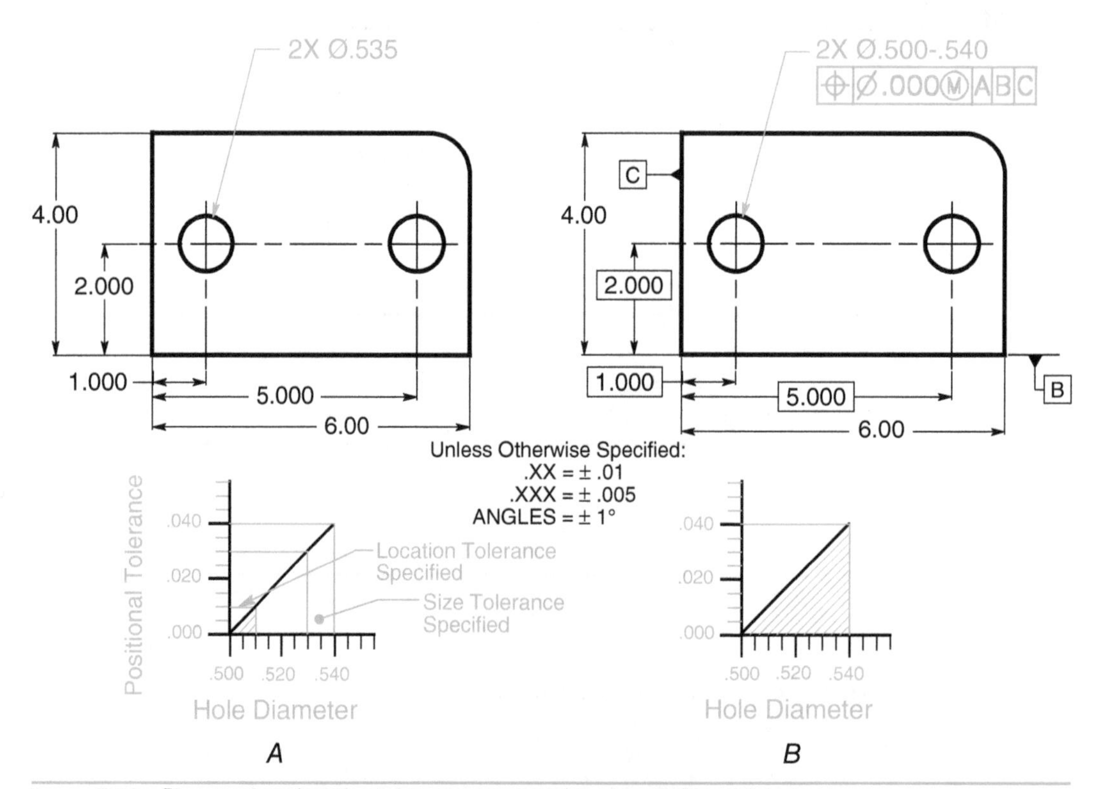

FIGURE 7-16 Plus or minus location tolerance compared to zero positional tolerance.

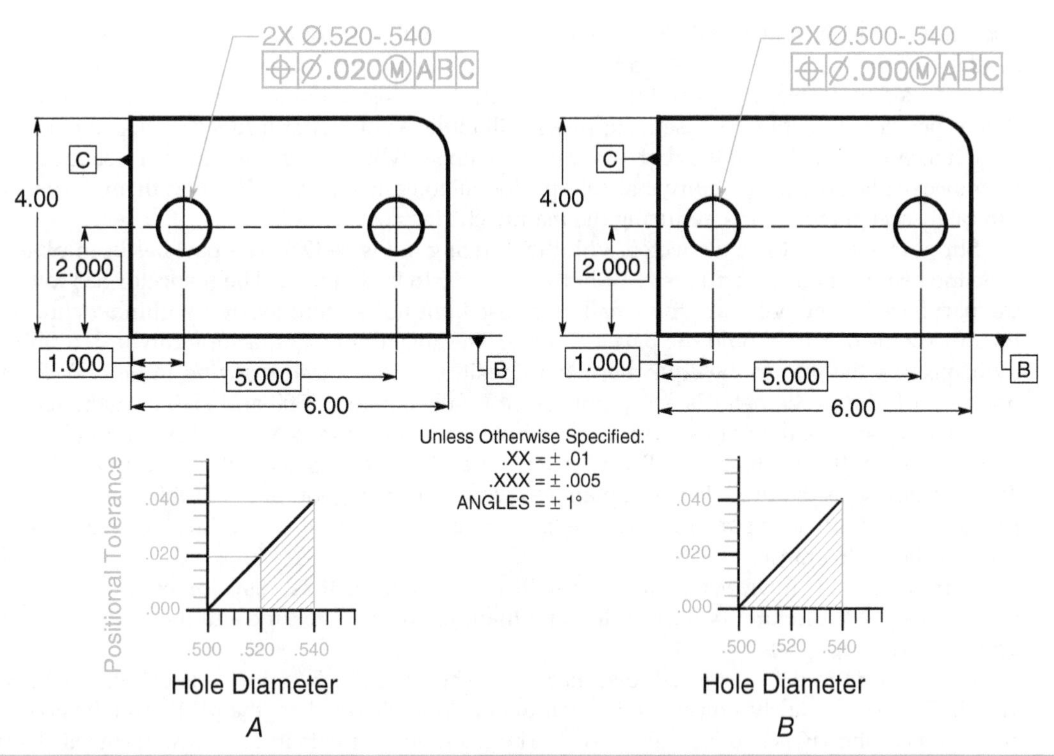

FIGURE 7-17 A specified position tolerance compared to zero positional tolerance.

The two parts in Fig. 7-17 are identical; they are just toleranced differently. If a part is made with the holes produced at .535 in diameter, what is the total location tolerance, and what is the virtual condition for these holes? In Table 7-3, the total tolerance and the virtual condition for the drawings in Fig. 7-17*A* and *B* are calculated.

The virtual condition and the total tolerance, for a given hole size at MMC, are the same no matter what tolerance is specified. But the range of the hole size has been increased when zero positional tolerance is used. Some engineers don't use zero positional tolerancing at MMC because they claim the manufacturing staff will not understand it. Consequently, they put some

	Total Positional Tolerance	
	Drawing *A*	**Drawing *B***
Actual Hole Size	.535	.535
MMC	– .520	– .500
Bonus	.015	.035
Geometric Tolerance	+ .020	+ .000
Total Positional Tolerance	.035	.035

	Virtual Condition	
	Drawing *A*	**Drawing *B***
MMC	.520	.500
Geometric Tolerance	– .020	– .000
Virtual Condition	.500	.500

TABLE 7-3 Both the Total Positional Tolerance and the Virtual Condition Are the Same Whether Controlled with a Numerical Tolerance or Zero Positional Tolerance at MMC

small number, such as .005, in the feature control frame with a possible .015 or .020 bonus tolerance available. If the machinist cannot read the bonus, he will produce the part within the tolerance of .005 specified in the feature control frame and charge the client company for the tighter tolerance. If zero positional tolerance is used, suppliers will either know what it means, ask someone else what it means, or not bid on the part. Actually, machinists who understand how to calculate bonus tolerance really like the flexibility that zero positional tolerancing at MMC gives them. Inspection can easily accept more parts, reducing the manufacturing costs.

Suppose a part is to be inspected with the drawing in Fig. 7-17*A*. The part has been plated a little too heavily, and the actual size of both holes is .518 in diameter. The inspector has to reject the part because the holes are too small. Suppose both holes were located within a cylindrical tolerance zone of .010. Would the part assemble? The answer to that question can be determined by inspecting the part to the equivalent, zero positional toleranced drawing in Fig. 7-17*B*. The hole size of .518 in diameter is acceptable since it falls between .500 and .540 in diameter. The location tolerance is the bonus, which is the hole size, .518, minus MMC, .500, or a cylindrical tolerance zone .018 in diameter. The part will fit and function since only a location tolerance .010 in diameter is required. Is it acceptable to scrap perfectly good parts? If this is a continuing problem for a particular part, submit an engineering change order converting the tolerance to a zero positional tolerance.

Zero positional tolerancing is a win-win-win situation. It is easy for engineers to use, machinists like the flexibility it provides, and manufacturers save time and money by reducing scrap and rework.

Converting the .020 positional tolerance for the holes in Fig. 7-18*A* to a zero positional tolerance in Fig. 7-18C is fairly simple. The only numbers to be changed are the MMC and the geometric tolerance, shown as blanks in Fig. 7-18*B*. The tolerance in the feature control frame is always converted to zero at MMC. A circle M symbol must follow the tolerance. Then convert the MMC of the feature to its virtual condition. In this case, the .520 MMC minus .020 geometric tolerance equals the virtual condition of .500 in diameter.

A zero tolerance is not used where the tolerance applies at RFS or where no bonus tolerance is available, as in a tolerance specified for threads or press fit pins.

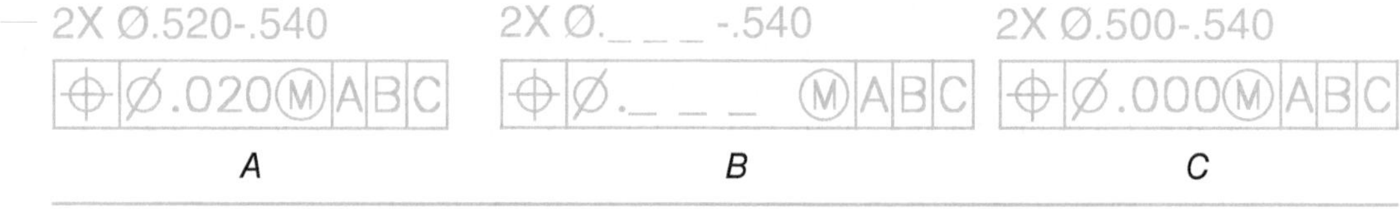

FIGURE 7-18 Converting the positional tolerance of a hole to zero positional tolerance.

Summary

- The tolerance of position may be viewed as a theoretical tolerance zone or a virtual condition boundary.
- A feature of size must be controlled for size, form, orientation, and location. Both size and form are controlled by the limits of size, and orientation and location are controlled by the tolerance of position.
- The orientation and location of true position is specified with basic dimensions from the datum features indicated. Once the feature control frame is assigned, an imaginary tolerance zone is defined and located about true position.

- RFS automatically applies to location and orientation tolerances of features of size if no material condition modifier is specified.
- When the MMC symbol modifies a position tolerance:
  1. The tolerance applies at the MMC size of the feature.
  2. As the size of the feature departs from MMC toward LMC, a bonus tolerance is achieved in the exact amount of such departure.
- Where datum features of size are specified at RMB, the datum is established by physical contact between the surface(s) of the datum feature and the surface(s) of the processing equipment. There is no shift tolerance for datum features specified at RMB.
- Where a datum feature of size is toleranced with a geometric tolerance and is referenced in a feature control frame at MMB, the resulting MMB for the datum feature is equal to its MMC or its virtual condition with respect to the preceding datum feature. It should be emphasized that when a shift tolerance applies to a pattern of features, it applies to the pattern as a group.
- Where the LMC symbol modifies a position tolerance:
  1. The tolerance applies at LMC.
  2. As the size of the feature departs from LMC toward MMC, a bonus tolerance is achieved in the exact amount of such departure.
- The worst-case inner and outer boundaries, or loci, are the virtual and resultant conditions; they are beneficial in performing a tolerance analysis.
- Zero positional tolerancing gives machinists more flexibility because manufacturing can easily accept more parts and charge less. For a given feature size, the total tolerance and the virtual condition are the same whether a numerical tolerance or a zero tolerance is specified.

Chapter Review

1. Position is a composite tolerance that controls both the ______________________________

   ______________________________ of features of size at the same time.

2. The tolerance of position may be viewed in either of two ways:

   - ____

     ____

     ____.

   - ____

     ____

     ____.

3. A feature of size has four geometric characteristics that must be controlled. These characteristics are ________________.

4. Since the position tolerance controls only features of sizes such as pins, holes, tabs, and slots, the feature control frame is always associated with a ________________.

5. The location of true position, the theoretically perfect location of an axis, is specified with ________________ from the datum features indicated.

6. Once the feature control frame is assigned, an imaginary ________________ ________________ is defined and located about true position.

7. Datum features are identified with ________________.

8. Datum features A, B, and C identify a ________________; consequently, they describe how the part is to be positioned for ________________.

9. If no material condition modifier is specified in the feature control frame, the ________________ modifier automatically applies to the tolerance of the feature.

10. To inspect a hole, the largest pin gage to fit inside the hole is used to simulate the ________________.

11. The measurement from the surface plate to the top edge of the pin gage minus half of the diameter of the pin gage equals the distance from ________________ ________________.

12. Where the MMC symbol is specified to modify the tolerance of a feature of size, the following two requirements apply:
    - The specified tolerance applies at ________________.
    - As the size of the feature departs from MMC toward LMC, ________________ ________________.

13. The difference between the actual mating envelope size and the MMC is the ________________.

14. The bonus plus the geometric tolerance equals ________________.

Ø.510 - .550

FIGURE 7-19 Geometric tolerance: Questions 15 through 18.

15. If the tolerance in Fig. 7-19 is for a pin .525 in diameter, what is the total positional tolerance?

    ____

16. What would be the size of the hole in a functional gage to inspect the pin above?

    ____

17. If the tolerance in Fig. 7-19 is for a hole .540 in diameter, what is the total positional tolerance? ______________________________

18. What would be the size of the pin on a functional gage to inspect the hole above?

    ____

19. Where a datum feature of size is toleranced with a geometric tolerance and is referenced in a feature control frame at MMB, the resulting MMB for the feature is equal to

    ______________________________.

20. A zero tolerance is not used where the tolerance applies at ______________

    or where no bonus tolerance is available as in a tolerance specified for ______________

    ______________________________.

Problems

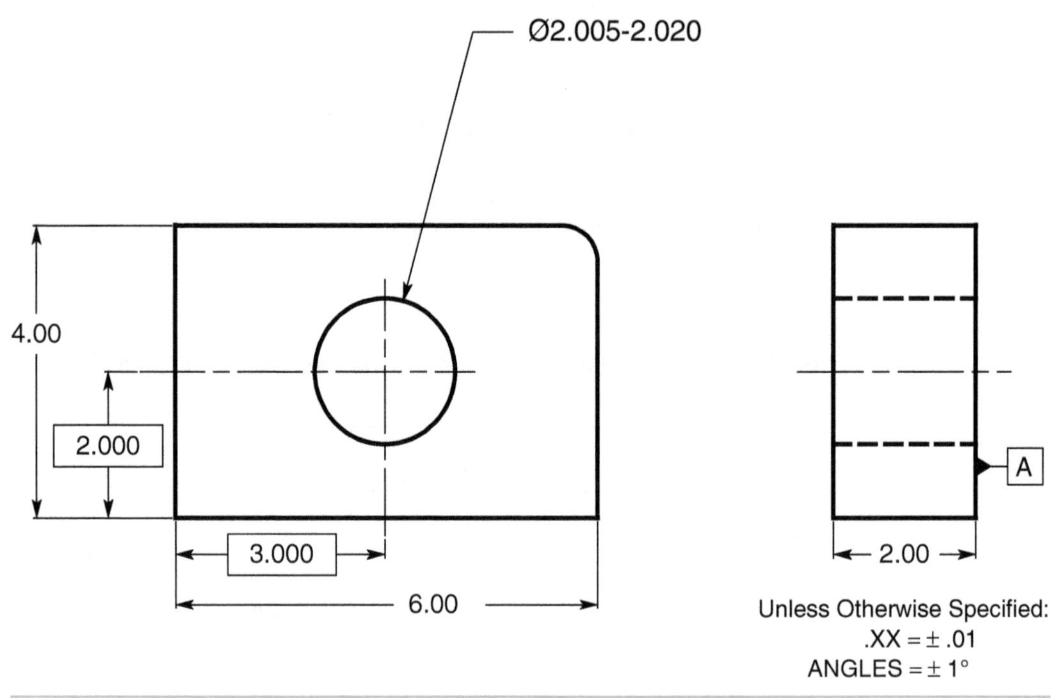

FIGURE 7-20 Locating a hole with the position control: Probs. 1 through 3.

1. Locate the hole in Fig. 7-20 with a positional tolerance of .005 at RFS.

2. Draw a feature control frame below to locate the hole in Fig. 7-20 with a positional tolerance of .005 at MMC.

____

____

3. If the actual mating envelope in the hole in Prob. 2 is produced at a diameter of 2.010 and the axis is located .003 over and .005 up from true position, is the part within tolerance? If not, can it be reworked to meet specifications?

____

____

____

____

____

____

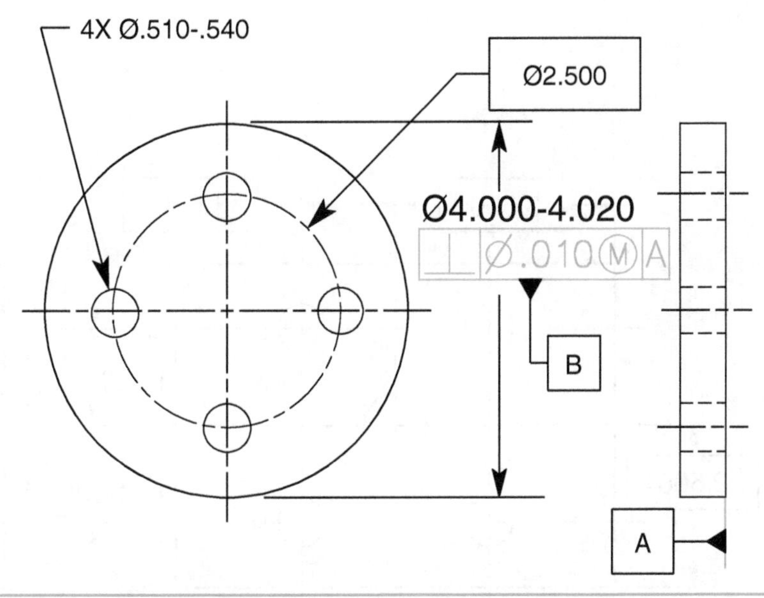

FIGURE 7-21 Locating features to a datum feature of size: Probs. 4 through 6.

4. Position the four-hole pattern with a tolerance of .010 at MMC in Fig. 7-21 perpendicular to datum feature A and located to datum feature B at RMB. If datum feature B is produced at a diameter of 4.010, how much shift tolerance is available?

5. Draw a feature control frame below to position the four-hole pattern in Fig. 7-21 perpendicular to datum feature A and located to datum feature B at MMB.

6. How much shift tolerance is available in Fig. 7-21 if datum feature B is specified at MMB and is produced at 4.015 in diameter?

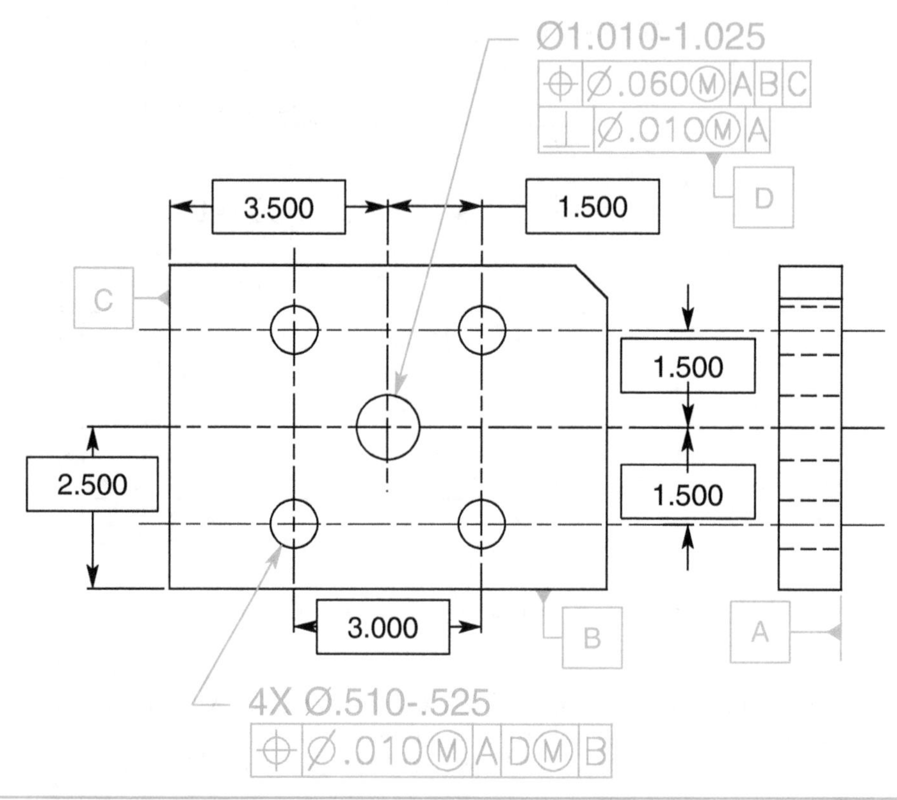

FIGURE 7-22 Design a gage to inspect for shift tolerance: Probs. 7 and 8.

7. On a gage designed to control the four-hole pattern in Fig. 7-22, what size pin must be produced to inspect the center hole (datum feature D)? ______________________

8. On the same gage, what is the diameter of the four pins locating the hole pattern?

____

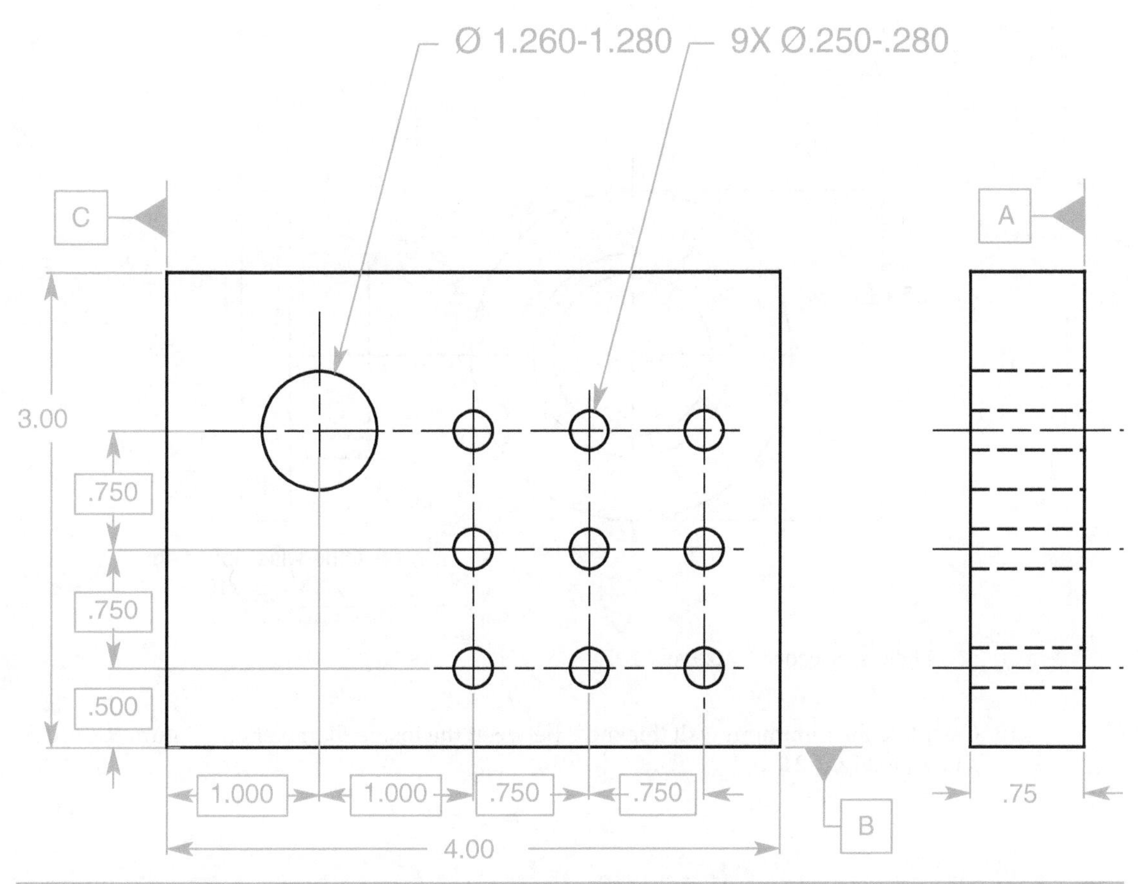

FIGURE 7-23 A pattern of holes located to a datum feature of size: Prob. 9.

9. In Fig. 7-23, locate the 1¼-inch-diameter hole to the edge datum features within a tolerance of .060 and refine its perpendicularity to datum feature A within a tolerance of .010. Locate the nine-hole pattern to the 1¼-inch-diameter hole and clock it to an edge datum feature with a zero positional tolerance. Use MMC and MMB wherever possible.

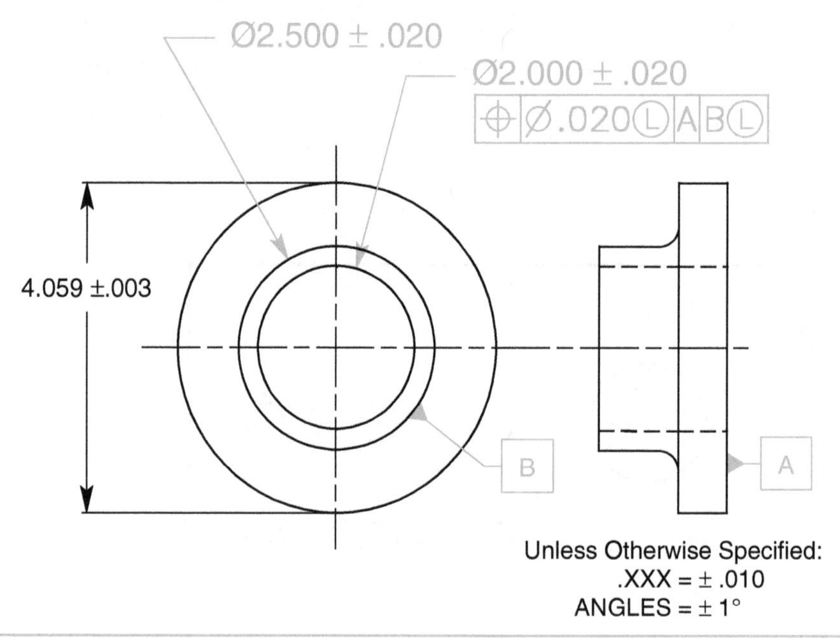

FIGURE 7-24 A hole specified at LMC: Prob. 10.

10. Calculate the minimum wall thickness between the inside diameter and datum feature B shown in Fig. 7-24.

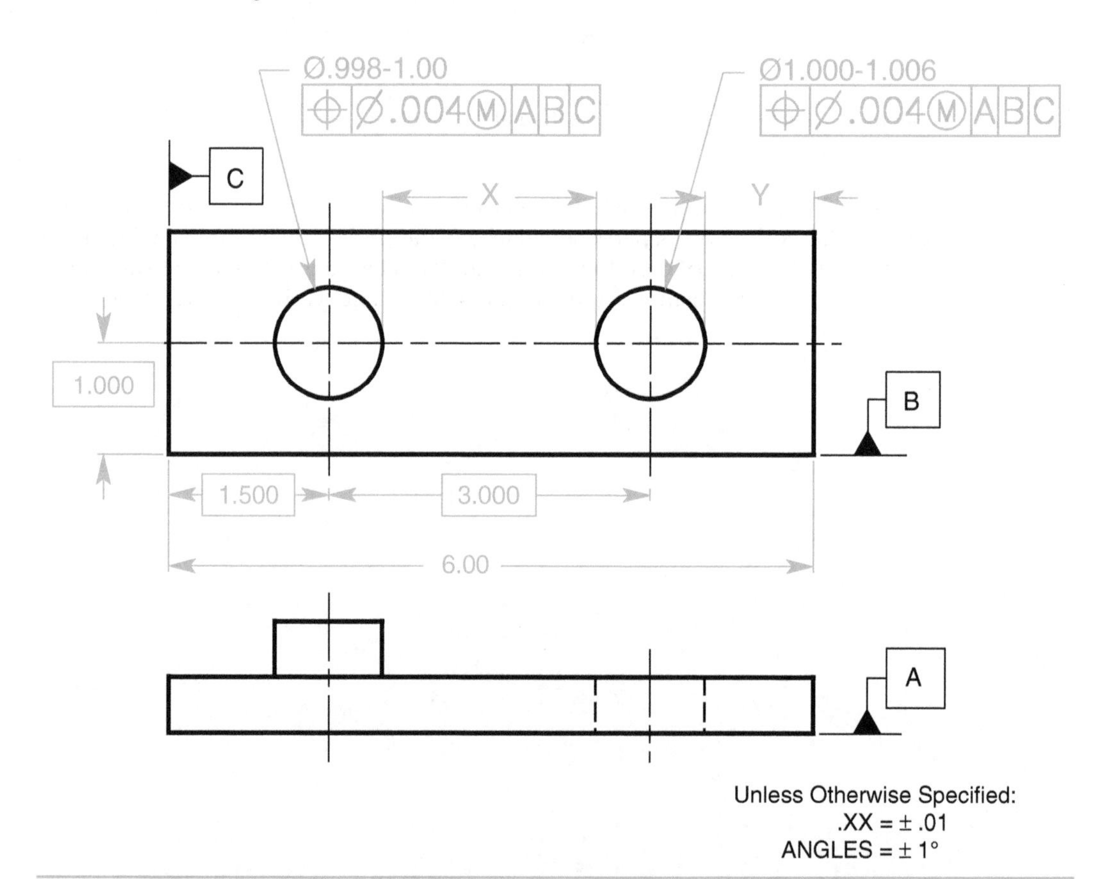

FIGURE 7-25 Boundary conditions: Prob. 11.

11. First calculate the virtual conditions and resultant conditions for the pin and hole in Fig. 7-25. Then calculate the maximum and minimum distances for dimensions X and Y.

Virtual Condition of the PIN

Virtual Condition of the HOLE

Resultant Condition of the PIN

Resultant Condition of the HOLE

The maximum and minimum distances for dimension **X**:

$X_{Max} =$

$X_{Min} =$

The maximum and minimum distances for dimension **Y**:

$Y_{Max} =$

$Y_{Min} =$

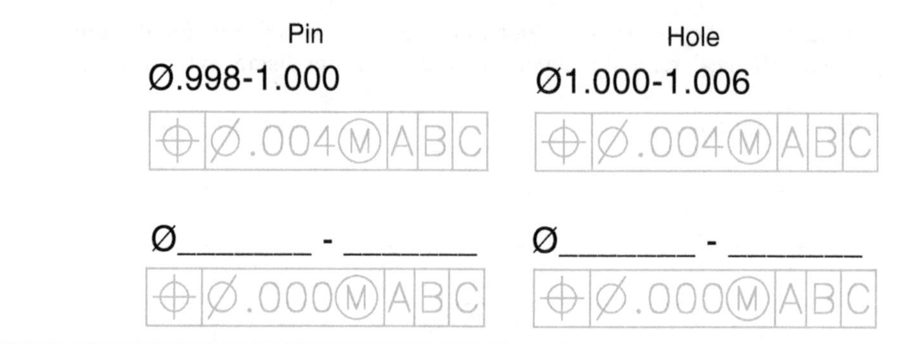

FIGURE 7-26 Zero positional tolerance conversion: Probs. 12 and 13.

12. Convert the tolerance in Fig. 7-26 to zero positional tolerances.

13. Zero tolerance is not used when the tolerance applies at ______________________ or when no bonus tolerance is available as in a tolerance specified for ______________________ ______________________.

CHAPTER 8

Position, Location

The most important function of the position control is to locate features relative to datum features and to one another. The position control is one of the most versatile of the 12 geometric controls. It controls both the location and the orientation of features of size and allows the application of maximum material condition (MMC), regardless of feature size (RFS), and least material condition (LMC) for location tolerances and maximum material boundary (MMB), regardless of material boundary (RMB), and least material boundary (LMB) for datum features of size. Most of the major applications of the position control are discussed in this chapter. Even though coaxial features may be located with the position control, coaxiality is a separate topic and will be discussed in the next chapter.

Chapter Objectives

After completing this chapter, the learner will be able to:

- *Calculate* tolerances for floating and fixed fasteners
- *Specify* projected tolerance zones
- *Apply* the concept of multiple patterns of features (simultaneous requirements)
- *Demonstrate* the proper application of composite positional tolerancing
- *Demonstrate* the proper application of multiple single-segment positional tolerancing
- *Tolerance* nonparallel holes
- *Tolerance* counterbores
- *Tolerance* noncircular features of size
- *Tolerance* spherical features
- *Tolerance* symmetrical features

Floating Fasteners

Because of the large number of fasteners used to hold parts together, tolerancing threaded and clearance holes may be among the most frequent tolerancing activities that an engineer performs. Often, due to ignorance, habit, or both, fasteners are toleranced too tightly. This section on fasteners attempts to provide the knowledge that allows engineers to make sound tolerancing decisions for floating and fixed fasteners.

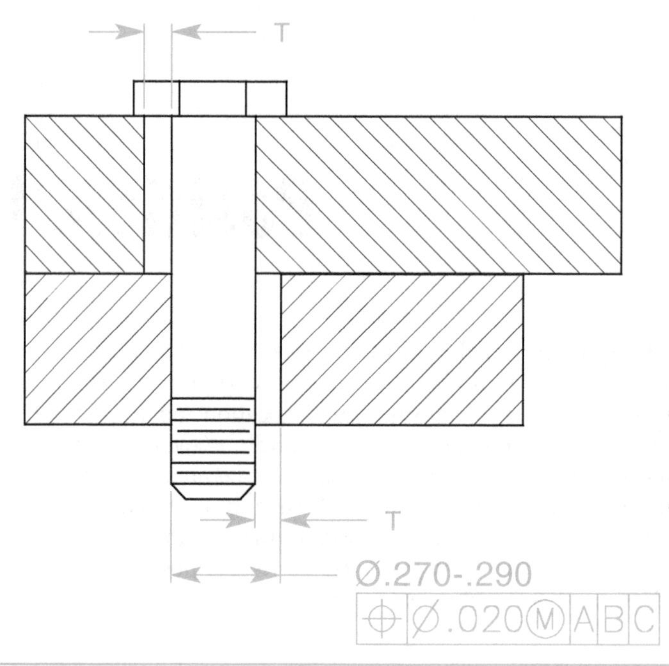

FIGURE 8-1 Floating fasteners.

The floating fastener got its name from the fact that the fastener is not restrained by any of the members being fastened. In other words, all parts being fastened together have clearance holes in which the fastener can float before the fastener is tightened demonstrated graphically in Fig. 8-1.

The floating fastener formula is:

$$T = H - F \qquad \text{or} \qquad H = F + T$$

T is the clearance hole location tolerance at MMC.

H is the clearance hole MMC diameter.

F is the fastener's MMC diameter, the nominal size.

The floating fastener tolerance applies to each hole in each part:

$$H = F + T$$
$$H = .250 + .020 = .270$$

Once the fastener size is determined, three pieces of information are needed to complete the clearance hole dimension and tolerances. They are illustrated in Fig. 8-2 and listed below:

1. Clearance hole LMC diameter (The numbers relate to the numbers in Fig. 8-2.)
2. Clearance hole location tolerance (T)
3. Clearance hole MMC diameter (H)

FIGURE 8-2 Floating fastener dimension and tolerances.

Clearance Hole LMC Diameter

The first step in calculating fastener tolerances is to determine the clearance hole LMC diameter, the largest possible clearance hole size. The LMC hole diameter is, essentially, arbitrary. Of course, the clearance hole must be at least large enough to include the fastener plus the stated positional tolerance, and it cannot be so large that the head of the fastener pulls through the hole. In a few cases where the fastener is highly torqued, the size of the clearance hole at LMC should be engineered.

It has been suggested that the clearance hole should not be larger than the largest hole that will fit under the head of the fastener. If a slotted clearance hole such as the one in Fig. 8-3*A* will fit and function for a ¼-inch fastener, then surely the .337-diameter hole in Fig. 8-3*B* will also fit and function especially if a washer is assembled. The largest hole that will fit under the head of a fastener is the sum of half of the diameter of the fastener plus half of the diameter of the fastener head, or half of the distance across the flats of the head, as shown in Fig. 8-3*C*. The LMC clearance hole can also be calculated by adding the diameter of the fastener and the diameter of the fastener head and then dividing the sum by two. The largest-diameter LMC clearance holes for some of the common fasteners are listed in the table "Machine and Cap Screw Sizes" shown in App. B:

Clearance hole LMC diameter = (F + F head)/2

Clearance hole LMC diameter = (.250 + .425)/2 = .337

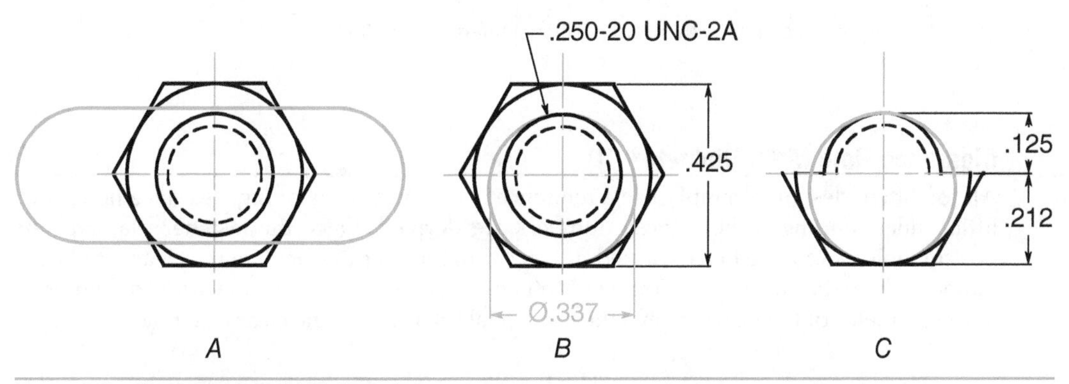

FIGURE 8-3 Clearance hole LMC diameter.

This method of selecting the LMC clearance hole size is a rule of thumb that will yield the largest hole that will fit under the head of the fastener. Engineers may select any size clearance hole that is required. For the ¼-inch fastener in Fig. 8-1, the .337 diameter might have been selected for the clearance hole in this example, but instead, a more conservative LMC clearance hole size of .290 in diameter was assigned. With the use of the above formula, engineers can make an informed decision and not have to blindly depend on an arbitrary clearance hole tolerance chart.

Clearance Hole Location Tolerance (T)

The positional tolerance at MMC for the clearance hole is also arbitrary since bonus tolerance is available. Zero positional tolerance at MMC is as good as any, but a positional tolerance of .020 was selected in the example in Fig. 8-1 to illustrate the relationship between the clearance hole MMC diameter and the location tolerance.

The location tolerance for a given hole diameter at MMC is the same no matter what tolerance is specified in the feature control frame. If the clearance hole actual mating envelope for Figs. 8-1 and 8-4 is produced at a diameter of .285, the total location tolerance is .035 no matter what positional tolerance is specified. The total positional tolerance equals the stated position tolerance plus the bonus. The bonus tolerance equals the actual mating envelope minus the MMC:

$$\text{Total positional tolerance} = \text{position tolerance} + \text{bonus}$$

For a .020 tolerance at MMC $\quad \text{Total positional tolerance} = .020 + (.285 - .270) = .035$

or

For a .000 tolerance at MMC $\quad \text{Total positional tolerance} = .000 + (.285 - .250) = .035$

If the machinist happens to produce the actual mating envelope size smaller than .270 for the clearance holes in Fig. 8-1, perhaps at a diameter .265, the hole size will be out of tolerance and will reject the part. If zero positional tolerance at MMC is specified, the hole size is acceptable, but the hole must be located within a positional tolerance zone .015 in diameter.

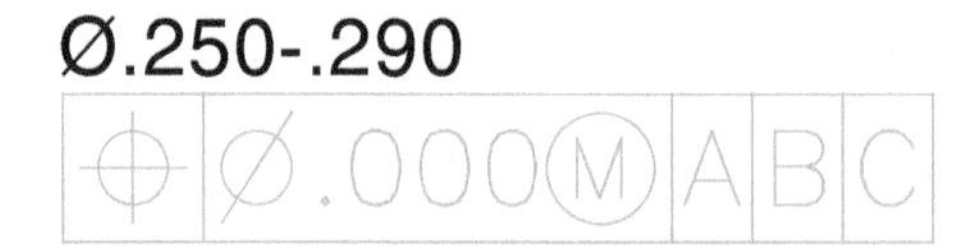

FIGURE 8-4 A clearance hole with a zero positional tolerance at MMC.

Clearance Hole MMC Diameter (H)

All too often, designers simply use a reference chart for dimensioning clearance holes and have little understanding of how these numbers are derived. Once the fastener diameter and the position tolerance have been selected, it is a simple matter to calculate the clearance hole MMC diameter. In reality, the calculations couldn't be easier. The clearance hole MMC diameter is equal to the diameter of the fastener plus the positional tolerance of the clearance hole:

$$H = F + T$$

$$H = .250 + .020 = .270$$

The floating fastener formula is a simple formula to remember. The hole has to be larger than the fastener. The clearance hole location tolerance is equal to the difference between the actual size of the clearance hole and the size of the fastener, as shown graphically in Fig. 8-1. No matter what tolerance is selected, it is important to use the formula to determine the correct MMC clearance hole diameter. If the clearance hole diameter is incorrect, either a possible no fit condition exists or tolerance is wasted.

Fixed Fasteners

The fixed fastener is fixed by one or more of the members being fastened. The formula for fixed fasteners is essentially the same as for floating fasteners except that the fixed fastener formula includes the tolerance for each hole demonstrated graphically in Fig. 8-5.

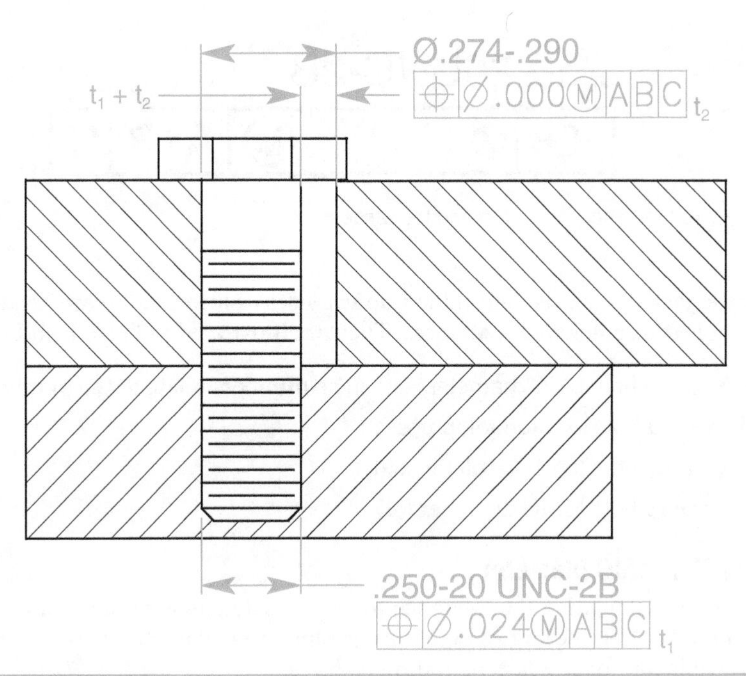

FIGURE 8-5 Fixed fasteners.

The fixed fastener formula is:

$$t_1 + t_2 = H - F \qquad \text{or} \qquad H = F + t_1 + t_2$$

t_1 is the threaded hole location tolerance at MMC

t_2 is the clearance hole location tolerance at MMC

H is the clearance hole MMC diameter

F is the fastener's MMC diameter, the nominal size

$$H = F + t_1 + t_2$$
$$H = .250 + .024 + .000 = .274$$

This formula is sometimes expressed in terms of 2T instead of $t_1 + t_2$; however, 2T implies that the tolerances for the threaded and clearance holes are the same. In most cases, it is desirable to assign more tolerance to the threaded hole than the clearance hole because the threaded hole is usually more difficult to manufacture.

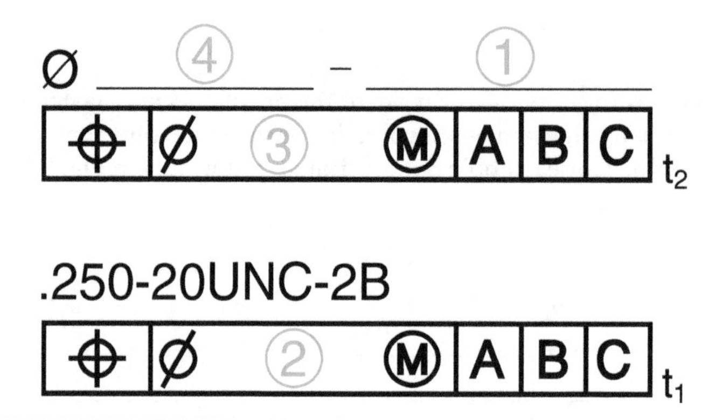

FIGURE 8-6 Fixed fastener dimension and tolerances.

Once the fastener size is determined, four pieces of information are needed to complete the clearance hole dimension and tolerances. They are illustrated in Fig. 8-6 and listed below.

1. Clearance hole LMC diameter (The numbers relate to the numbers in Fig. 8-6.)
2. Threaded hole location tolerance (t_1)
3. Clearance hole location tolerance (t_2)
4. Clearance hole MMC diameter (H)

Clearance Hole LMC Diameter

The first step in calculating the tolerance for fixed fasteners is to determine the diameter of the clearance hole at LMC, the largest possible clearance hole diameter. This step is the same for fixed fasteners as it is for floating fasteners shown above.

$$\text{Clearance hole LMC diameter} = (F + F\ \text{head})/2$$
$$\text{Clearance hole LMC diameter} = (.250 + .425)/2 = .337$$

For this example, the engineer might have specified the largest hole that will fit under the head of the ¼-inch fastener, a diameter of .337, but instead, the engineer selected the more conservative .290 diameter, as shown in Fig. 8-5.

Threaded Hole Location Tolerance (t_1)

The location tolerance for both the threaded hole and the clearance hole must come from the difference between the actual mating envelope of the clearance hole and the diameter of the fastener, the total tolerance available:

$$\text{Total location tolerance} = \text{Clearance hole LMC diameter} - \text{fastener diameter}$$
$$\text{Total location tolerance} = .290 - .250 = .040$$

Since drilling and taping a hole involves two operations and cutting threads in a hole is more problematic than just drilling the hole, it is common practice to assign a larger portion of the location tolerance to the threaded hole. In this example, 60% of the tolerance is assigned to the threaded hole; the remaining tolerance applies to the clearance hole:

$$\text{Thread location tolerance} = 60\% \times \text{the total location tolerance}$$
$$\text{Thread location tolerance} = 60\% \times .040 = .024$$

The position control locating the threaded hole has a cylindrical tolerance zone .024 in diameter at MMC. Zero positional tolerance is not appropriate for a threaded hole since there is almost no bonus tolerance between mating threaded features. The tolerance is specified at MMC because there is some movement, however small, between the assembled threaded features, and a very small bonus tolerance may be available. Those who are tempted to specify a location tolerance at RFS should be aware that costly inspection equipment, such as a spring thread gage, is required to inspect the thread, and a more restrictive tolerance is imposed on the thread. Parts should be toleranced and inspected the way they fit and function in assembly.

Clearance Hole Location Tolerance (t_2)

There are those who like to assign a position tolerance of .005 or .010 at MMC for the clearance hole location. However, a tolerance at MMC is arbitrary since bonus tolerance is available. If there is doubt about which location tolerance to use, specify zero positional tolerance at MMC. Zero positional tolerance at MMC will provide all of the tolerance available and give the machinist the most size flexibility in producing the clearance hole. Zero positional tolerance at MMC has been assigned to the clearance hole in this example and will be used to calculate the MMC hole diameter.

Clearance Hole MMC Diameter (H)

Once the fastener and the position tolerances for the threaded and clearance holes have been selected, it is a simple matter to calculate the clearance hole MMC diameter. The positional tolerance for the threaded hole, t_1, is .024 in diameter, and the positional tolerance for the clearance hole, t_2, is .000. Whichever tolerance is selected, it is important to use this formula to calculate the correct clearance hole MMC diameter. If the MMC clearance hole diameter is incorrect, either a possible no fit condition exists or tolerance is wasted:

$$H = F + t_1 + t_2$$

$$H = .250 + .024 + .000 = .274$$

At this point, the engineer may wish to check a drill chart (Table 8-1) to determine the actual tolerance available. A drill chart and a chart of oversize diameters in drilling are located in App. B. The letter L drill would not be used since the drill will probably produce a hole .002 or .003 oversize. If the letter K drill were used and drilled only .002 oversize, the actual mating envelope of the clearance hole would be .283 in diameter:

$$\text{Clearance hole location tolerance} = \text{actual mating envelope} - \text{MMC}$$

$$\text{Clearance hole location tolerance} = .283 - .274 = .009$$

Because of the drill size used, the total available clearance hole tolerance is not .040 but .033, and the amount of tolerance assigned to the threaded hole is more than 70% of the total tolerance. At this point, the designer may want to increase the clearance hole LMC diameter or reduce the threaded hole location tolerance.

Letter	Fraction	Decimal
	17/64	.266
H	–	.266
I	–	.272
J	–	.277
	9/32	.281
K	–	.281
L	–	.290

TABLE 8-1 A Partial Drill Chart (a complete drill chart is located in App. B)

A fixed fastener is fixed by one or more of the members being fastened. Both fasteners in Fig. 8-7 are fixed; the fastener heads are fixed by their countersunk holes. The fastener in Fig. 8-7*B* is also fixed in the threaded hole at the other end of the screw. This screw is considered to be a double fixed fastener. Double fixed fasteners should be avoided. It is not always possible to avoid double fixed fastener conditions where flat head fasteners are required, but a misaligned, double fixed fastener with a high torque may cause the fastener to fail. Cap screws in counterbored holes will produce single fixed fasteners that are flush with the surface of the part.

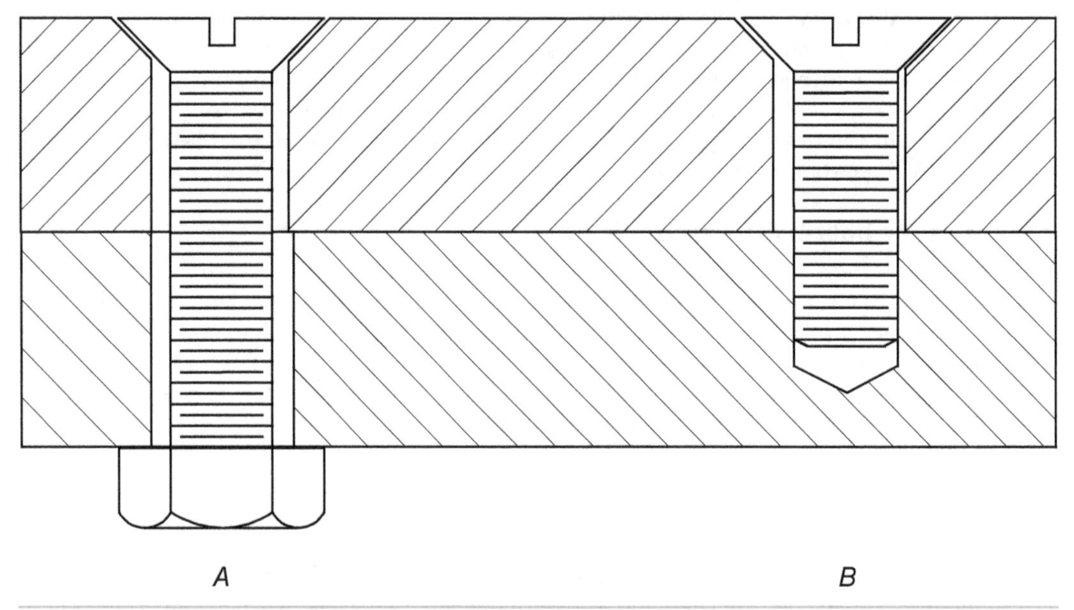

FIGURE 8-7 A fixed fastener and a double fixed fastener.

Projected Tolerance Zones

Where specifying a threaded hole or a hole for a press fit pin, the orientation of the hole determines the orientation of the mating pin. Although the location and orientation of the hole and the location of the pin will be controlled by the tolerance zone of the hole, the orientation of the pin outside the hole cannot be guaranteed, as shown in Fig. 8-8*A*. The most convenient way to control the orientation of the pin outside the hole is to project the tolerance zone into the mating part. The tolerance zone must be projected in the direction of and at the greatest thickness of the mating part as shown in Fig. 8-8*B*. The specified height of the tolerance zone is equal to or greater than the thickest mating part or tallest stud or pin after installation. In other words, the tolerance zone height is specified to be at least as tall as the MMC thickness of the mating part or the maximum height of the installed stud or pin.

Through Holes

When specifying a projected tolerance zone for a through hole, place a circle P in the feature control frame after the tolerance and any material condition modifier, then specify both maximum height and direction by drawing and dimensioning a thick chain line next to an extension of the centerline. The chain line is the MMC height of the mating part and located on the side where the mating part assembles. If the mating part is 1.500 ± .030 thick and assembles on top of the plate over the through hole shown in Fig. 8-9, the chain line is extended up above the hole and dimensioned with the MMC thickness of the mating part, 1.530. The projected height is a minimum.

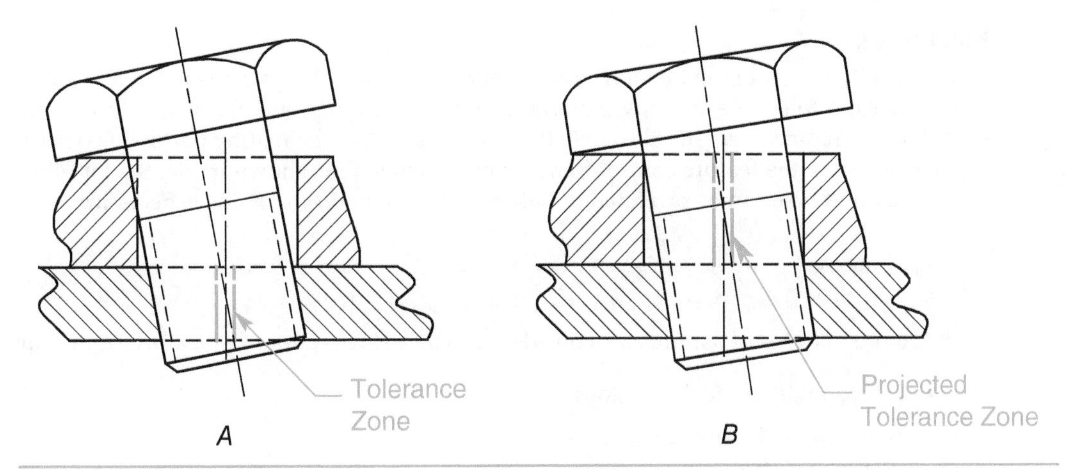

FIGURE 8-8 A standard tolerance zone compared to a projected tolerance zone.

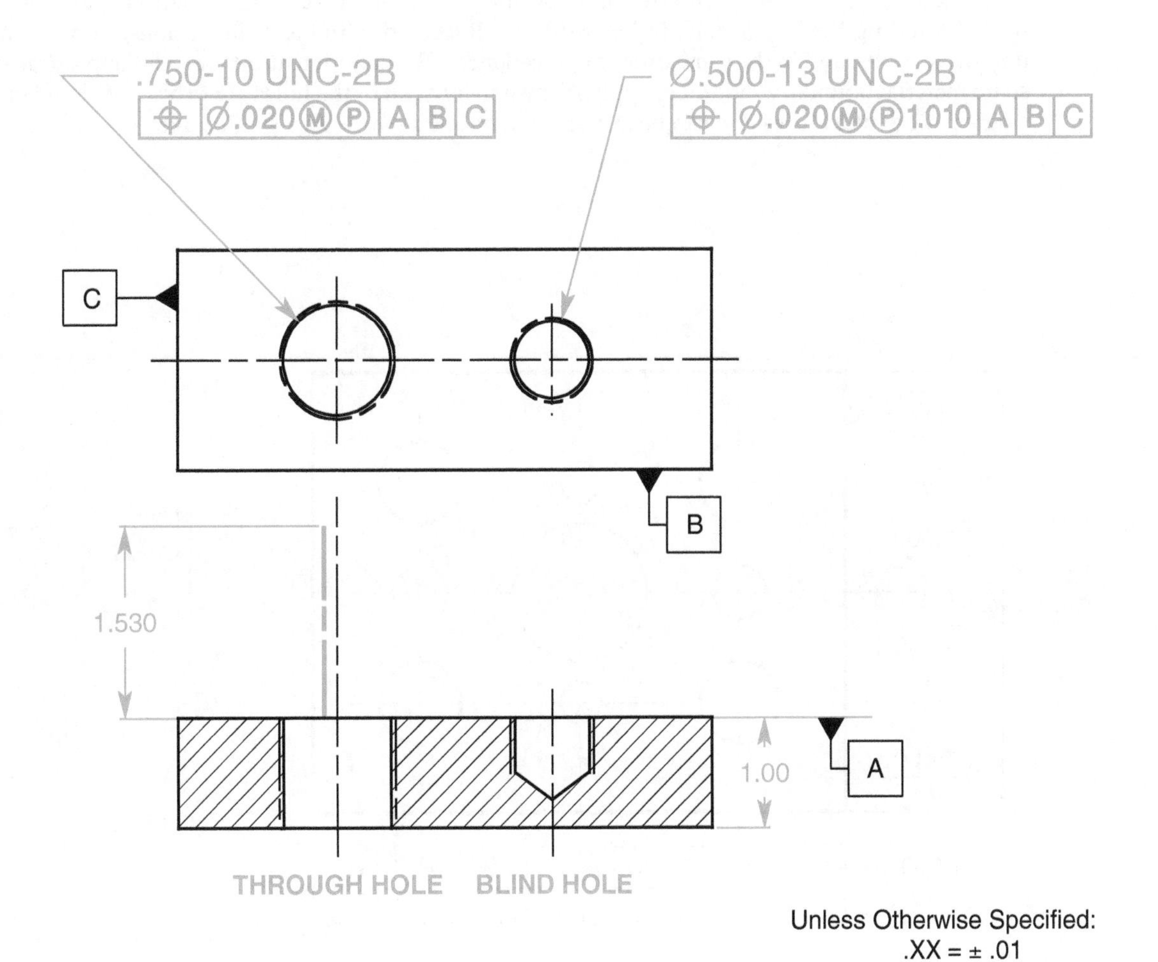

FIGURE 8-9 Specifying projected tolerance zones for through and blind holes.

Blind Holes

Where specifying a projected tolerance zone for a blind hole, place a circle P in the feature control frame after the tolerance and any material condition modifier and specify the projected MMC height of the mating part after the circle P. If the thickness of the mating part is 1.000 ± .010, then 1.010 is placed in the feature control frame after the circle P, as shown in Fig. 8-9 for blind holes. Because a blind hole can be projected in only one direction, no chain line is necessary.

Multiple Patterns of Features, Simultaneous Requirements

Two or more patterns of features are considered to be a single pattern of features if they are:

- Located with basic dimensions
- To the same datums features
- In the same order of precedence
- At the same material boundary modifier

Even though they are different sizes and specified at different tolerances, the four patterns of holes (including the center hole) in Fig. 8-10 are all located with basic dimensions, to the same datums features, and in the same order of precedence. (The datum features are all plane surfaces, so no material boundary modifiers apply.) Consequently, all of the holes are to be considered one single pattern of features and can be inspected in one setup or with a single gage.

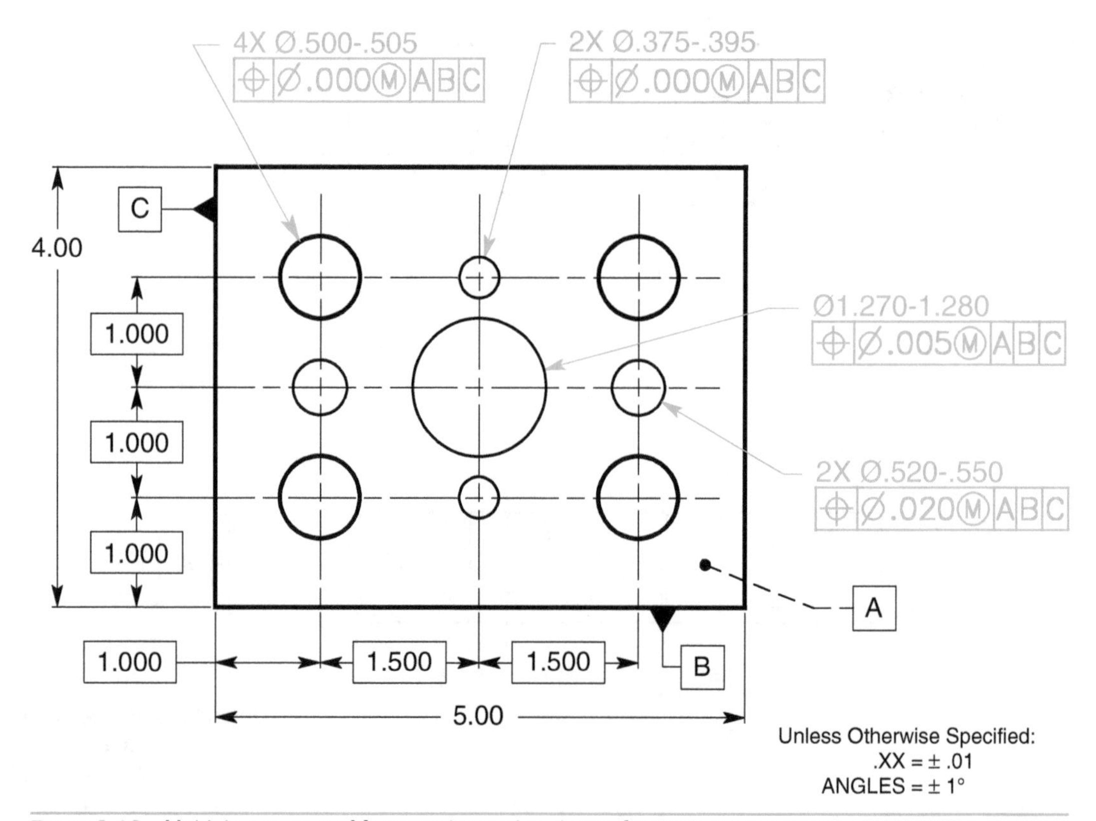

FIGURE 8-10 Multiple patterns of features located to datum features not subject to size variation (plane surfaces).

In spite of the fact that they are different sizes and specified at different tolerances, the four-hole patterns in Fig. 8-11 are all located with basic dimensions, to the same datum features, in the same order of precedence, and with the same material boundary modifiers. The outside diameter, datum feature B, is a feature of size specified at RMB. Datum features of size specified at RMB require physical contact between the gaging element and the datum feature. Consequently, the part cannot shift inside a gage or open setup. The four patterns of holes are to be considered one single pattern of features and can be inspected in one setup or with a single gage.

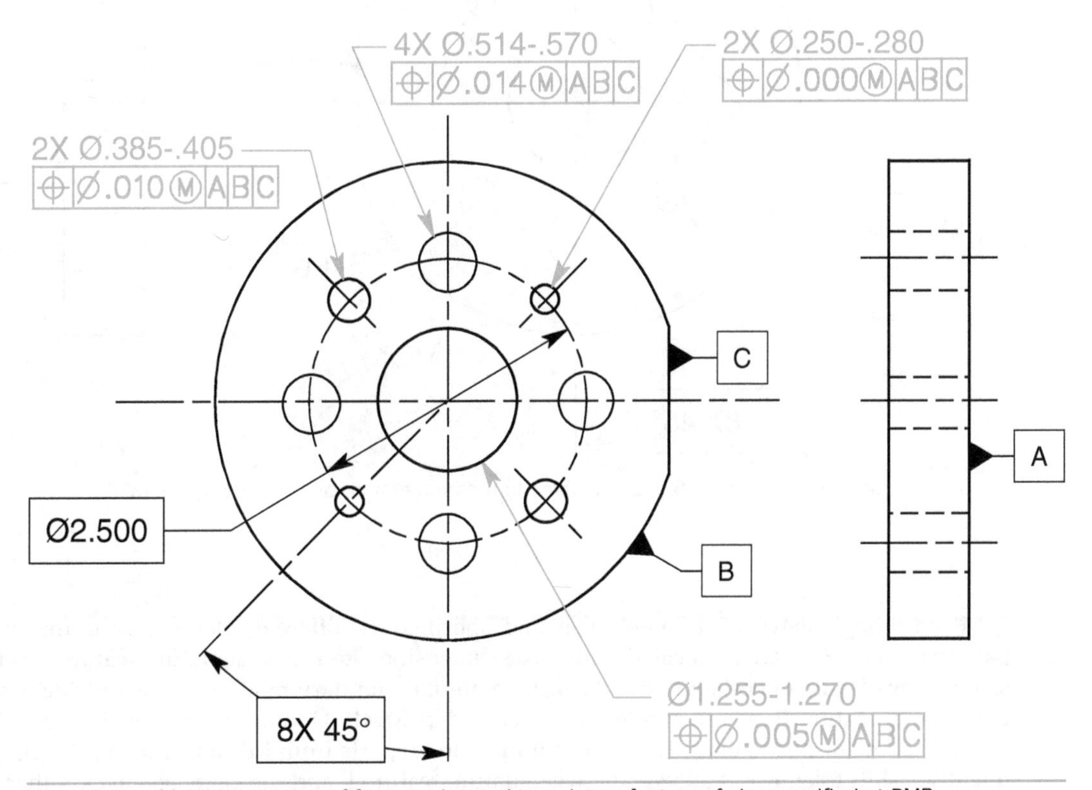

FIGURE 8-11 Multiple patterns of features located to a datum feature of size specified at RMB.

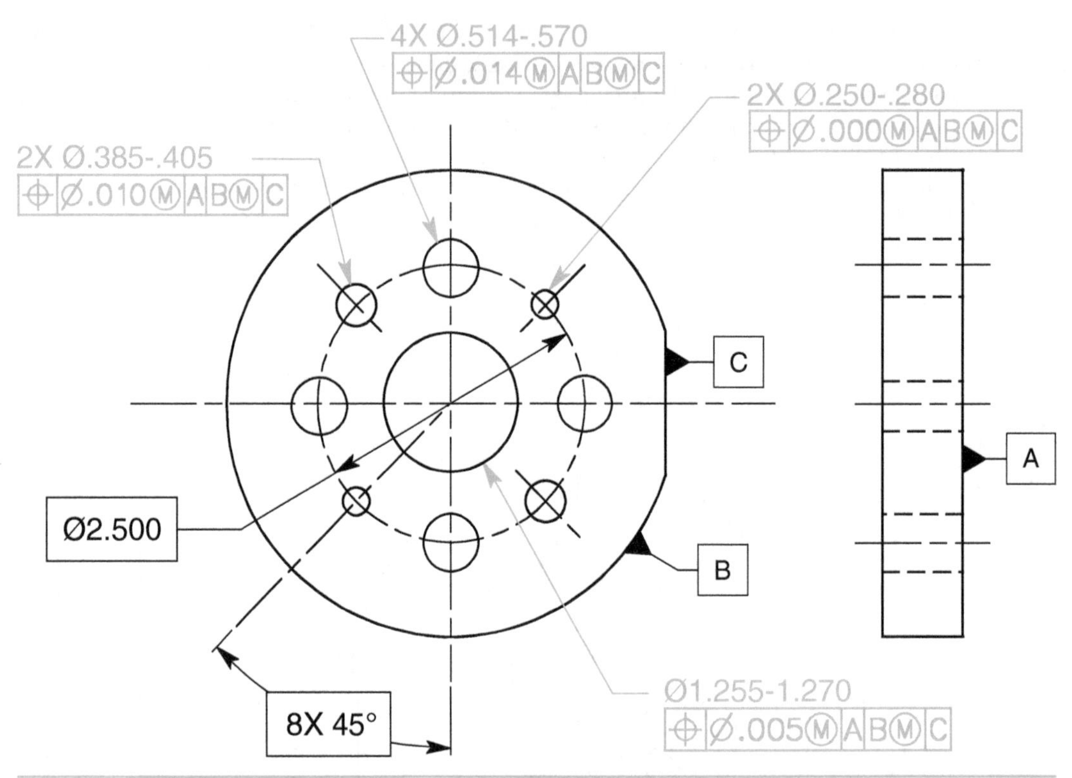

FIGURE 8-12 Multiple patterns of features located to a datum feature of size specified at MMB.

Even though they are different sizes and specified at different tolerances, the four-hole patterns in Fig. 8-12 are all located with basic dimensions, to the same datum features, in the same order of precedence, and with the same material boundary modifiers. The outside diameter, datum feature B, is a feature of size specified at MMB. Datum features of size specified at MMB allow a shift tolerance as the datum feature departs from MMB toward LMC. Consequently, a shift tolerance is allowed between datum feature B and the gage; however, if there is no note, the four patterns of holes are to be considered one single pattern of features and must be inspected in one setup or with a single gage. No matter how the features are specified, as long as they are located with basic dimensions, to the same datum features, in the same order of precedence, and with the same material boundary modifiers, the default condition is that patterns of features are to be treated as one single pattern of features.

If patterns of features controlled to datum features of size specified at MMB have no relationship to each other, a note such as SEP REQT may be placed under each feature control frame, allowing each pattern to be inspected separately. If some patterns are to be inspected separately and some simultaneously, a local note is required to clearly communicate the desired specifications.

Composite Positional Tolerancing

When locating patterns of features, there are situations where the relationship from feature to feature must be kept to a certain tight tolerance and the relationship between the pattern and its datum features is not as critical and may be held to a looser tolerance. These situations often occur when combining technologies that are typically held to different tolerances. For example, composite tolerancing is recommended if a hole pattern on a sheet metal part must be held to a tight tolerance from feature to feature and located from a datum feature that has several bends between the datum feature and the pattern requiring a larger tolerance. Since many industries make machined components that are mounted to a welded frame, the location of the components may be able to float within a tolerance of 1/8 of an inch to the welded frame, but the mounting hole pattern might require a .030 tolerance from feature to feature. Both of these tolerancing arrangements can easily be achieved with composite positional tolerancing as shown in Fig. 8-13.

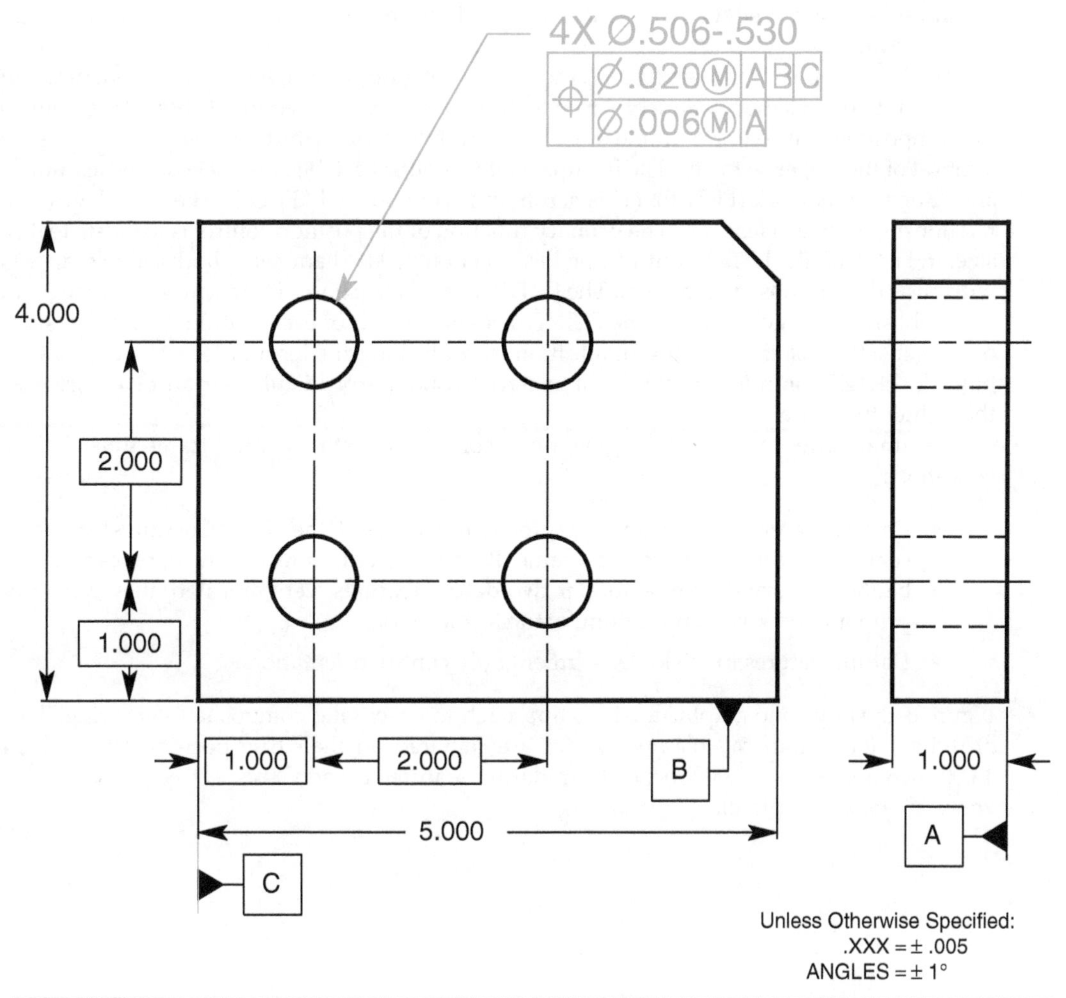

FIGURE 8-13 A composite positional tolerance controlling a four-hole pattern to its datum features with one tolerance and a feature-to-feature relationship with a smaller tolerance.

FIGURE 8-14 The composite feature control frame.

A composite feature control frame, shown in Fig. 8-14, has one position symbol that applies to the two horizontal segments that follow. The upper segment is called the pattern-locating control. It acts like any other position control. The pattern-locating control establishes the Pattern-Locating Tolerance Zone Framework (PLTZF). (This acronym is pronounced "Plahtz.") The PLTZF governs the relationship between the datum features and the pattern. Letter A in the upper segment of the feature control frame orients the .020-diameter cylindrical tolerance zones perpendicular to datum feature A. Letters B and C locate the tolerance zones with basic dimensions to datum features B and C.

The lower segment is referred to as the feature-relating control. It is a refinement of the upper control and governs the relationship from feature to feature. Each complete horizontal segment in the composite feature control frame must be separately verified, but the lower segment is always a subset of the upper segment. The feature-relating control establishes the Feature-Relating Tolerance Zone Framework (FRTZF). (This acronym is pronounced "Fritz.") The FRTZF governs the relationship between features. The primary function of the position control is to locate features of size. In Fig. 8-15, the FRTZF controls the location of the .006-diameter cylindrical tolerance zones with basic dimensions to each other. The FRTZF is free to rotate and translate within the boundaries established and governed by the PLTZF. That is, the axis of each feature must fall inside each of its respective tolerance zones. In addition to controlling the location of features from hole to hole, the FRTZF controls the cylindrical tolerance zones perpendicular to datum feature A within the tighter tolerance.

Datum features in the lower segment of a composite feature control frame must satisfy two conditions:

- Datum features in the lower segment must repeat the datum features in the upper segment of the feature control frame. If only one datum feature were repeated, it would be the primary datum feature; if two datum features were repeated, they would be the primary and secondary datum features, and so on.
- Datum features in the lower segment only control orientation.

Figure 8-15 shows the graphic analysis approach to specifying composite tolerancing. The four .020-diameter cylindrical tolerance zones are centered on their true positions located a basic 1.000 inch and a basic 3.000 inches from datum features B and C, respectively. These tolerance zones are locked in place.

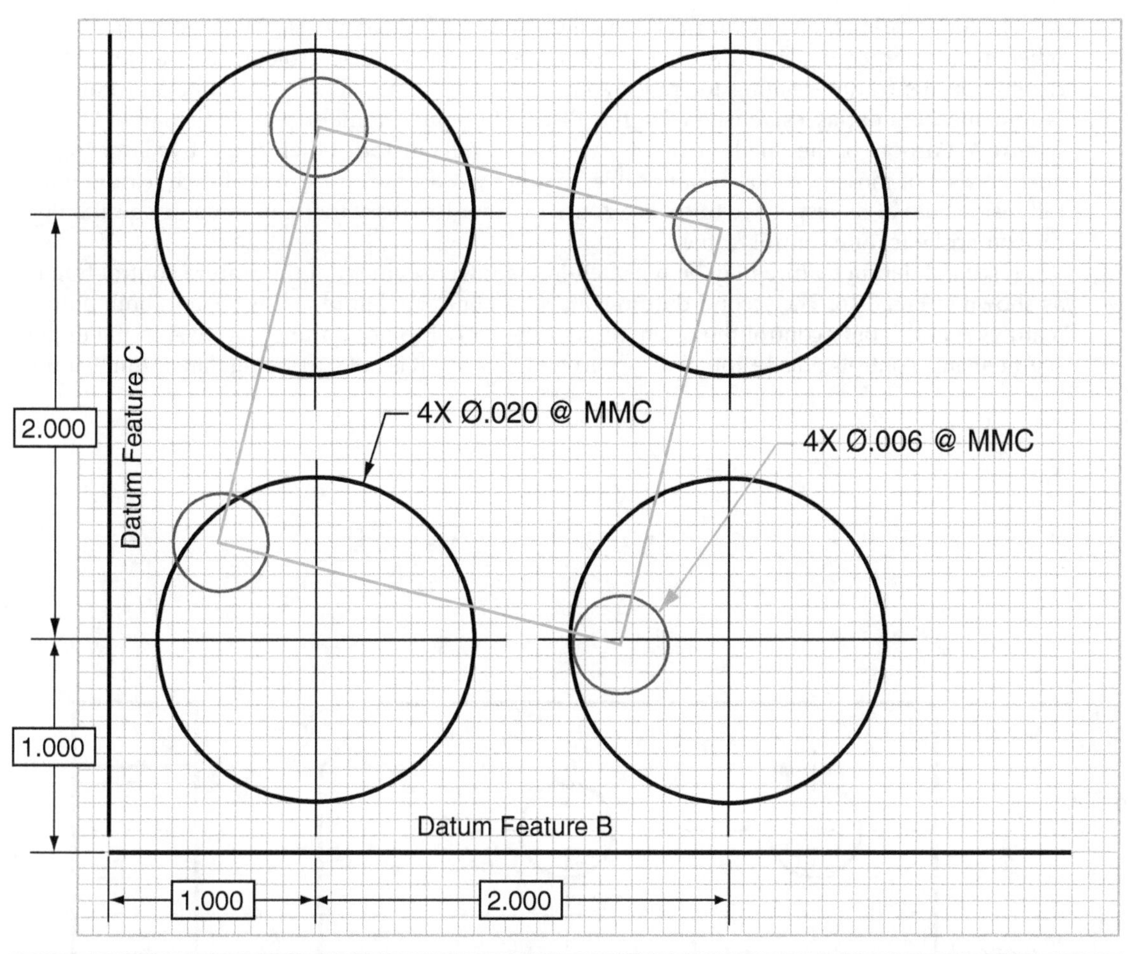

FIGURE 8-15 A graphic analysis approach to specifying the relationship between the tolerance zones of the PLTZF and the FRTZF for the drawing in Fig. 8-13.

The four .006-diameter cylindrical tolerance zones are located a basic 2.000 inches from each other and are perpendicular to datum feature A. These four cylindrical tolerance zones are locked together in the feature-relating tolerance zone framework. The FRTZF can rotate and translate within the boundaries established by the PLTZF but must remain perpendicular to datum feature A. Portions of the smaller tolerance zones may fall outside of their respective larger tolerance zones, but those portions are unusable. In other words, the entire axis of each feature must fall inside both of its respective tolerance zones in order to satisfy the requirements specified by the composite feature control frame.

A second datum feature may be repeated in the lower segment of the feature control frame, as shown in Fig. 8-16. The second datum feature can only be datum feature B, and both datum features A and B only control the orientation of the FRTZF. Since datum feature A in the upper segment only controls orientation, that is, perpendicularity, it is not surprising that datum feature

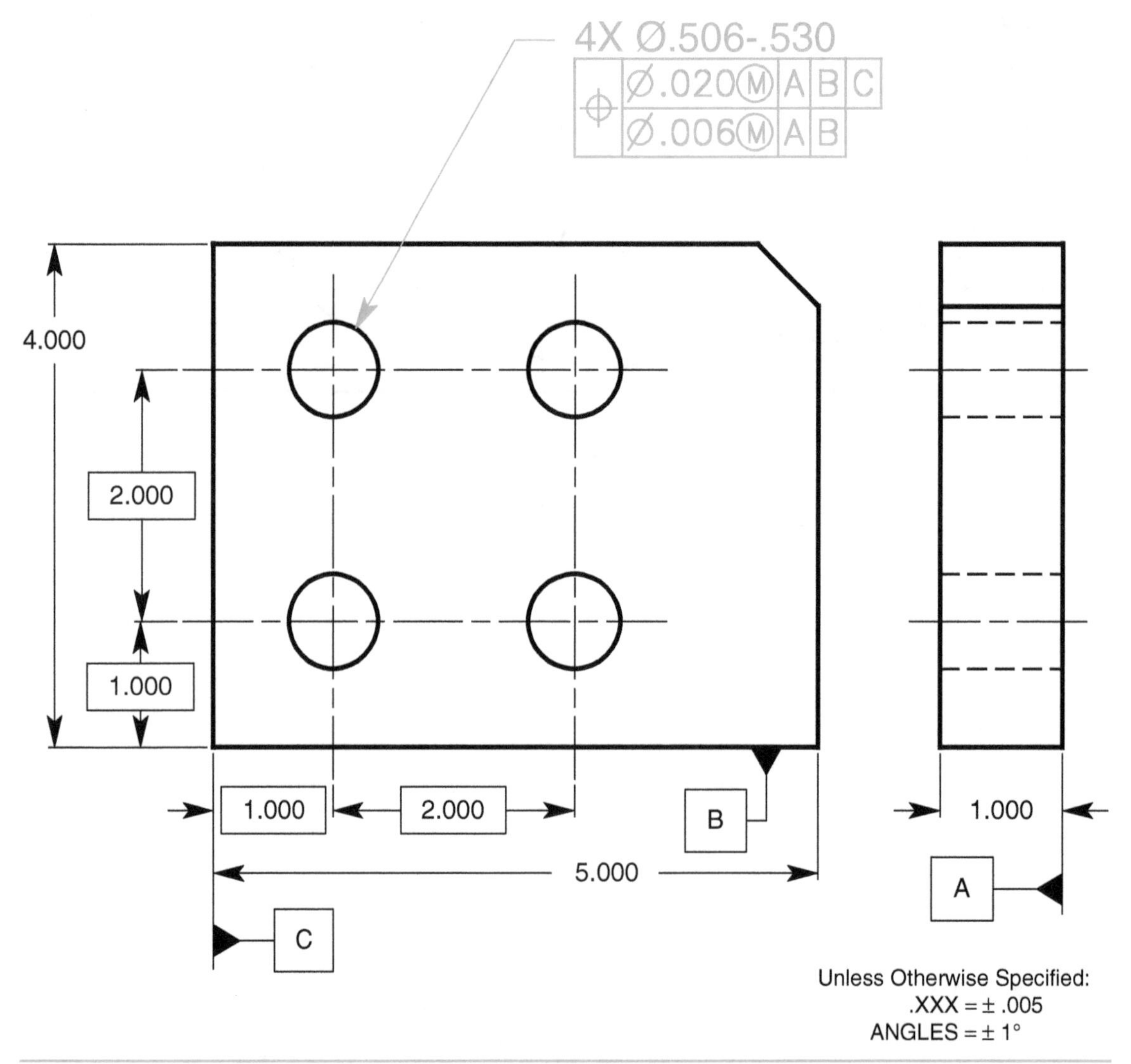

FIGURE 8-16 A composite positional tolerance with datum features A and B repeated in the lower segment of the feature control frame.

A in the lower segment is a refinement of perpendicularity to a tighter tolerance. Where datum feature B is included in the lower segment of the composite feature control frame, the .006-diameter cylindrical tolerance zone framework must remain parallel to datum feature B. That means that the FRTZF is allowed to translate up and down and left and right but may not rotate about an axis perpendicular to datum feature A. The tolerance zone framework must remain parallel to datum plane B at all times, as shown in Fig. 8-17.

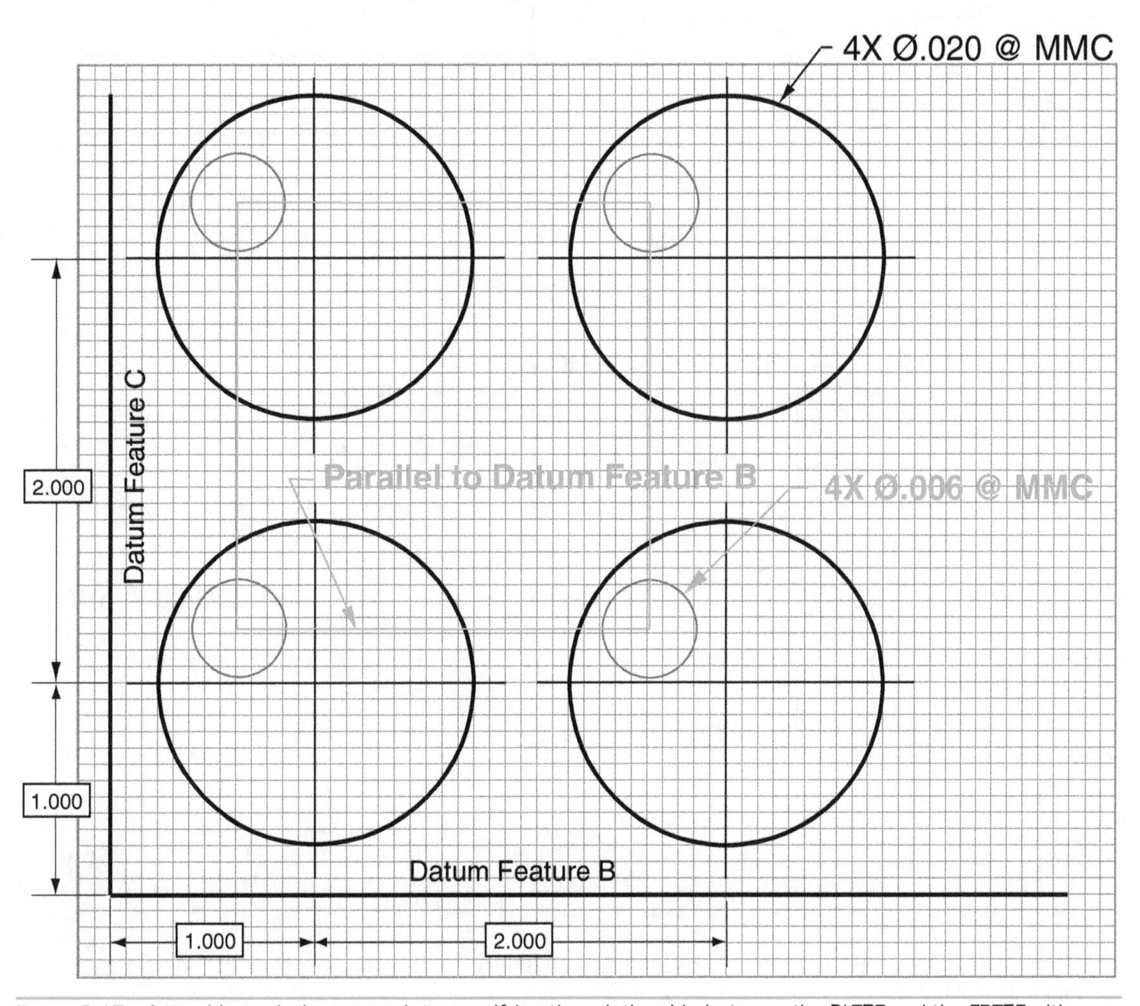

FIGURE 8-17 A graphic analysis approach to specifying the relationship between the PLTZF and the FRTZF with datum features A and B repeated in the lower segment of the composite feature control frame specified in the drawing in Fig. 8-16.

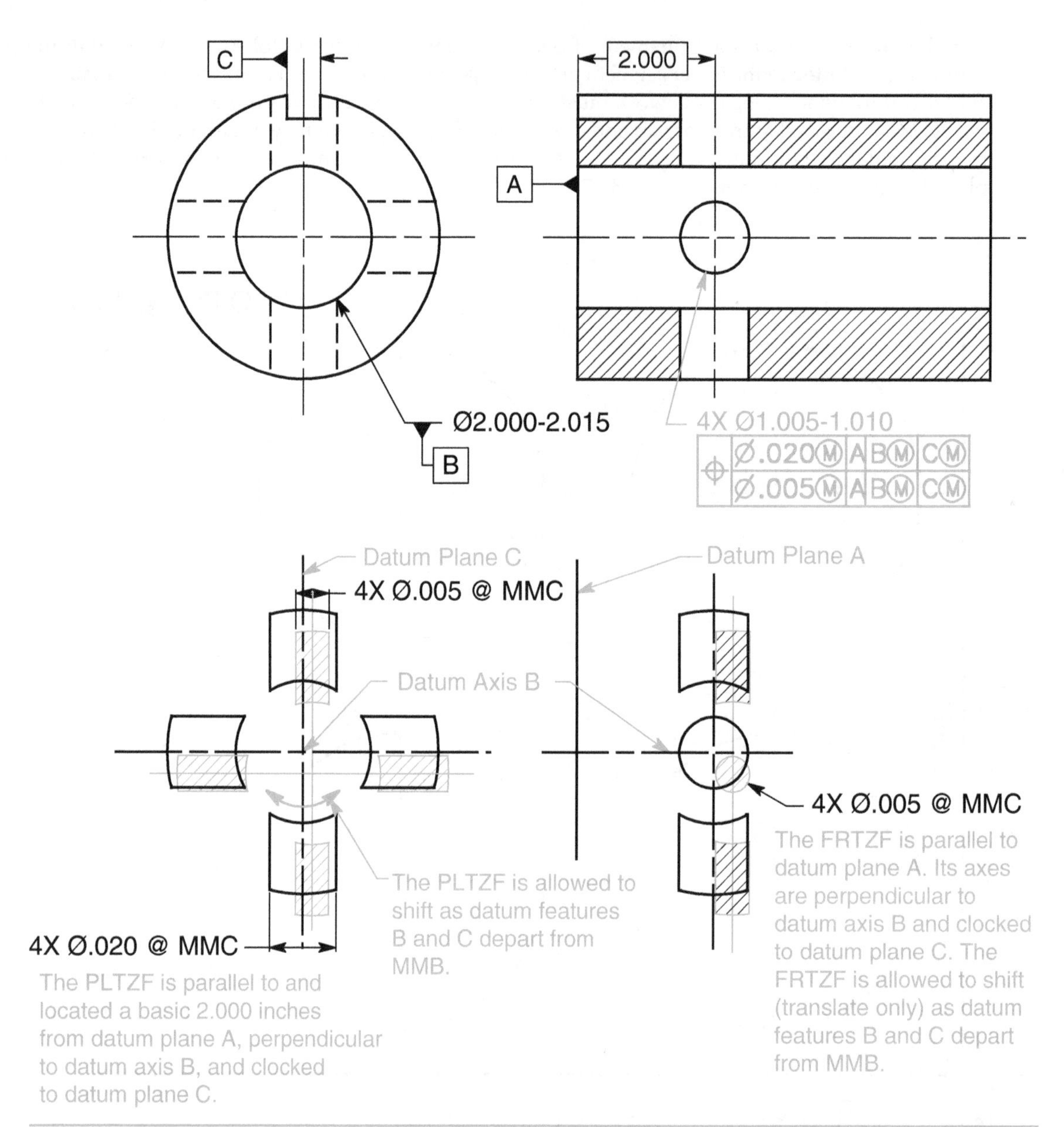

FIGURE 8-18 A composite positional tolerance with three datum features in both the upper and lower segments of the feature control frame.

In a more complex geometry, Fig. 8-18, the four holes are located by the .020-diameter PLTZF held parallel to and located with a basic dimension to datum plane A, centered on and perpendicular to datum axis B, and clocked to datum center plane C. Since datum features B and C are feature of sizes and specified at MMB, a shift tolerance is allowed. As the datum features depart from MMB toward LMC, the PLTZF, as a group, can shift with respect to datum axis B and rotate about datum axis B as permitted by datum feature C. The pattern location is further refined by the FRTZF within the .005-diameter cylindrical tolerance zones, which may translate in all directions, but held parallel to datum plane A, perpendicular to datum axis B, and clocked to datum center plane C. As datum features B and C depart from MMB toward LMC, a shift tolerance with respect to translation is allowed for the .005-diameter FRTZF. Because both datum features B and C are repeated in the lower segment of the feature control frame, the pattern-locating tolerance zone framework may not rotate in any direction. Each feature axis must fall inside both of its respective tolerance zones.

Multiple Single-Segment Positional Tolerancing

The four-hole pattern in Fig. 8-19 is toleranced with a control called the multiple single-segment feature control frame. The position symbol is entered in each of the single segments. In this case, the lower segment refines the feature-to-feature relationship just as does the lower segment of the composite feature control frame, but the datums behave differently. The lower segment of the multiple single-segment feature control frame acts just like any other position control. If datum feature C had been included in the lower segment, the upper segment would be meaningless, and the entire pattern would be controlled to the tighter cylindrical tolerance of .006 in diameter. In Fig. 8-19, the lower segment of the multiple single-segment feature control frame refines the feature-to-feature relationship oriented perpendicular to datum feature A and located with a

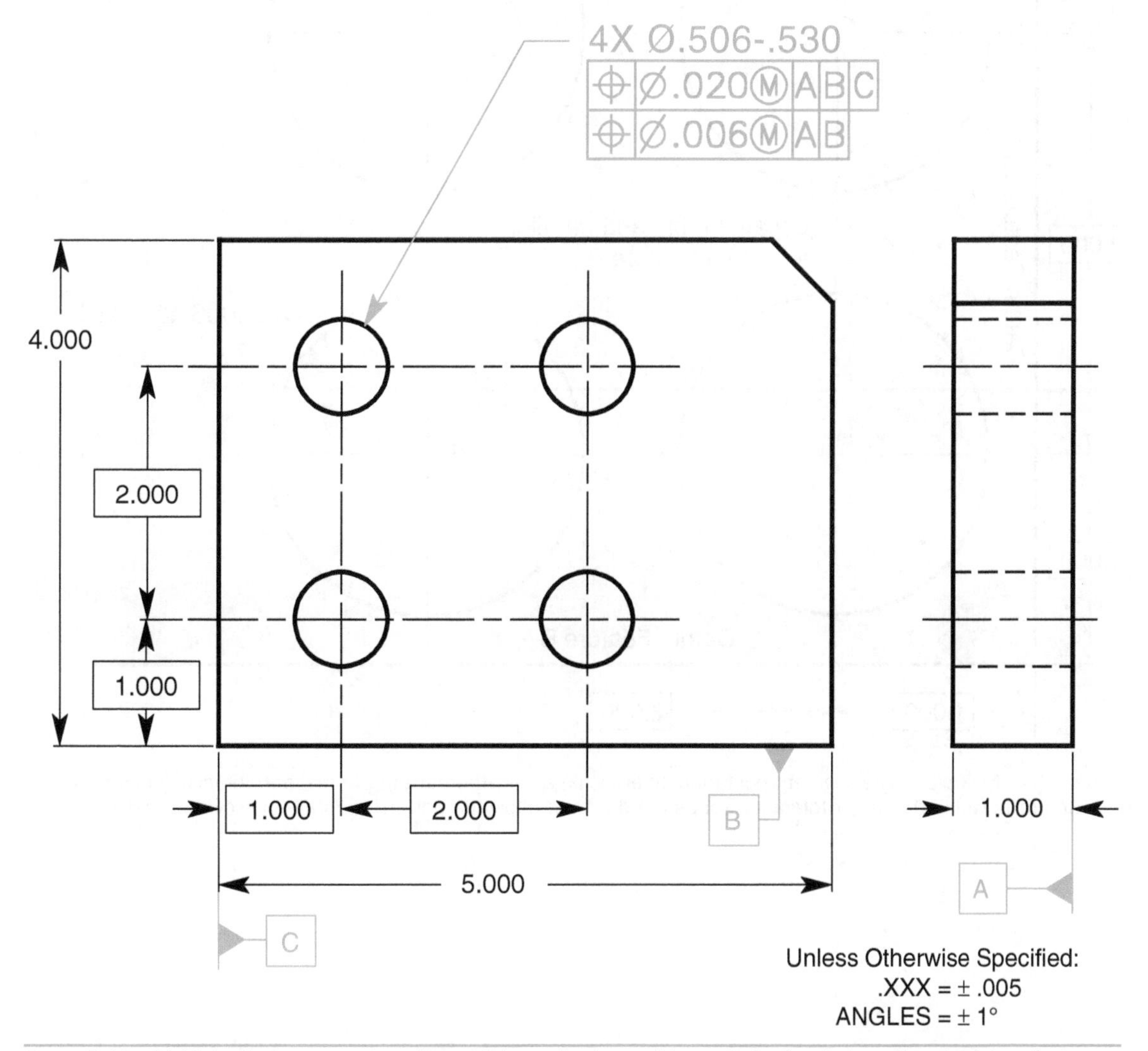

FIGURE 8-19 Multiple single-segment positional tolerancing is used to control a four-hole pattern to datum features A and B with a .006-diameter cylindrical tolerance and to datum feature C with a .020-diameter cylindrical tolerance.

basic dimension to datum feature B within a .006-diameter cylindrical tolerance zone. The upper segment allows the feature relating tolerance zone framework to translate back and forth relative to datum feature C within the .020-diameter cylindrical tolerance zone. In other words, the smaller tolerance zone framework is locked to datum feature B by a basic 1.000-inch dimension and cannot rotate or move up or down. This control only allows the .006-diameter cylindrical tolerance zone framework to shift back and forth relative to datum feature C within the larger tolerance zone of .020 in diameter, as shown in Fig. 8-20.

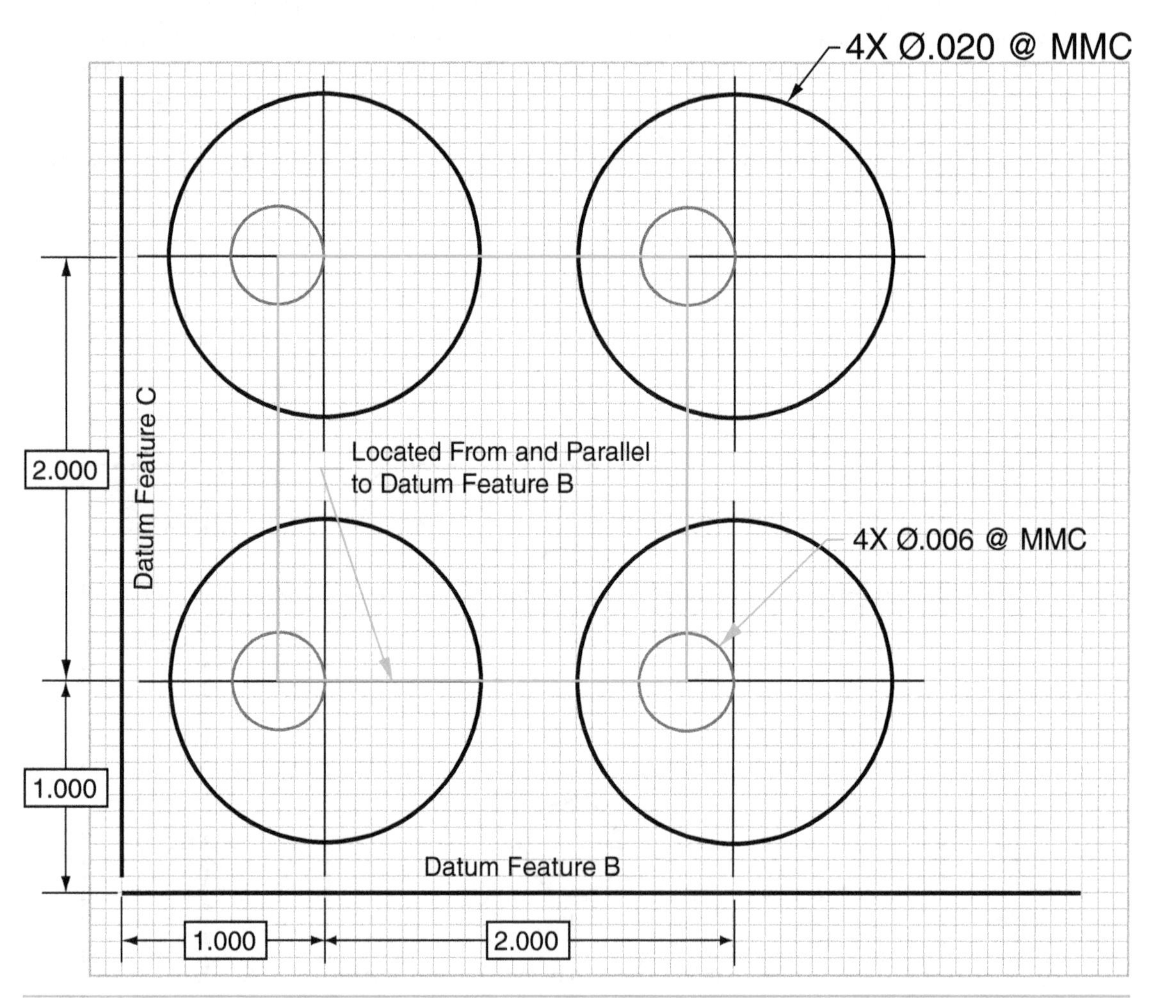

FIGURE 8-20 Multiple single-segment positional tolerancing allows the pattern of smaller tolerance zones to move back and forth within the larger tolerance zones but does not allow the pattern to rotate or move up or down.

Nonparallel Holes

The position control is so versatile that it can control a radial pattern of holes at an angle to a primary datum plane. As shown in Fig. 8-21, the radial pattern of 8 holes is dimensioned with a basic 45° to each other about datum axis B and at a basic 30° angle to datum plane A.

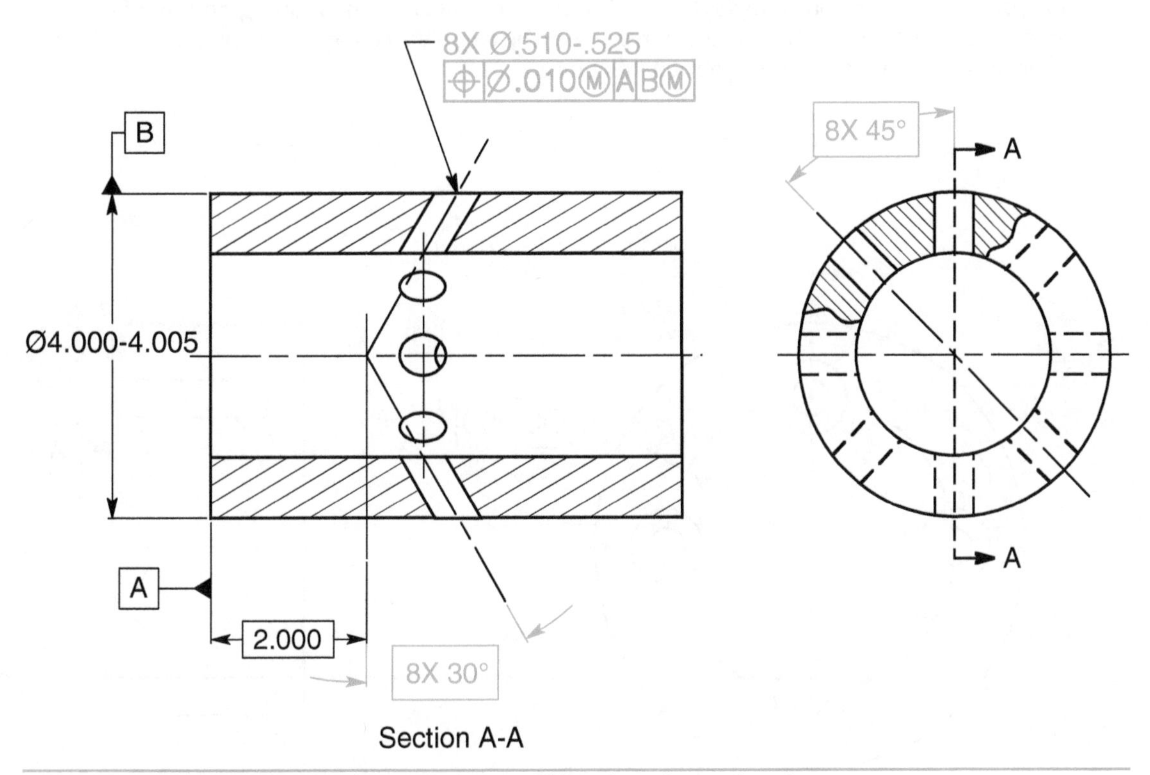

FIGURE 8-21 Eight radial holes about a cylinder at a 30° angle to datum plane A.

Counterbored Holes

Counterbores that have the same location tolerance as their respective holes are specified by indicating the hole and the counterbore callouts followed by the geometric tolerance for both. The counterbore callout includes the dimension and tolerance for the hole diameter, the counterbore diameter, and the counterbore depth. The depth is specified using the depth symbol followed by the depth dimension and tolerance. The counterbore and depth symbols shall precede, with no space, the dimension of the symbols. The feature control frame locating both the hole and counterbore patterns is placed below the callout. The complete callout is shown in Fig. 8-22. This is the most frequently used counterbore tolerance.

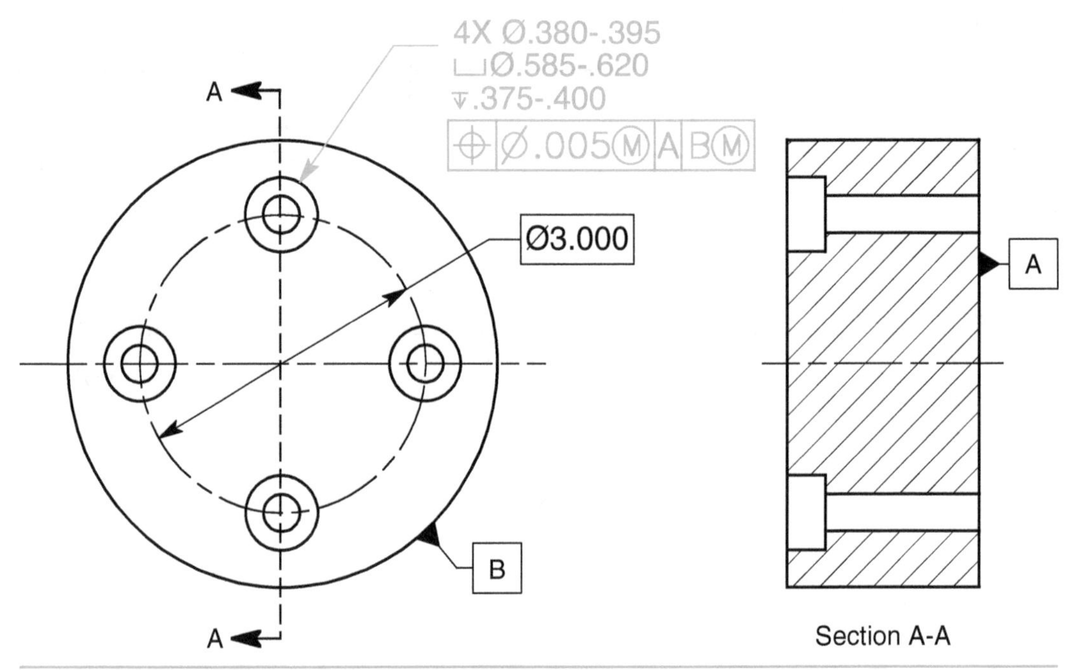

FIGURE 8-22 Specifications for holes and counterbores with the same tolerances for both.

Counterbores with a larger location tolerance than their respective holes are specified by separating the hole callout from the counterbore callout. After specifying the hole pattern callout and its geometric tolerance, state the complete counterbore callout followed by its larger geometric tolerance, as shown in Fig. 8-23. Note that the 4X is repeated before the counterbore callout.

If a hole has a larger tolerance than its counterbore and is controlled with the tolerance callout in Fig. 8-23, the hole may move out of position more than the counterbore and create a no fit condition with the head of the fastener. In this case, the counterbore must be controlled to the hole. Consequently, counterbores with a smaller location tolerance than their respective holes are toleranced by first specifying the hole callout followed by the geometric tolerance. Then each counterbore is located to its respective hole by identifying one of the holes as datum feature C, including the note 4X INDIVIDUALLY next to the datum feature symbol. A second note, 4X INDIVIDUALLY, is placed beneath the feature control frame locating each counterbore to its respective datum feature C, as shown in Fig. 8-24.

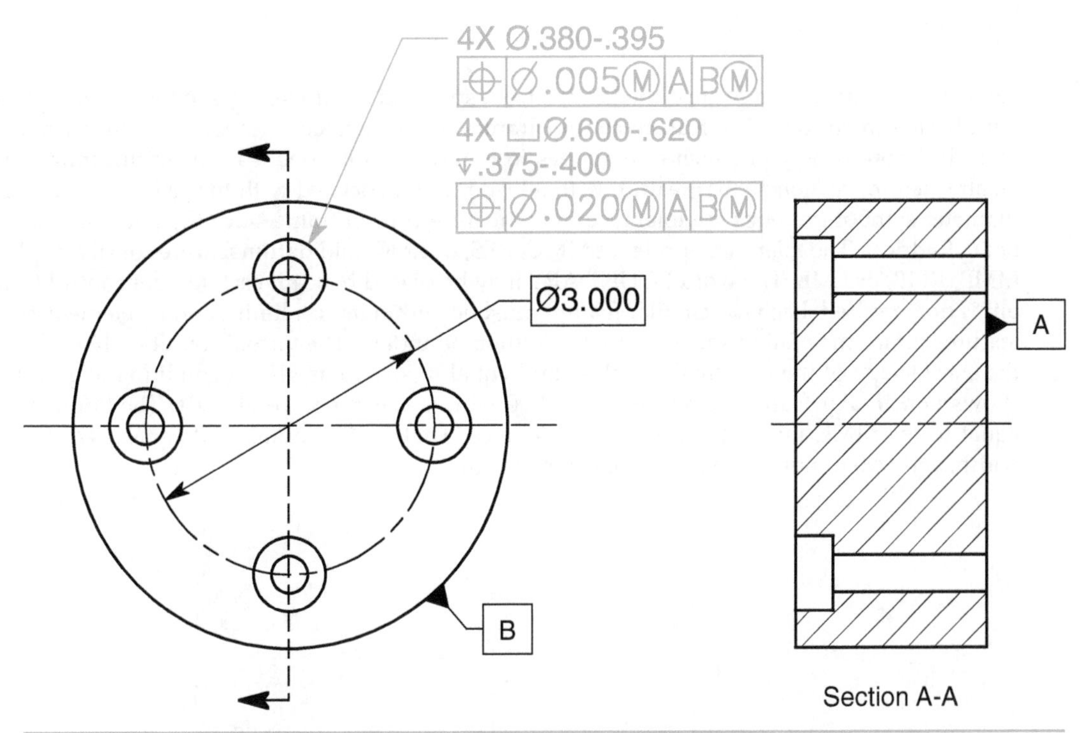

FIGURE 8-23 Specifications for holes and counterbores with a larger tolerance for the counterbores.

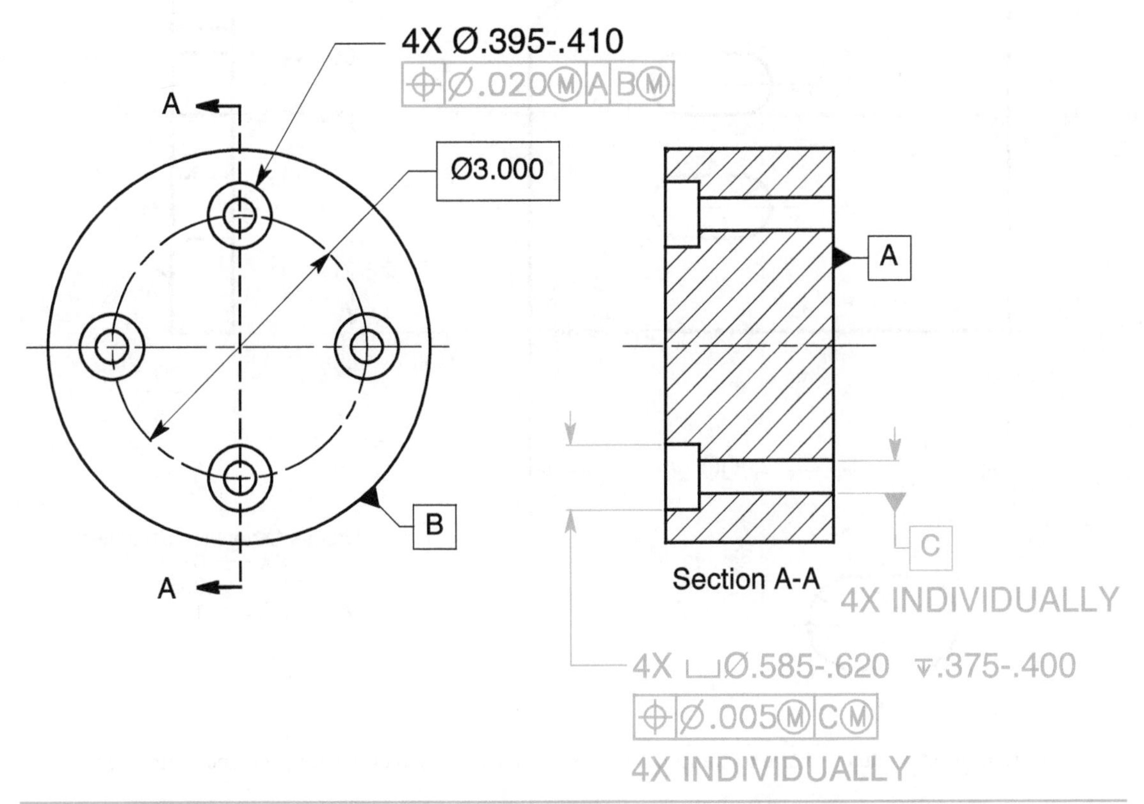

FIGURE 8-24 Specifications for holes and counterbores with a smaller tolerance for the counterbores.

Noncircular Features of Size

Noncircular features of size are dimensioned from specified datum features to their center planes with basic dimensions. The feature control frames are associated with the size dimensions in each direction. If only one tolerance applies in both directions, one feature control frame may be attached to the noncircular feature with a leader not associated with the size dimension. No diameter symbol precedes the tolerance in the feature control frame since the tolerance zone is not cylindrical. The tolerance applies at MMC, RFS, or LMC, and datum features of size apply at MMB, RMB, or LMB. The word BOUNDARY may be placed beneath each feature control frame but is not required. Each noncircular feature must be within its size limits, and no element of the feature surface may fall inside its virtual condition boundary. The virtual condition boundary is the exact shape of the noncircular feature and equal in size to its virtual condition. Figure 8-25 shows noncircular features that are .50 by 1.00 with a size tolerance of ± .01. The boundary is equal to the MMC minus the geometric tolerance. That is .490 minus .020 equals .470 for the width, and .990 minus .060 equals .930 for the length.

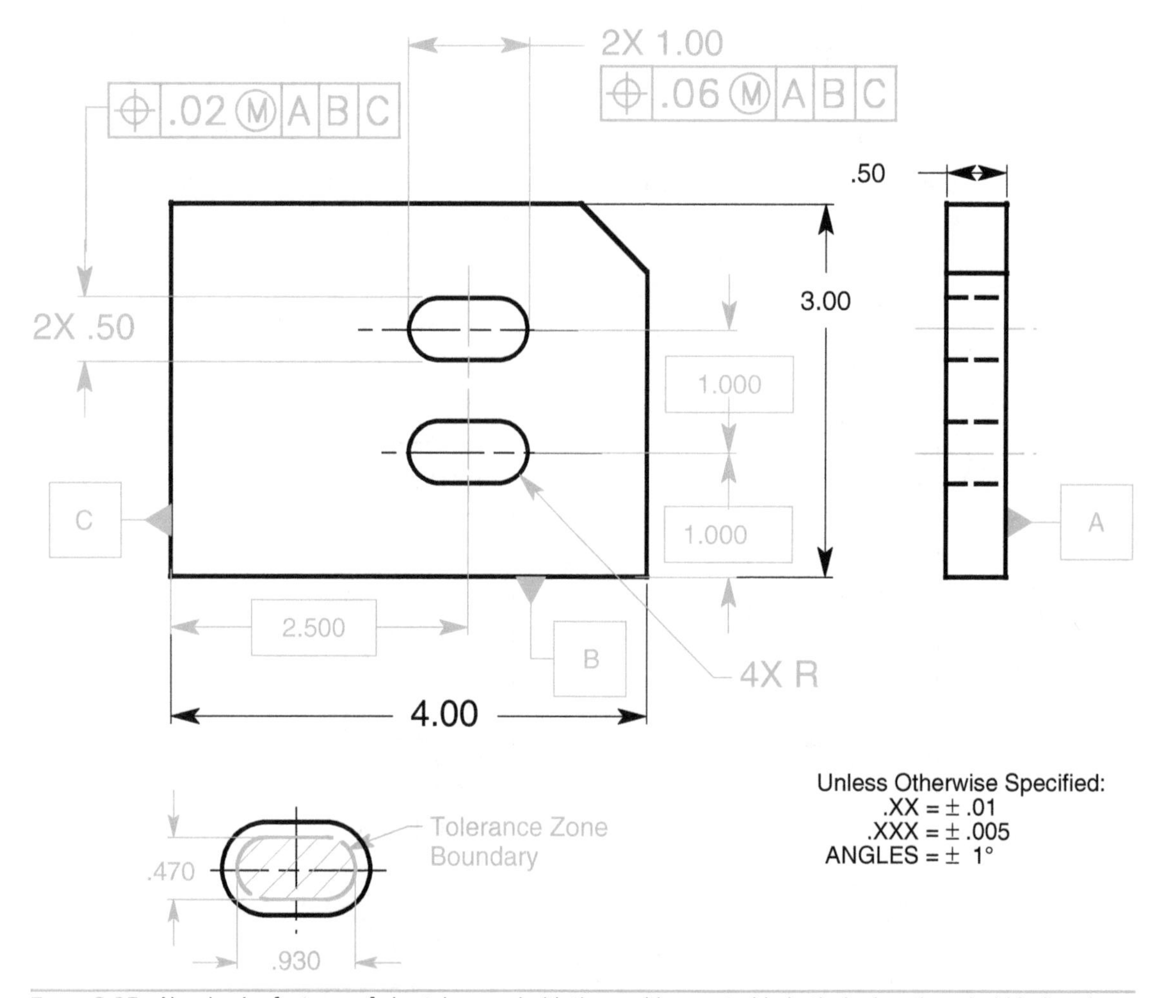

FIGURE 8-25 Noncircular features of size toleranced with the position control in both the length and width directions.

Spherical Features Located with the Position Control

As shown in Fig. 8-26, a spherical feature may be located to datum features with the position control. The spherical diameter symbol precedes the size dimension of the feature and the positional tolerance value in the feature control frame.

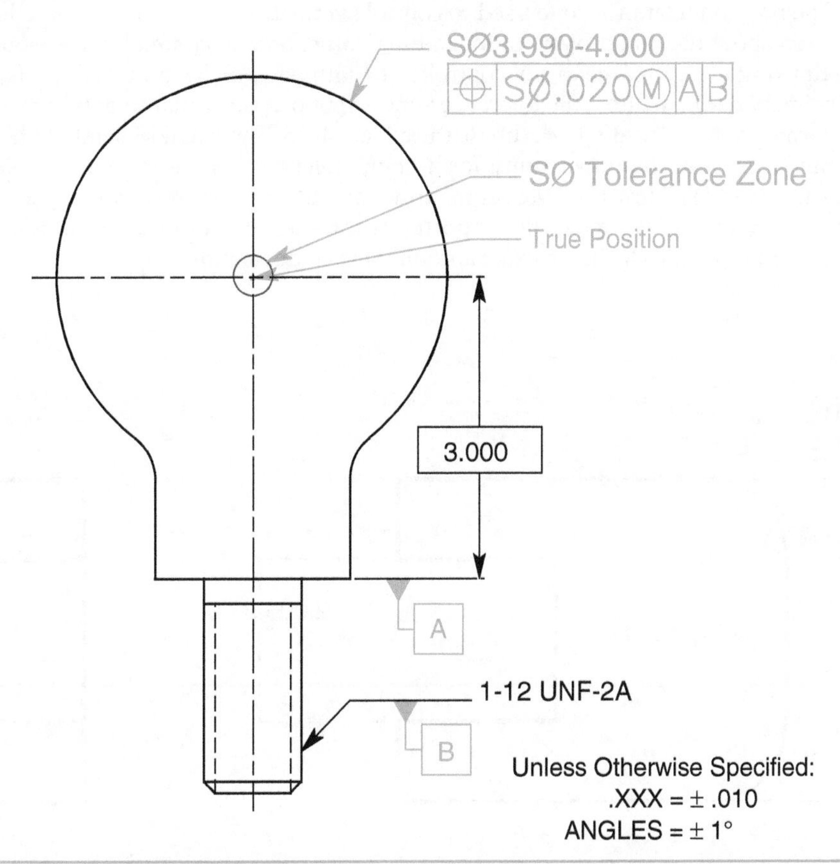

FIGURE 8-26 The center point of a sphere is controlled with a spherical tolerance zone.

Symmetrical Features Located with the Position Control

The position control may be used to orient and locate features of size symmetrical to datum features. The feature control frame is associated with the size dimension. No diameter symbol precedes the tolerance in the feature control frame since the tolerance zone is not a cylinder. The tolerance applies at MMC, RFS, or LMC, and datum features of size apply at MMB, RMB, or LMB.

The position tolerance zone used to control symmetry consists of two parallel planes evenly disposed about the center plane of the datum feature and separated by the geometric tolerance. The drawing in Fig. 8-27 has a slot controlled to datum features A and B. Since datum feature A is the primary datum feature, the tolerance zone is first perpendicular to datum feature A, and then it is located symmetrically to datum feature B at MMB. The circle M symbol after the geometric tolerance provides the opportunity for a bonus tolerance as the size of the toleranced feature departs from MMC toward LMC in the exact amount of such departure. The circle M symbol after datum feature B provides the opportunity for a shift tolerance as the datum feature departs from MMB toward LMC in the exact amount of such departure.

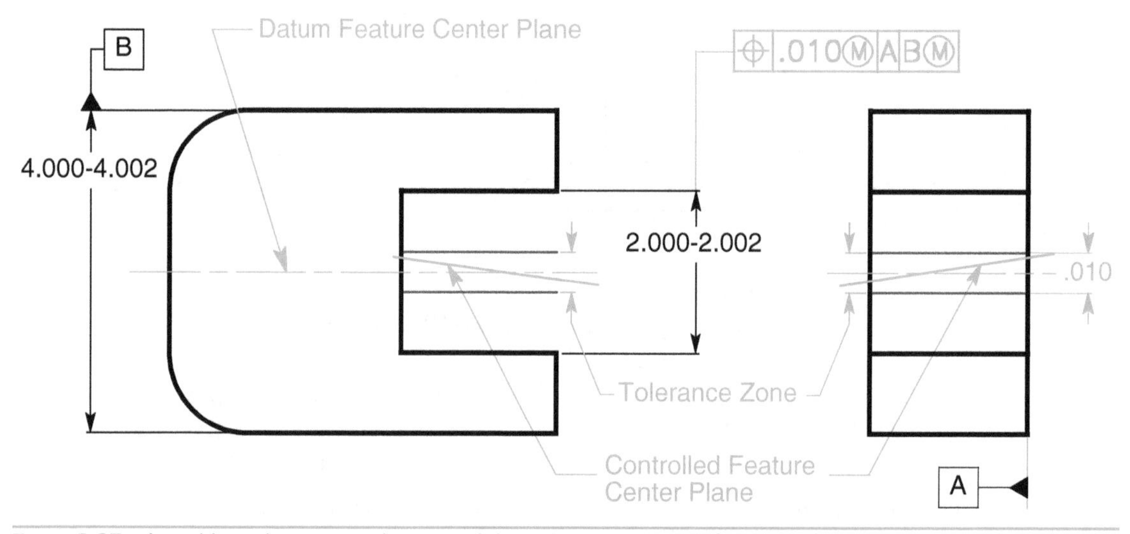

FIGURE 8-27 A position tolerance used to control the symmetry between features of sizes.

Summary

- The floating fastener formula is: $T = H - F$ or $H = F + T$
- The fixed fasteners formula is: $t_1 + t_2 = H - F$ or $H = F + t_1 + t_2$
- The clearance hole at LMC formula is: H at LMC = (F + F head)/2
- The most convenient way to control the orientation of a pin outside a threaded or press fit hole is to project the tolerance zone into the mating part.
- The default condition for multiple patterns of features is that they be treated as one single pattern of features as long as they are:
  - Located with basic dimensions
  - To the same datums features
  - In the same order of precedence
  - At the same material conditions
- If patterns of features are controlled to datum features of size specified at MMB and have no relationship to each other, a note such as SEP REQT may be placed under each feature control frame, allowing each pattern to be inspected separately.
- A composite positional tolerance controls a pattern of features to its datum features with one tolerance and a feature-to-feature relationship with a smaller tolerance.
- When the secondary datum feature is included in the lower segment of a composite feature control frame, the smaller tolerance zone framework must remain parallel to that secondary datum feature.
- The lower segment of multiple single-segment positional tolerancing acts just like any other position control. The lower segment only refines the feature-to-feature tolerance zone framework by orienting it to the primary datum feature and locating it to the secondary datum feature with basic dimensions to a smaller tolerance.
- Nonparallel holes: The position control is so versatile that it can control patterns of nonparallel holes at a basic angle to a principal plane.
- Counterbores can be toleranced in one of three ways—with the same tolerance, more tolerance, or less tolerance—than their respective holes.
- Elongated holes are dimensioned and toleranced in both directions. The feature control frames do not have cylindrical tolerance zones.
- A feature of size may be located symmetrically to a datum feature of size and toleranced with a position control associated with the size dimension of the feature being controlled.

Chapter Review

1. The floating fastener formula is:

   ____.

2. T = ____.

   H = ____.

   F = ____.

3. The clearance hole LMC diameter formula is ____________________.

4. The fixed fastener is fixed by one or more of the ____________________.

5. The formula for fixed fasteners is:

   ____.

6. The formula for fixed fasteners is essentially the same as for floating fasteners except that the fixed fastener formula includes the tolerance for ____________________.

7. It is common practice to assign a larger portion of the location tolerance to the __________ hole.

8. A fastener fixed at its head in a countersunk hole and in a threaded hole at the other end is called what? ____________________

9. Where specifying a threaded hole or a hole for a press fit pin, the orientation of the ____________________ determines the orientation of the mating pin.

10. The most convenient way to control the orientation of a pin outside the hole is to ____________________ the tolerance zone into the mating part.

11. The height of the projected tolerance zone is equal to or greater than the thickest ____________________ or tallest ____________________ after installation.

12. Two or more patterns of features are considered to be a single pattern of features if they are:

    ____

    ____

    ____

    ______________________________.

13. Datum features of size specified at RMB require ______________ between the gagging element and the datum feature.

14. If patterns of features controlled to datum features of size specified at MMB have no relationship to each other, a note such as ______________ may be placed under each feature control frame allowing each pattern to be inspected separately.

15. When locating patterns of features, there are situations where the relationship from ______________ must be kept to a certain tight tolerance and the relationship between the ______________ and its datum features is not as critical and may be held to a looser tolerance.

16. A composite feature control frame has one ______________ symbol that applies to the two horizontal ______________ that follow.

17. The upper segment of a composite feature control frame is called the ______________ control, it governs the relationship between the datum features and the ______________.

18. The lower segment of a composite feature control frame is called the ______________ control; it governs the relationship from ______________.

19. The primary function of the position control is to control ______________.

20. Datum features in the lower segment of a composite feature control frame must satisfy what two conditions?

____

____

(Assume plane surface datum features for Questions 21 and 22.)

21. Where the secondary datum feature is included in the lower segment of a composite feature control frame, the tolerance zone framework must remain ______________ to the secondary datum plane.

22. The lower segment of a multiple single-segment feature control frame acts just like any other ______________________________.

23. Counterbores that have the same location tolerance as their respective holes are specified by indicating the ______________________________

______________________________.

24. Counterbores that have a larger location tolerance than their respective holes are specified by ______________________________.

25. When tolerancing elongated holes, no ______________ precedes the tolerance in the feature control frame since the tolerance zone is not ______________.

26. The virtual condition boundary is the ______________ of the noncircular feature and equal in size to its ______________.

27. The ______________ is used to locate a feature of size symmetrically at MMC to a datum feature of size specified at MMB.

Problems

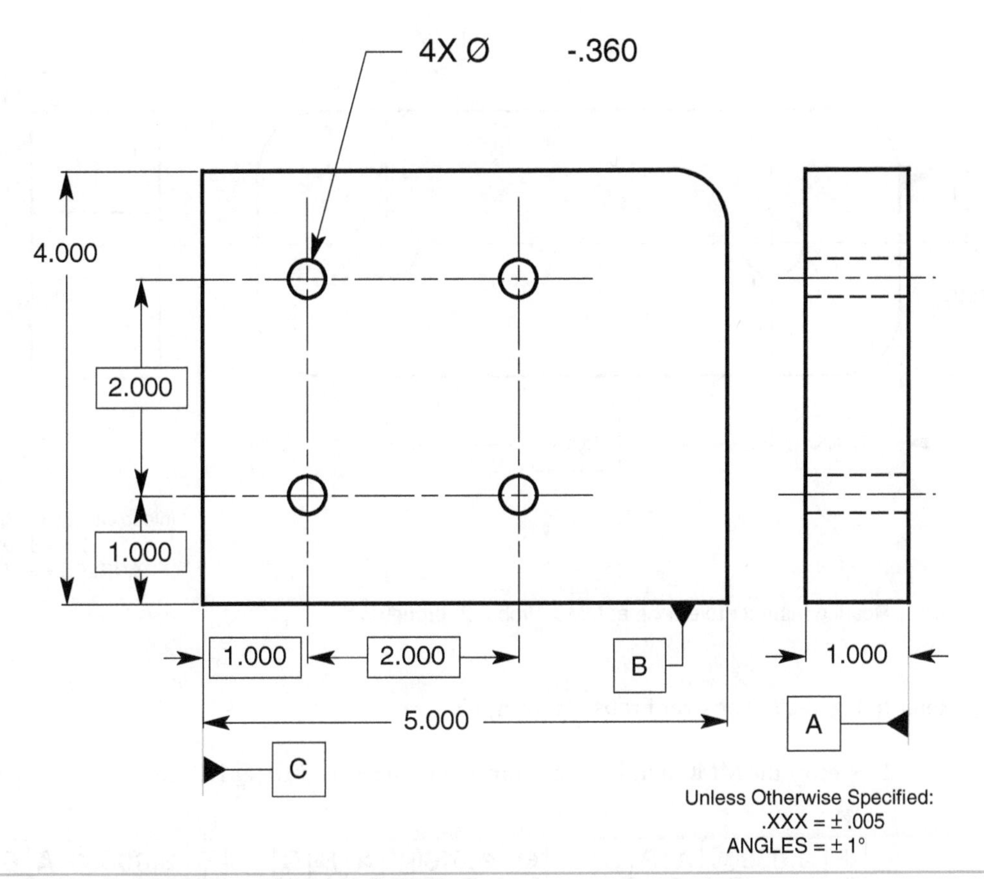

FIGURE 8-28 Floating fastener tolerance: Prob. 1.

1. Tolerance the clearance holes on the plate in Fig. 8-28 to be fastened with 5/16–18 UNC hex head bolts (.313 in diameter) and nuts with a .010-diameter positional tolerance at MMC.

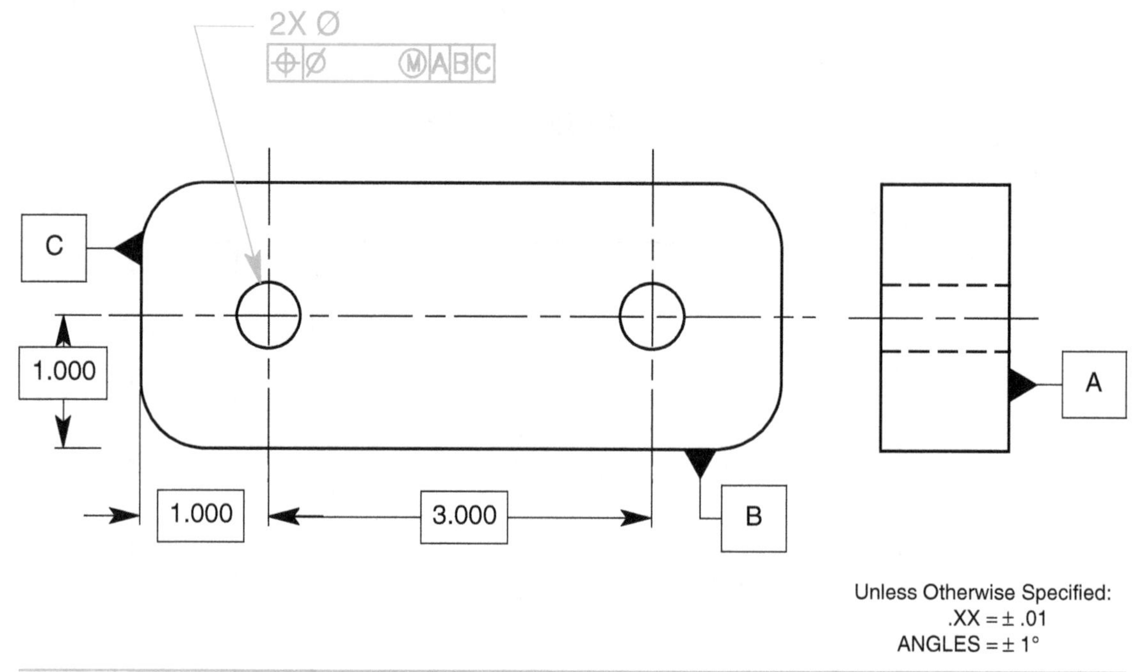

FIGURE 8-29 Floating fastener tolerance at MMC: Probs. 2 through 5.

Refer to Fig. 8-29 to answer Probs. 2 through 5.

2. Specify the MMC and LMC clearance hole sizes for #10 (Ø.190) socket head cap screws.

Ø	Ø	Ø
⊕ | Ø.030Ⓜ | A | B | C	⊕ | Ø.010Ⓜ | A | B | C	⊕ | Ø.000Ⓜ | A | B | C

3. If the actual mating envelope size of the clearance holes in Prob. 2 is .230 in diameter, calculate the total positional tolerance for each callout.

Actual Size	.230	.230	.230
MMC	______	______	______
Bonus			
Geo. Tolerance	______	______	______
Total Tolerance	______	______	______

4. Specify the MMC and LMC clearance hole sizes for 3/8 (.375 in diameter) hex head bolts.

Ø	Ø	Ø
⊕ | Ø.025Ⓜ | A | B | C	⊕ | Ø.015Ⓜ | A | B | C	⊕ | Ø.000Ⓜ | A | B | C

5. If the actual mating envelope size of the clearance holes in Prob. 4 measures .440 in diameter, calculate the total positional tolerance for each callout.

Actual Size	.440	.440	.440
MMC	______	______	______
Bonus			
Geo. Tolerance	______	______	______
Total Tolerance	______	______	______

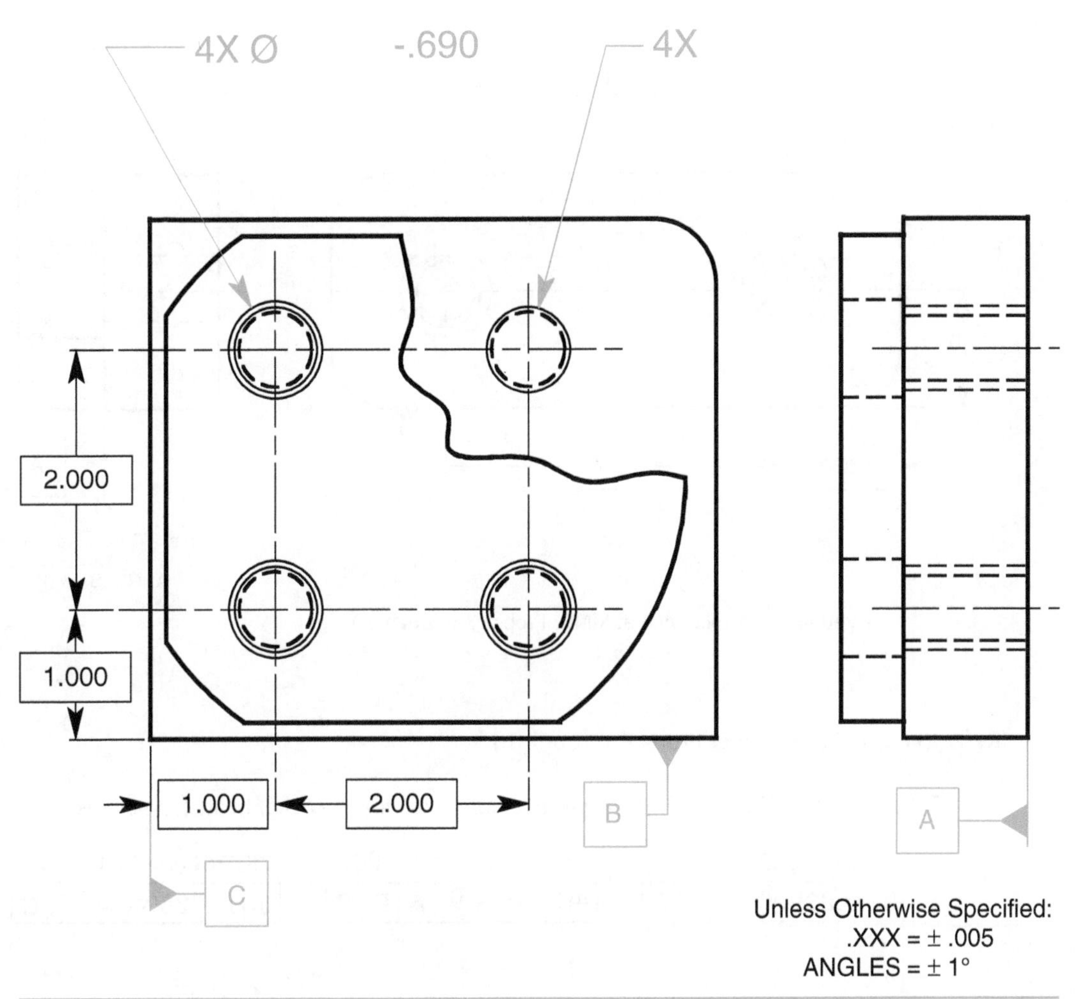

FIGURE 8-30 Fixed fastener tolerance: Prob. 6.

6. Tolerance the clearance and threaded holes in the plates in Fig. 8-30 to be fastened with 5/8–11 UNC hex head bolts (.625 in diameter). Use a .000 positional tolerance at MMC wherever possible and calculate a 60% location tolerance for the threaded holes.

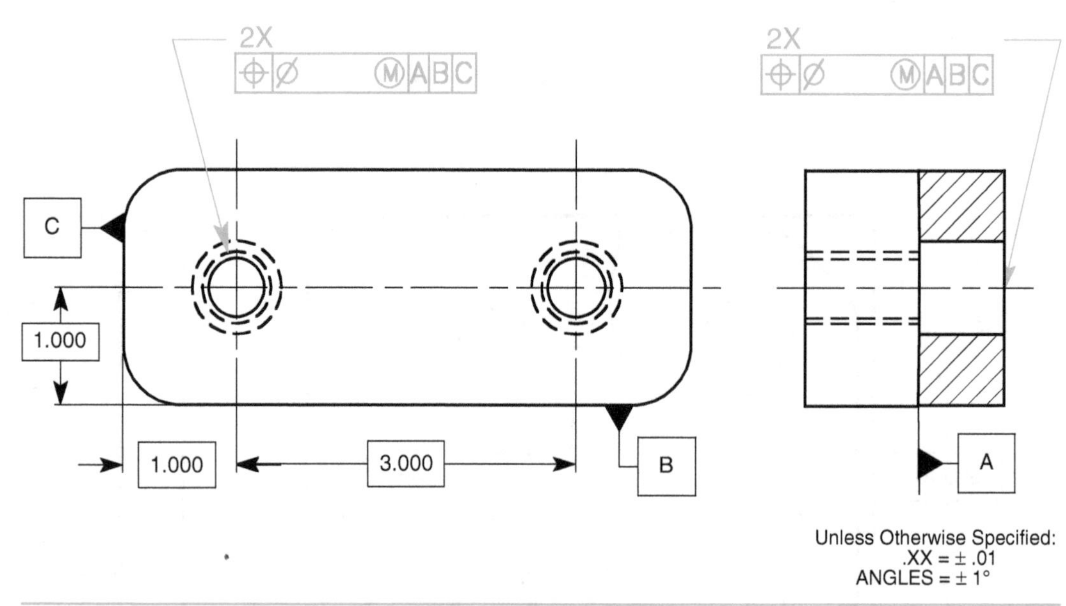

FIGURE 8-31 Fixed fastener tolerance at MMC: Probs. 7 through 10.

Refer to Fig. 8-31 to answer Probs. 7 through 10.

7. Specify the MMC and LMC clearance hole sizes for #8 socket head cap screws.

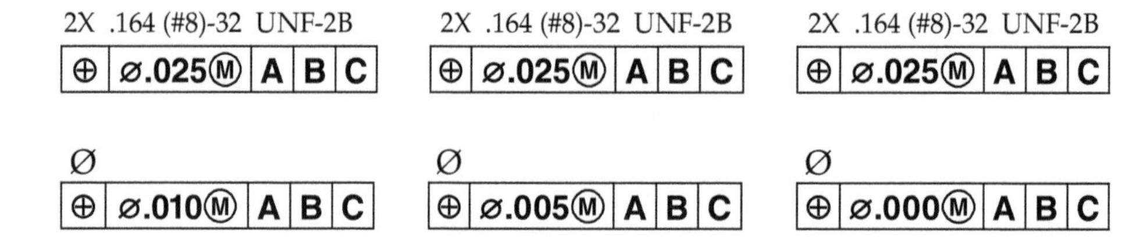

8. If the actual mating envelope size of the clearance holes in Prob. 7 measures .205 in diameter, calculate the total positional tolerance for each callout.

Actual Size	.205	.205	.205
MMC	______	______	______
Bonus			
Geo. Tolerance	______	______	______
Total Tolerance	______	______	______

9. Specify the MMC and LMC clearance hole sizes for the ½-inch hex head bolts.

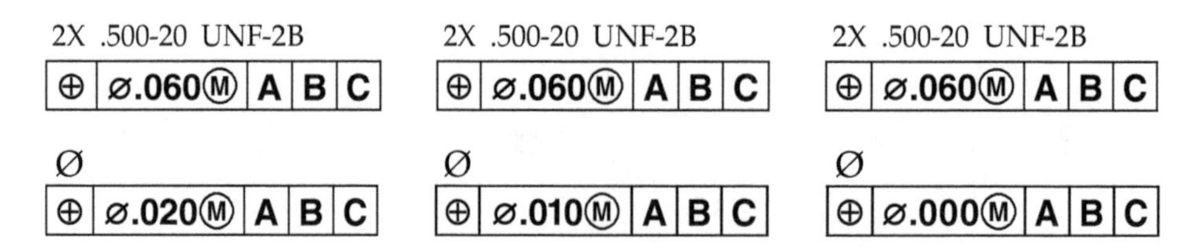

10. If the actual mating envelope size of the clearance holes in Prob. 9 measures .585 in diameter, calculate the total positional tolerance for each callout.

Actual Size	.585	.585	.585
MMC	______	______	______
Bonus			
Geo. Tolerance	______	______	______
Total Tolerance	______	______	______

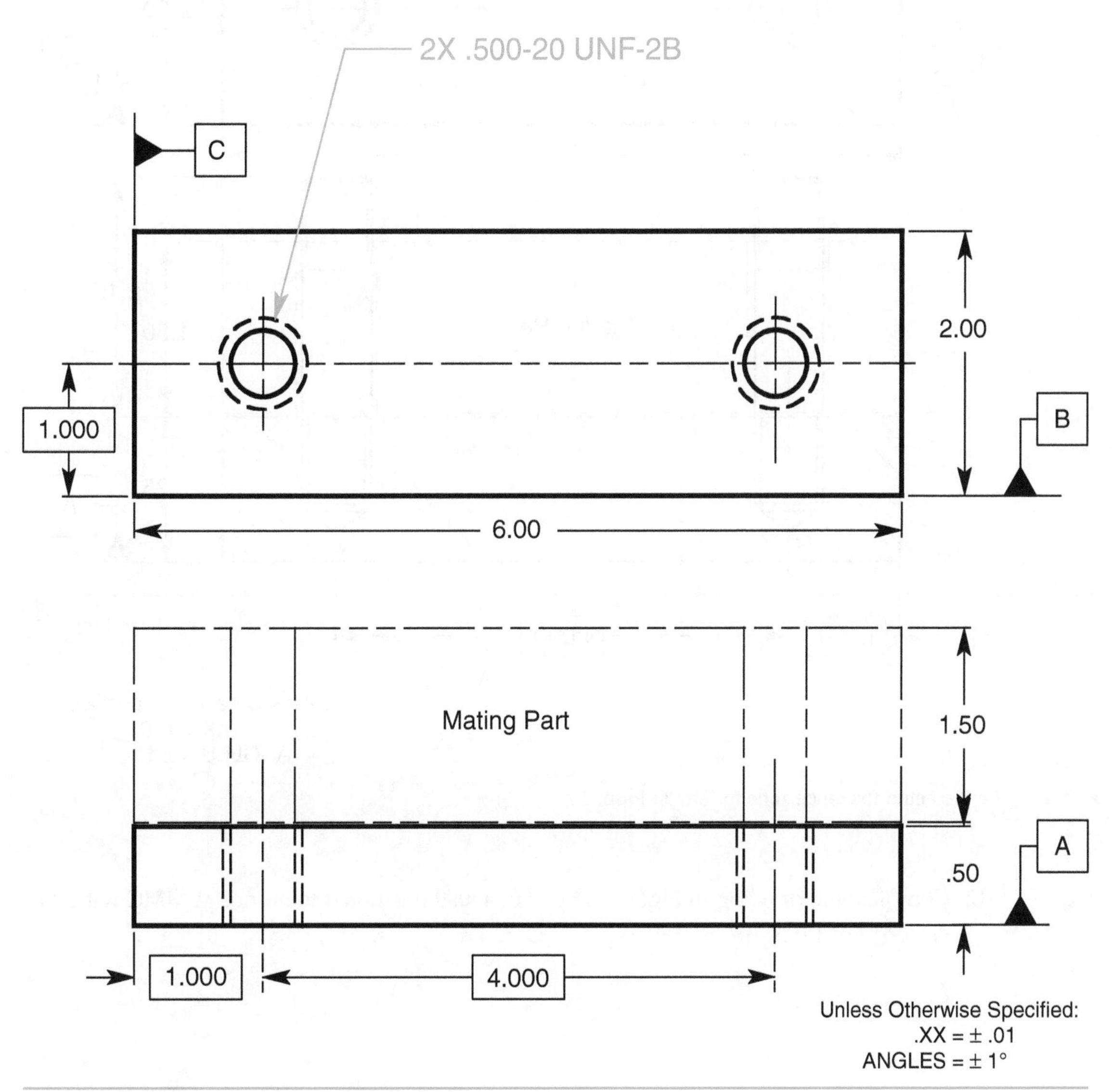

FIGURE 8-32 Projected tolerance zone: Prob. 11.

11. Complete the drawing in Fig. 8-32. Specify a .040 positional tolerance at MMC with the appropriate projected tolerance.

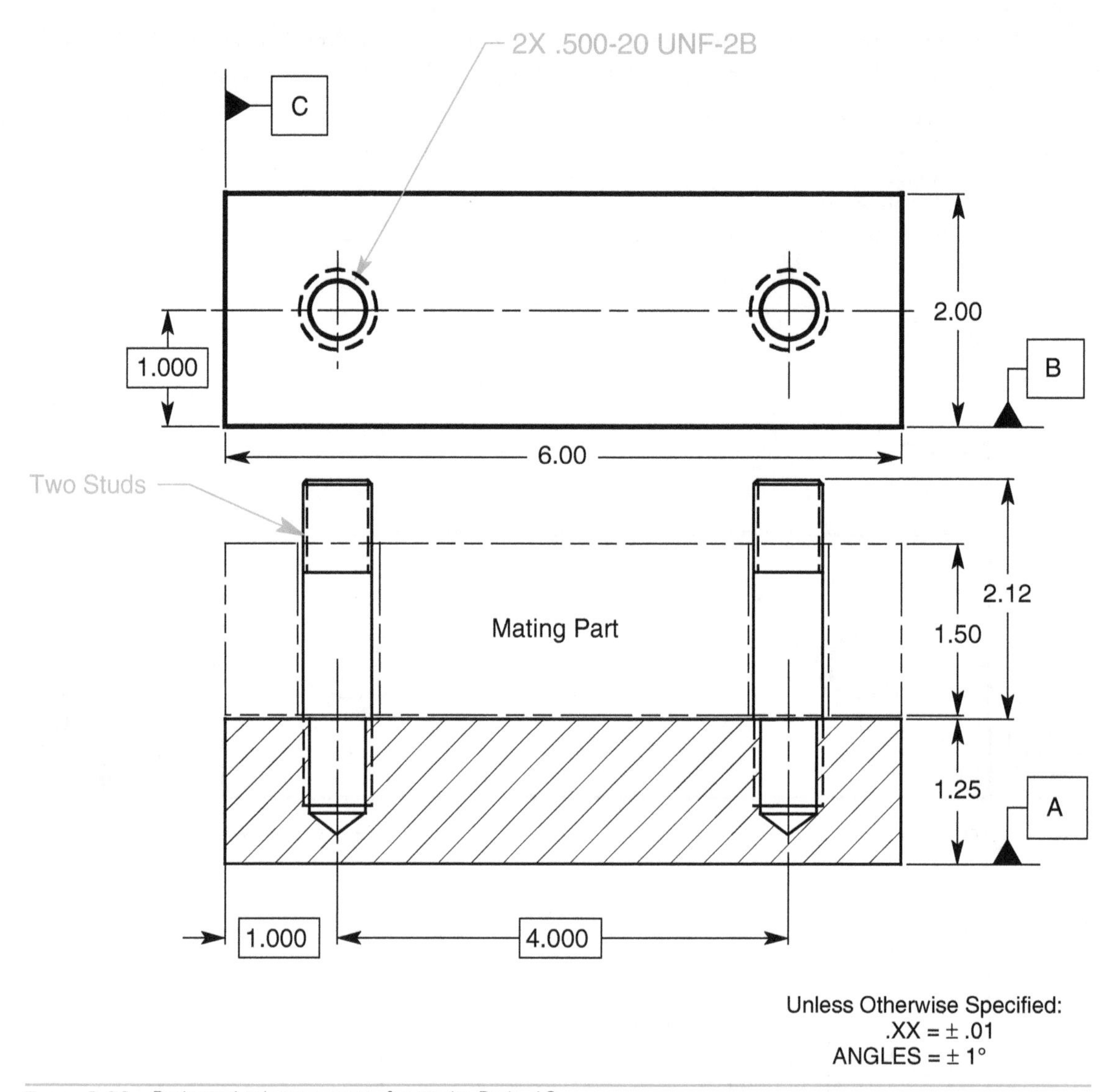

FIGURE 8-33 Projected tolerance zone for studs: Prob. 12.

12. Complete the drawing in Fig. 8-33. Specify a .050 positional tolerance at MMC with the appropriate projected tolerance.

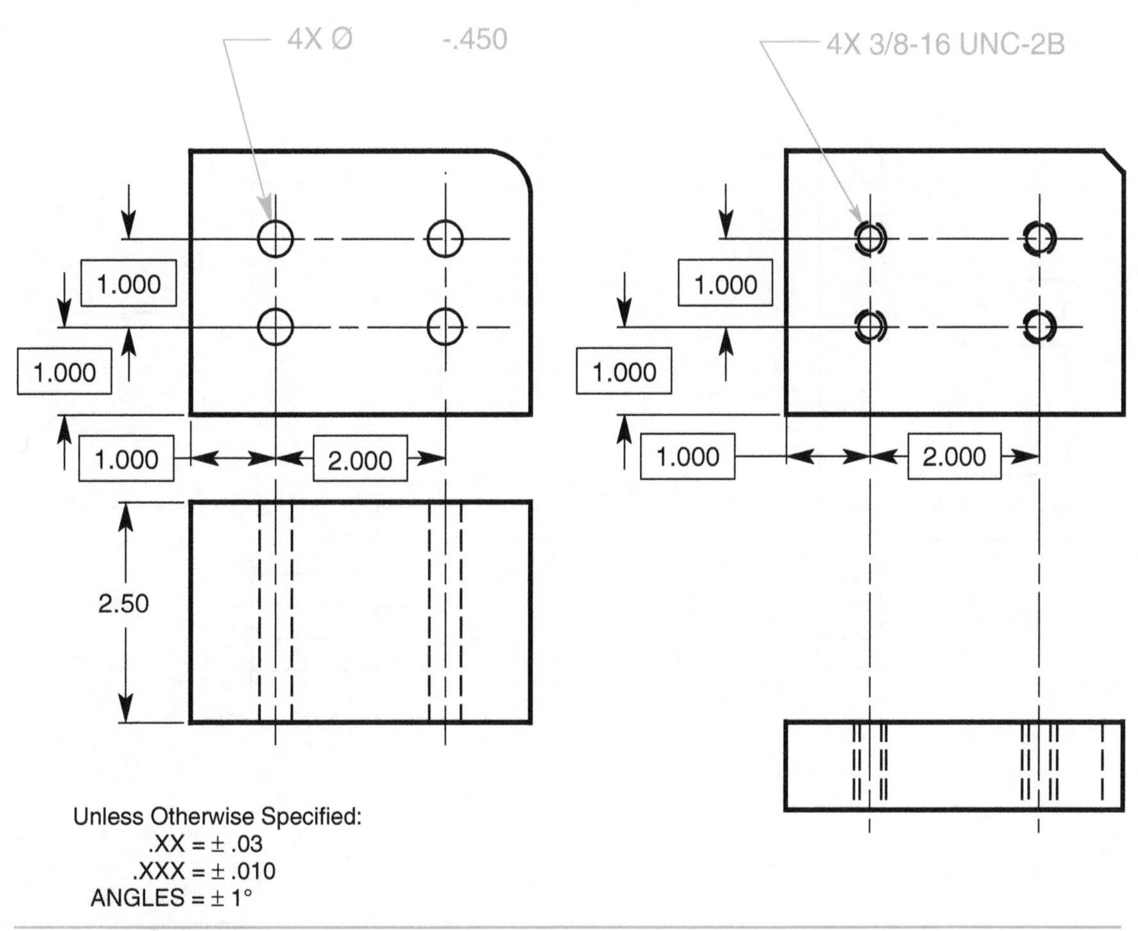

FIGURE 8-34 Fixed fastener assembly: Prob. 13.

13. The part with clearance holes in Fig. 8-34 assembles on top of the part with threaded holes and is fastened with cap screws. Allow a tolerance of at least .030 on both threaded and clearance holes, use zero positional tolerance at MMC, and specify projected tolerance zones.

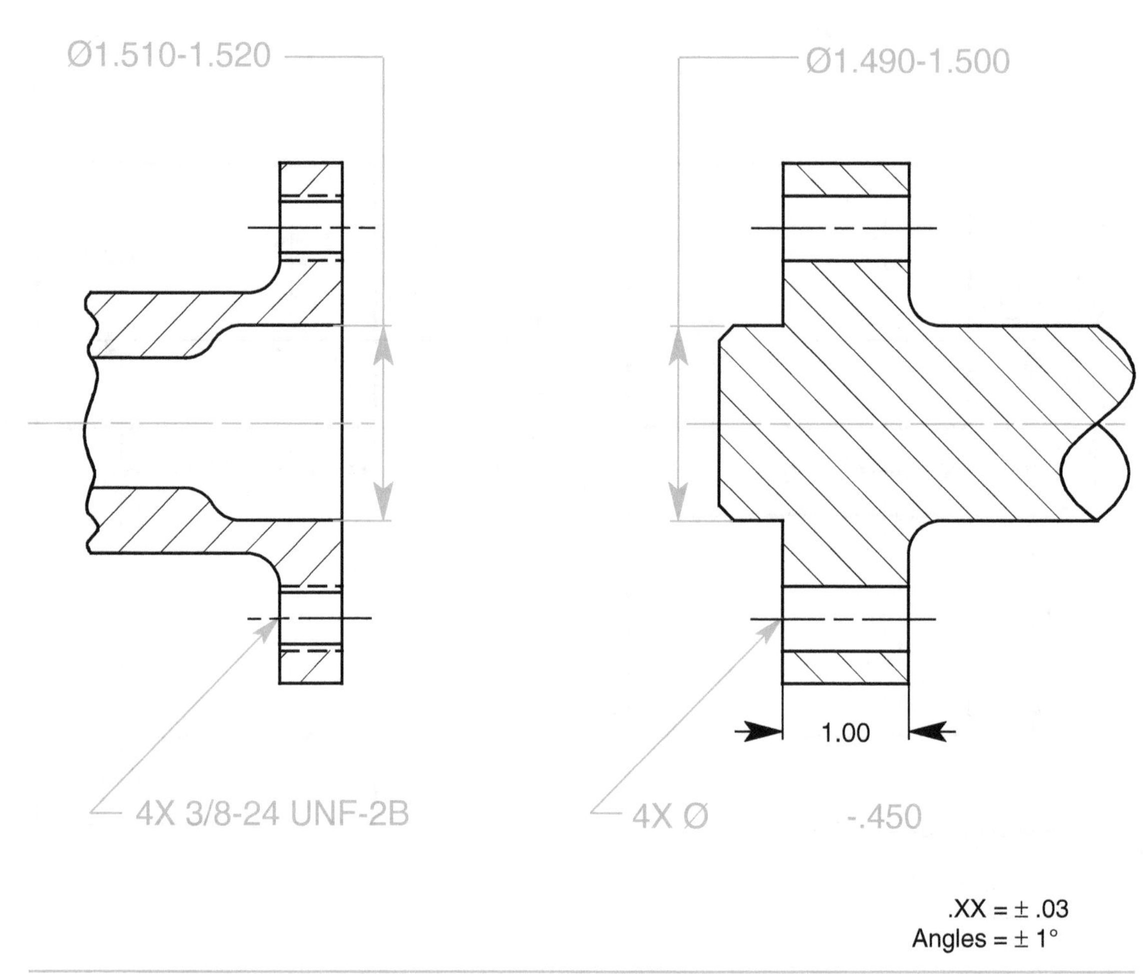

FIGURE 8-35 A coupling assembly: Prob. 14.

14. Tolerance the two parts in Fig. 8-35. Specify a flatness control of .002 on each of the primary datum features. Specify the appropriate orientation tolerance to control the relationships between the primary and secondary datum features. Finally, complete the location tolerances for the hole patterns for 3/8-inch cap screws using a .010 positional tolerance for the clearance holes. Specify MMC and MMB wherever possible.

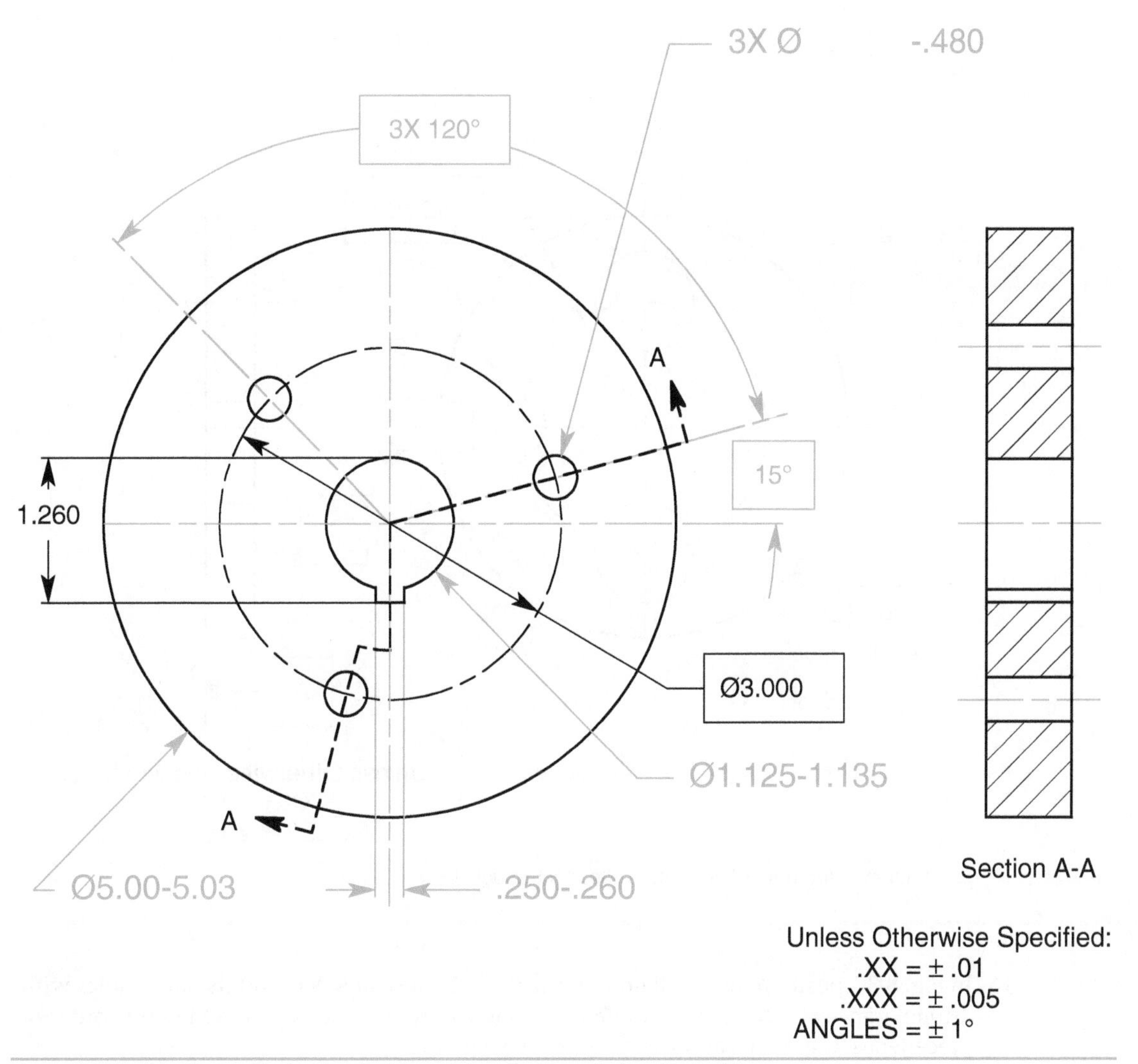

FIGURE 8-36 A pattern of holes located to a datum feature of size: Prob. 15.

15. In Fig. 8-36, the inside diameter and the back are mating features. Select the primary datum feature. (Consider a form control.) The virtual condition of the mating shaft is 1.125 in diameter. Locate the keyway for a ¼-inch key. Locate the three-hole pattern for 7/16-inch (Ø.438) cap screws as floating fasteners with a zero positional tolerance at MMC. Specify MMC and MMB wherever possible.

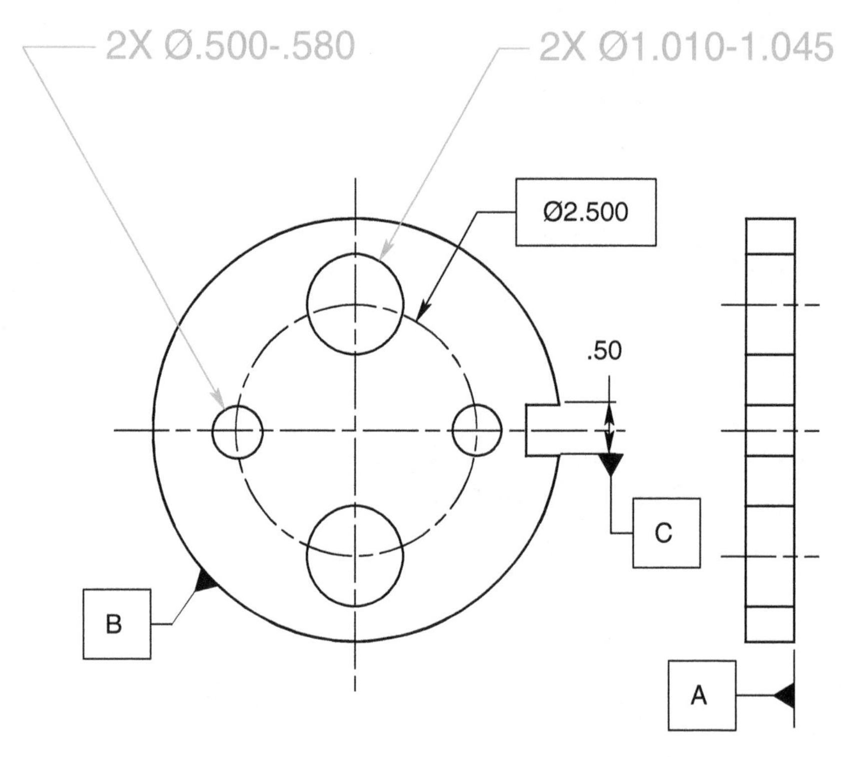

Unless Otherwise Specified:
.XX = ± .01
ANGLES = ± 1°

FIGURE 8-37 Multiple patterns of features: Probs. 16 through 18.

16. In Fig. 8-37, position the small holes with .000 tolerance at MMC and the large holes with .010 tolerance at MMC; locate them to the same datum features and in the same order of precedence. Use MMC and MMB wherever possible.
17. Must the hole patterns be inspected in the same setup or in the same gage? Explain?

18. Can the requirement be changed? If so, how?

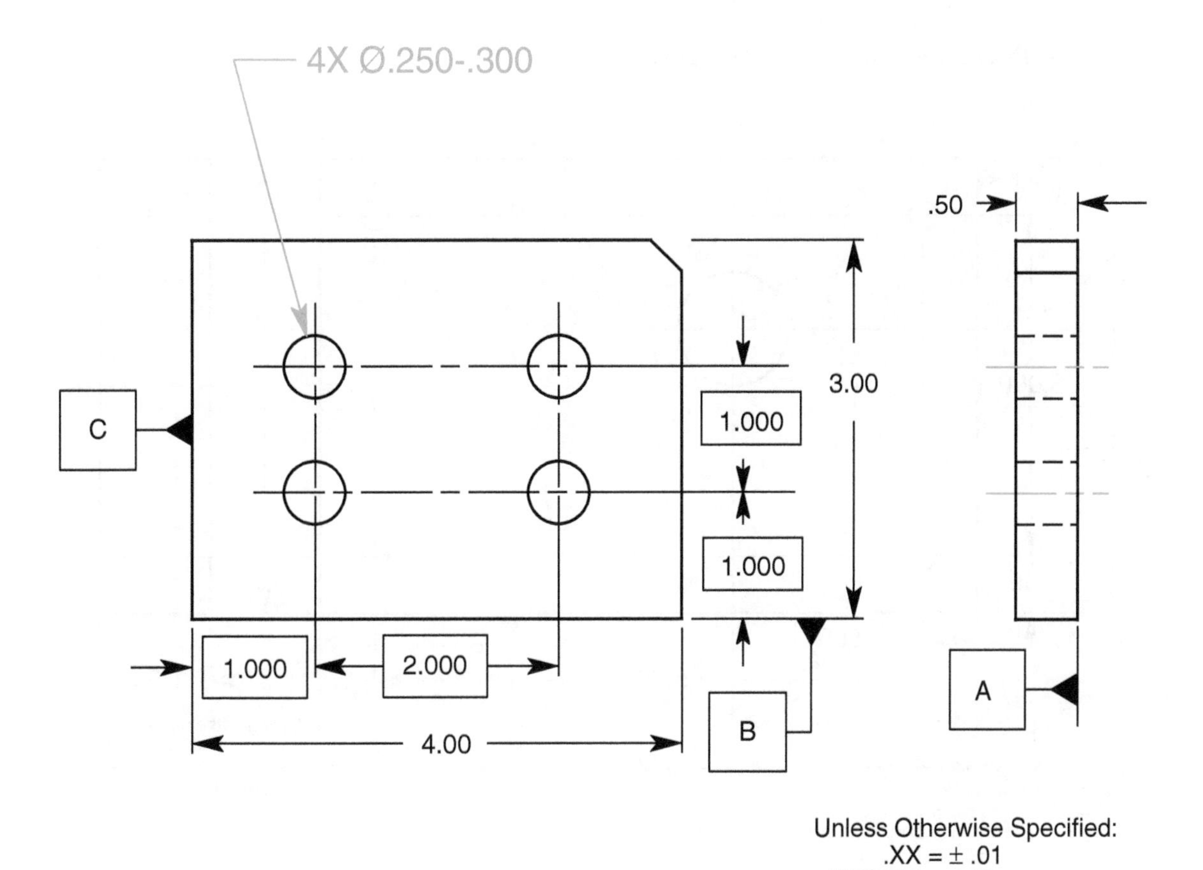

FIGURE 8-38 Composite positional tolerancing: Probs. 19 and 20.

19. Locate the pattern of clearance holes on the part in Fig. 8-38 with a tolerance of at least .060 in diameter at MMC to the datum features specified. This plate is required to assemble to the mating part with ¼-inch hex bolts as floating fasteners. Complete the geometric tolerance.

20. It has been determined that the hole pattern in Fig. 8-38 is required to remain parallel to datum feature B within the smaller tolerance. Draw the feature control frame that will satisfy this requirement.

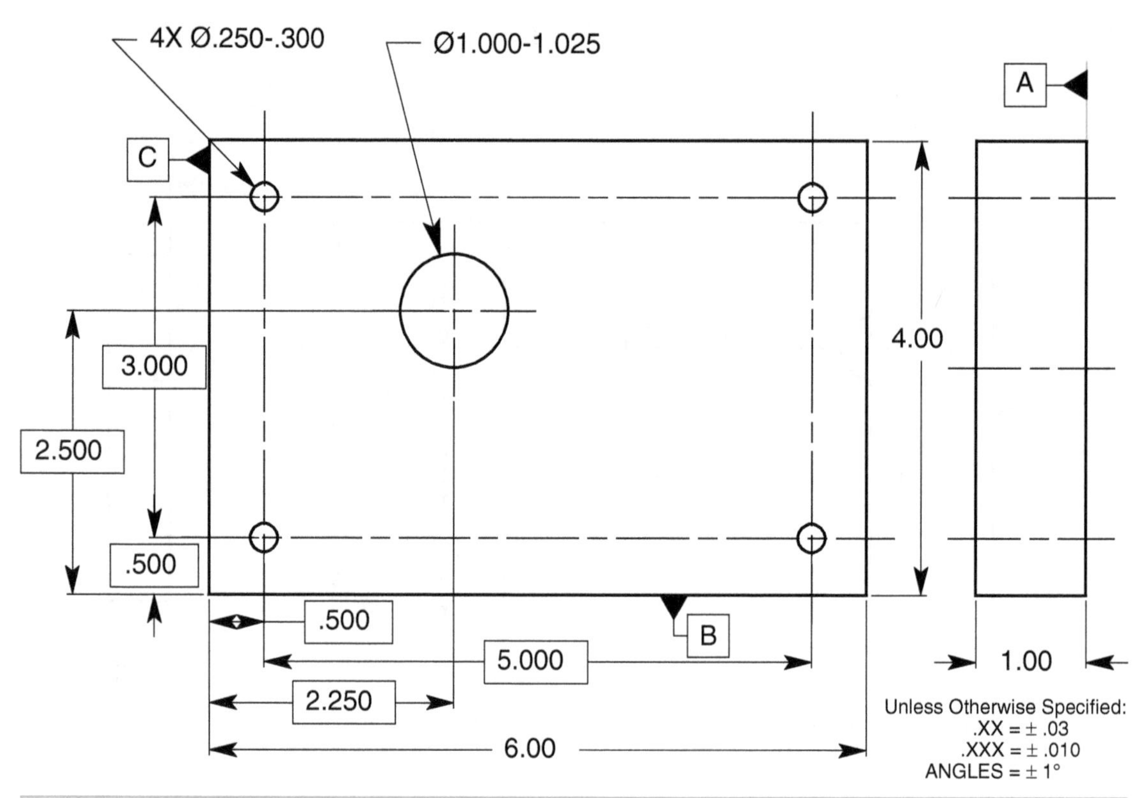

FIGURE 8-39 Composite positional tolerancing of holes with different sizes: Prob. 21.

21. Locate all five holes in Fig. 8-39 within a tolerance of .060 to the datum features specified. Also, locate all five holes to each other and perpendicular to datum feature A within a positional tolerance of .010. Use MMC and MMB wherever applicable.

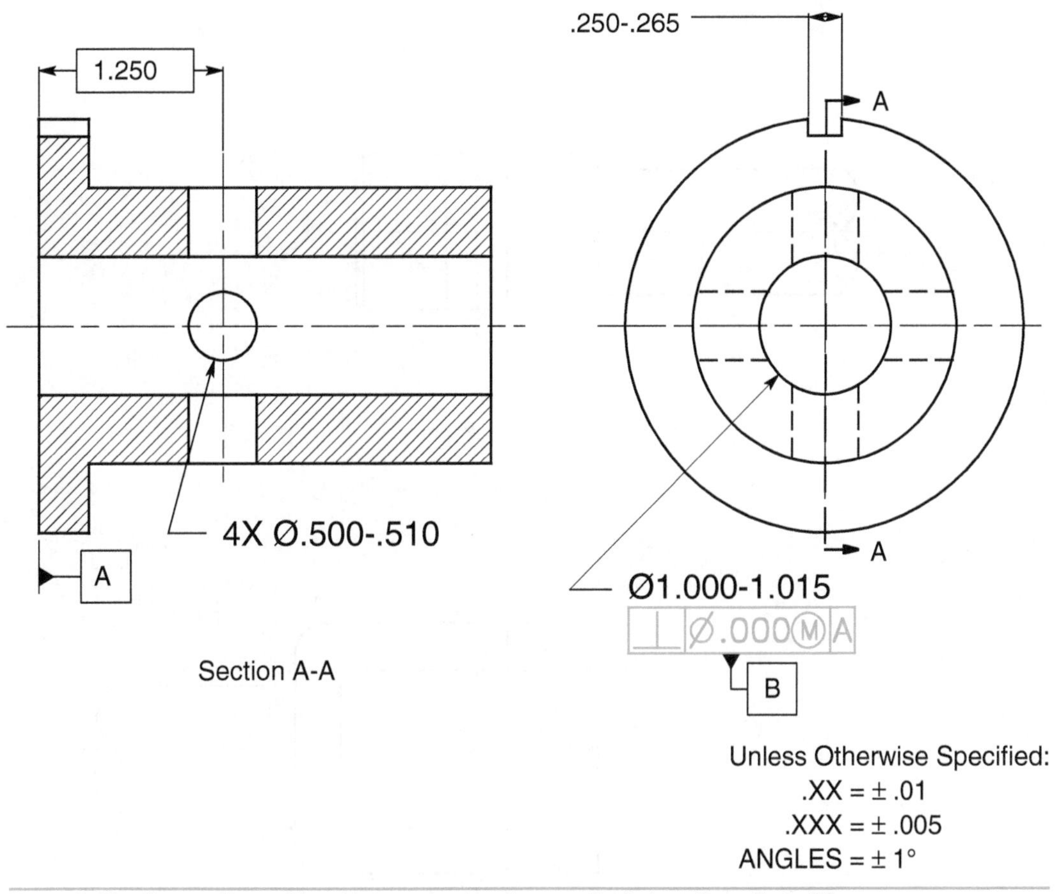

FIGURE 8-40 Composite positional tolerancing locating a radial hole pattern: Prob. 22.

22. Locate the ¼-inch keyway to datum features A and B on Fig. 8-40. Locate the four-hole pattern within a tolerance of .020. Refine the orientation of the four-hole pattern parallel to datum feature A, perpendicular to datum feature B at MMB, and parallel and perpendicular to the center plane of the keyseat within zero tolerance at MMC. Specify MMC and MMB wherever possible.

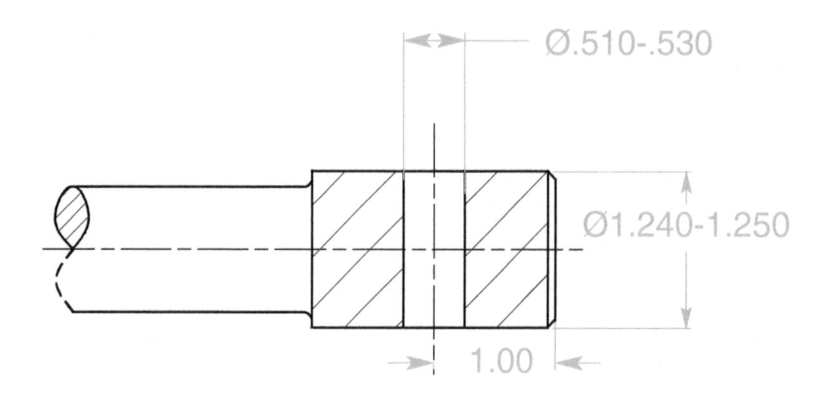

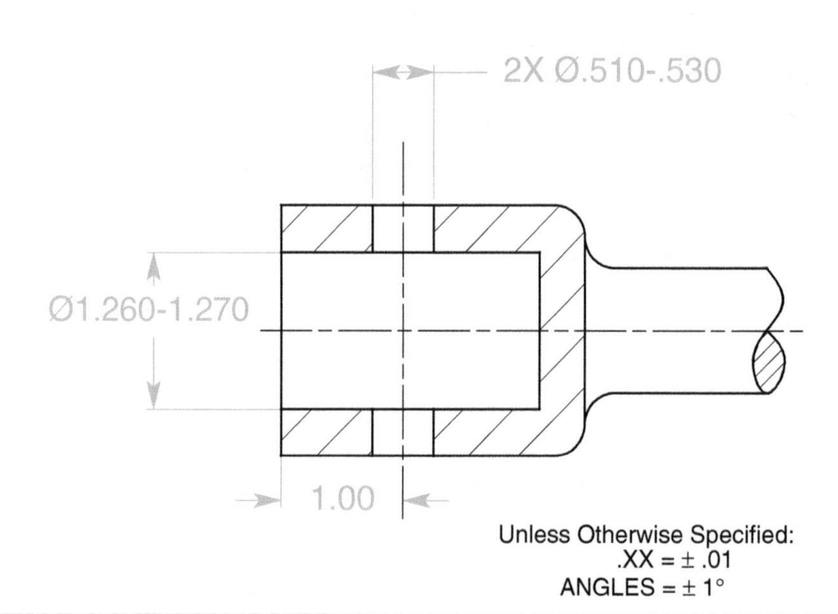

FIGURE 8-41 Multiple single-segment positional tolerancing to control holes: Prob. 23.

23. The inner and outer shafts in Fig. 8-41 will assemble every time. Control the location of the clearance holes for a ½-inch fastener with a multiple single-segment positional tolerance. Locate the holes to the end of each shaft with a tolerance of .040 Locate the holes to the axis of each shaft using the floating fastener formula. Use MMC and MMB wherever applicable.

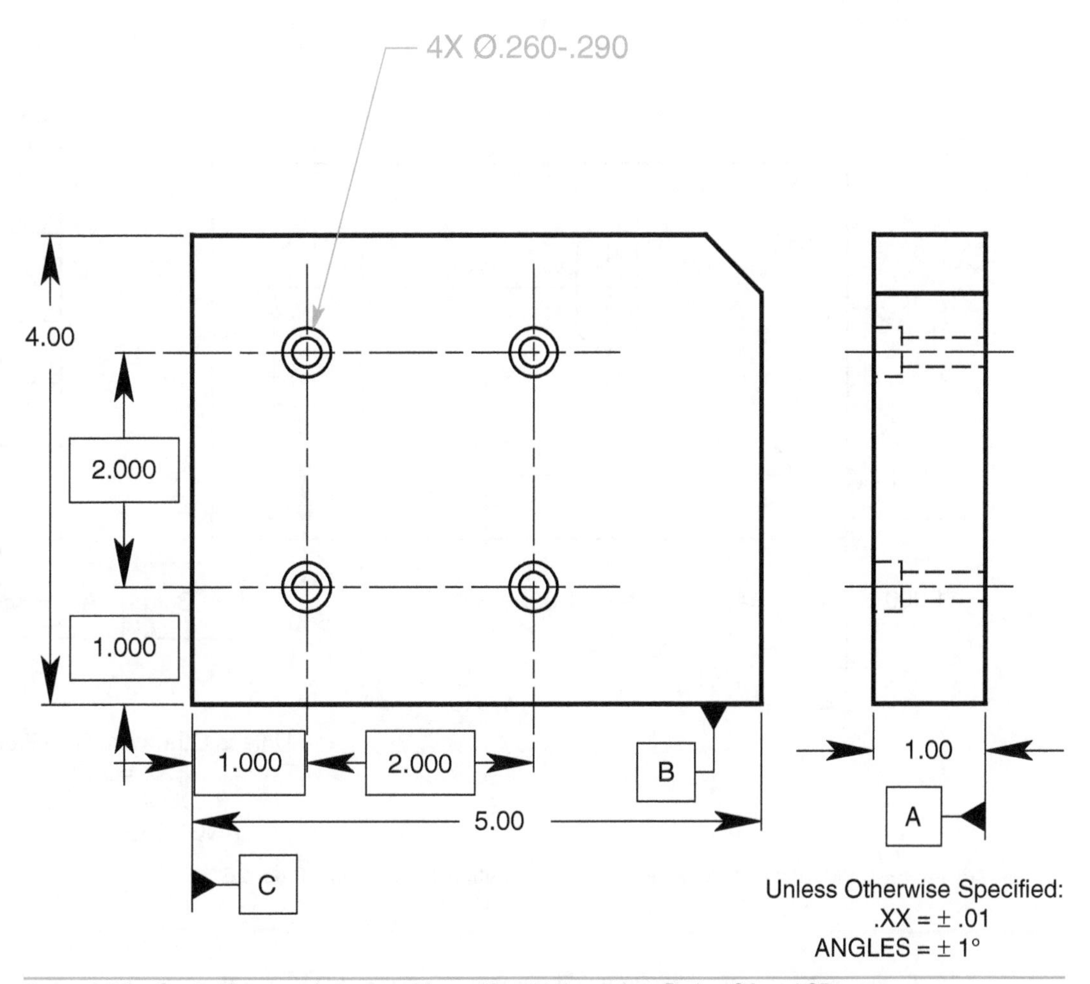

FIGURE 8-42 Controlling counterbores with positional tolerancing: Probs. 24 and 25.

24. Tolerance the holes and counterbores in Fig. 8-42 for four ¼-inch socket head cap screws. The cap screw head is a diameter of .365-.375, the height is .244-.250. Specify MMC and MMB wherever possible.
25. If the geometric tolerance for just the counterbores in Fig. 8-42 can be loosened to .020 at MMC instead of .010, draw the entire callout below.

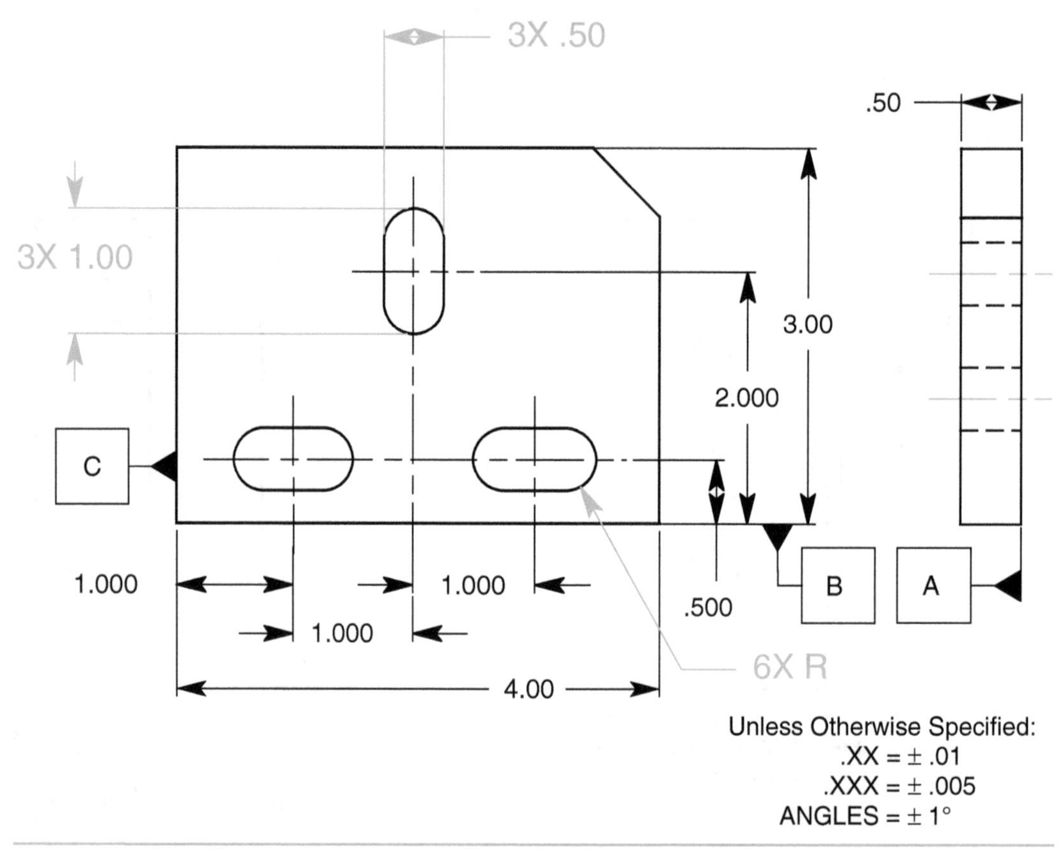

FIGURE 8-43 Controlling noncircular features with positional tolerancing: Prob. 26.

26. In Fig. 8-43, specify a geometric tolerance of .040 at MMC for the ½-inch direction and .060 at MMC for the 1-inch direction for the noncircular features.

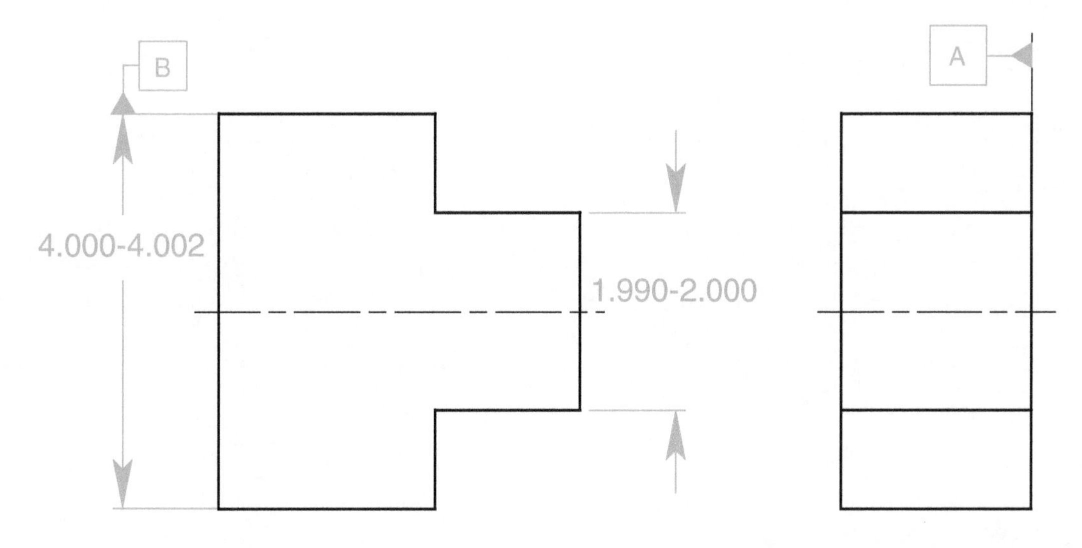

FIGURE 8-44 Controlling symmetrical features with positional tolerancing: Probs. 27 and 28.

27. In Fig. 8-44, control the symmetry of the 2-inch feature with respect to the 4-inch feature and perpendicular to datum feature A within a tolerance of .020. Use MMC and MMB wherever possible.
28. If the controlled feature in Fig. 8-44 happened to be produced at 1.995 and the datum feature produced at 4.000, what would the total positional tolerance be? ____________

CHAPTER 9

Position, Coaxiality

One of the most common errors on drawing is the failure to specify a coaxiality tolerance for coaxial features. Many practitioners think that coaxiality tolerance is unnecessary; they may not even be aware that a coaxiality tolerance is required. The tolerance of position is a very effective and versatile coaxiality control. The applications of the positional tolerance to control coaxiality will be discussed in this chapter.

Chapter Objectives

After completing this chapter, the learner will be able to:

- *Explain* the difference between coaxiality controls
- *Specify* position tolerance for coaxiality
- *Specify* coaxiality on a material condition basis
- *Specify* composite positional control of coaxial features
- *Tolerance* a plug and socket

Definition

Coaxiality is that condition where the axes of two or more surfaces of revolution are coincident.

Many engineers produce drawings similar to the one in Fig. 9-1 showing two or more cylinders on the same axis. This is an incomplete drawing because there is no coaxiality tolerance between the two cylinders. There is a misconception that centerlines or title block tolerances control the coaxiality between two cylindrical features. Centerlines indicate that the cylinders are dimensioned to the same axis. That is, in Fig. 9-1, the distance between the axes of the 1.000-inch-diameter and the 2.000-inch-diameter cylinders is zero. Of course, zero dimensions are implied and never placed on drawings. Even though the dimension is implied, the tolerance is not; there is no tolerance indicating how far out of coaxiality the axis of an acceptable part may be. Many practitioners incorrectly think that the title block tolerance controls coaxiality. It does not. See Rule #1 in Chap. 3, "The relationship between individual features…" for a more complete discussion of the coaxiality tolerance between individual features.

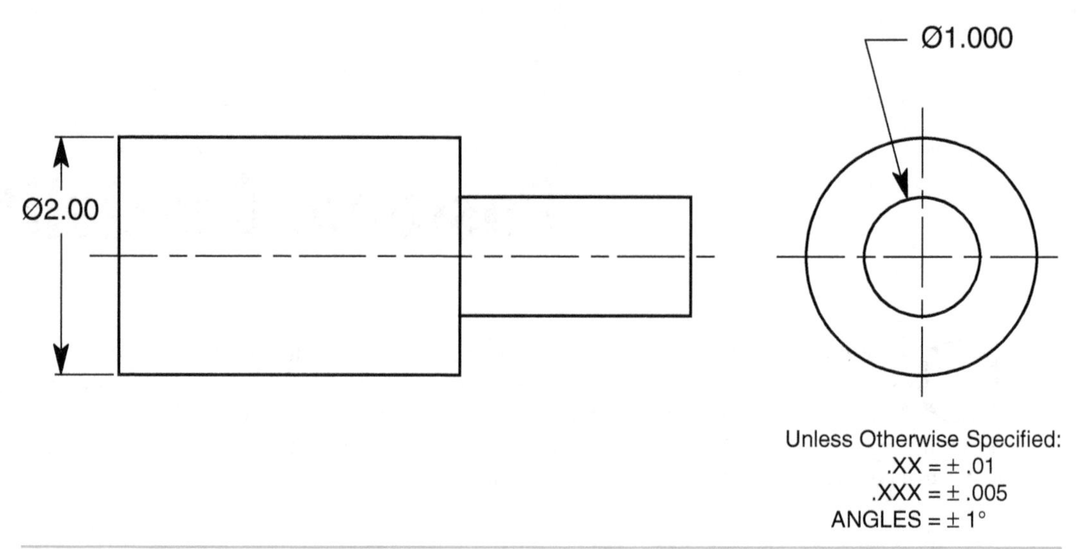

FIGURE 9-1 Definition: two coaxial cylinders.

There are other methods of controlling coaxiality, such as a note or a dimension and tolerance between diameters, but a geometric tolerance, such as the one in Fig. 9-2, is preferable. The position control is the appropriate tolerance for coaxial cylindrical surfaces of revolution that require a maximum material condition (MMC) or least material condition (LMC). The position control provides the most tolerancing flexibility.

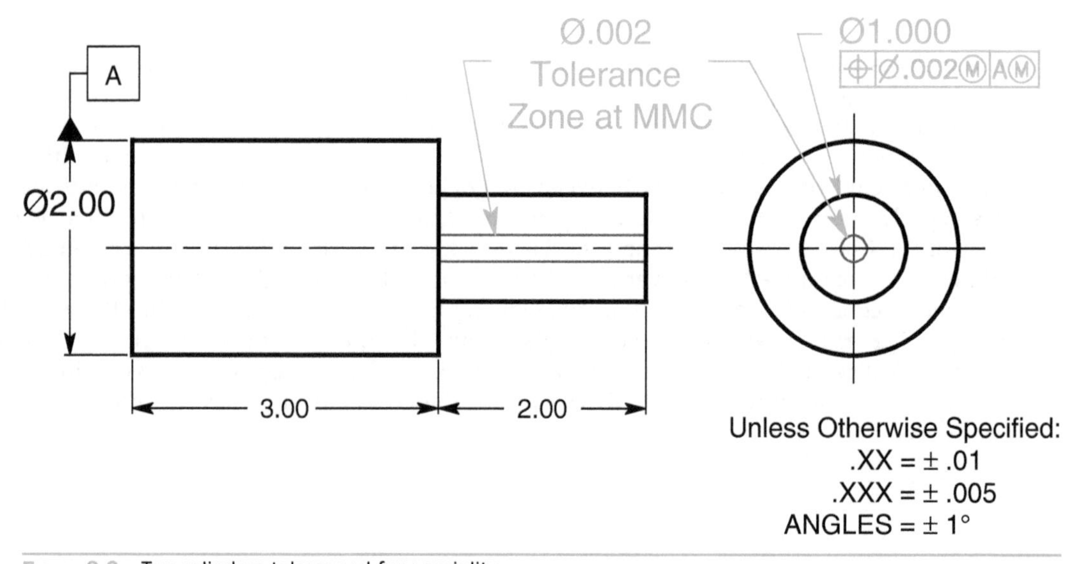

FIGURE 9-2 Two cylinders toleranced for coaxiality.

Comparison between Coaxiality Controls

The amount of permissible variation from coaxiality may be expressed by a tolerance of position, runout, or profile of a surface. Comparisons between these controls is shown in Table 9-1. In general:

- The position control is used when coaxial features are mated in a static assembly and apply at MMC, regardless of feature size (RFS), or LMC.
- Runout is specified for high-speed rotating assemblies where the surfaces of a feature must be controlled relative to the datum axis.
- Profile of a surface is specified to achieve the combined control of size, form, orientation, and coaxiality of a feature within the stated tolerance.

Many people automatically specify a concentricity tolerance for the control of coaxiality, perhaps because they use the terms *coaxial* and *concentric* interchangeably. *Coaxial* means that the axes of two or more features of size are coincident. *Concentric* means that the center points of two or more plane geometric figures are coincident. The concentricity tolerance controls all of the median points of a figure of revolution within a cylindrical tolerance zone. The concentricity control requires an expensive inspection process and is appropriate in only a few unique applications where precise balance is required.

NOTE *The discussion above about concentricity applies to all previous standards. The concentricity and symmetry controls have been deleted from the ASME Y 14.5-2018 standard. The chapter in the previous (second) edition on concentricity and symmetry appears in its entirety as App. A of this text.*

Characteristic Symbol	Tolerance Zone	Material Condition	Surface Error
⌖	Ø	Ⓜ Ⓛ	Includes
↗	Two concentric circles	None	Includes
⌰	Two coaxial cylinders	None	Includes
◎	Ø	None	Independent
NOTE: *The concentricity control has been deleted from the ASME Y 14.5-2018 standard.*			
⌓	The shape of the profile	None	Includes

TABLE 9-1 A Comparison between Position, Runout, Profile of a Surface, and Concentricity

Specifying Coaxiality at MMC

Where specifying coaxiality with the position control, the feature control frame is associated with the size dimension of the feature being controlled. The axis of the toleranced feature is controlled with a cylindrical tolerance zone. The tolerance may apply at MMC, RFS, and LMC, and the datum feature(s) may apply at maximum material boundary (MMB), regardless of material boundary (RMB), and least material boundary (LMB). At least one datum feature is specified in the feature control frame.

In Fig. 9-3, a coaxiality tolerance is specified at MMC and a datum feature of size is specified at MMB, and bonus and shift tolerances are available. The circle M symbol after the geometric tolerance provides the opportunity for a bonus tolerance as the feature departs from MMC toward LMC. The circle M symbol after the datum feature provides the opportunity for a shift tolerance as the datum feature departs from its MMB toward LMC. It should be emphasized here that the bonus tolerance and the shift tolerance are not the same. Bonus tolerance increases the size of the tolerance zone. Shift tolerance does not increase the size of the tolerance zone; it allows the datum feature to shift within its MMB.

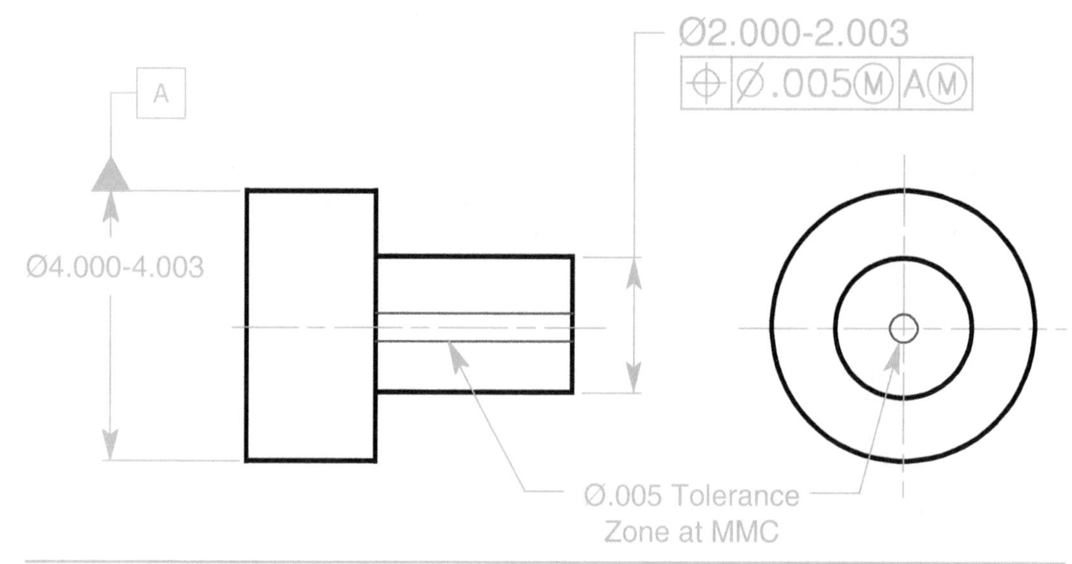

FIGURE 9-3 Specifying a coaxiality tolerance at MMC to a datum feature of size at MMB.

Composite Positional Control of Coaxial Features

A composite positional tolerance may be applied to a pattern of coaxial features, such as those in Fig. 9-4. The upper segment of the feature control frame controls the location of the hole pattern to the datum features. The lower segment of the feature control frame controls the coaxiality of the holes to one another within the tighter tolerance. The smaller tolerance zone framework may float up and down, in and out, and at any angle within the larger tolerance zones. Portions of the smaller tolerance zone may fall outside the larger tolerance zone, but these portions are unusable. The axes of the holes must fall inside both of their respective tolerance zones at the same time.

Datum features in the lower segment of a composite feature control frame must repeat the datum feature(s) that appear in the upper segment, and they only control orientation. Since both of the datum features in Fig. 9-5 are repeated in the lower segment of the feature control frame, the smaller tolerance zone framework may float up and down and in and out but must remain parallel to datum features A and B. The axes of the holes must fall inside both of their respective tolerance zones at the same time.

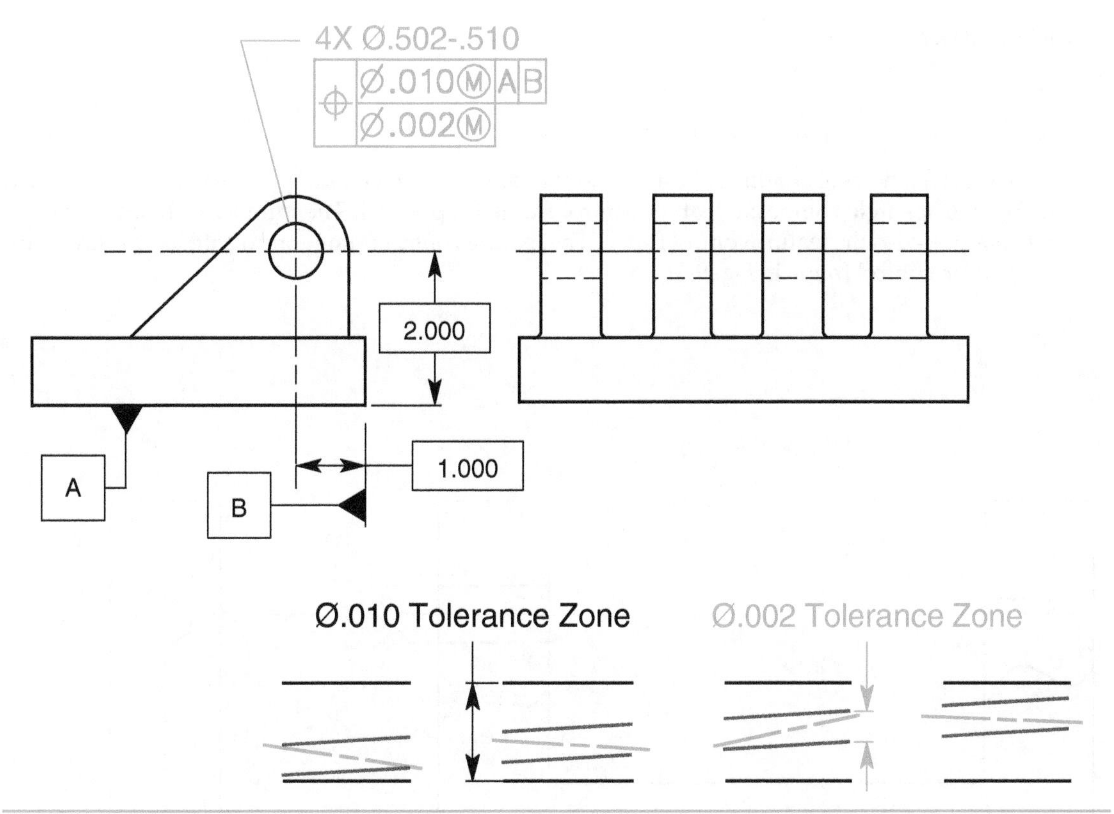

FIGURE 9-4 Composite control of coaxial features.

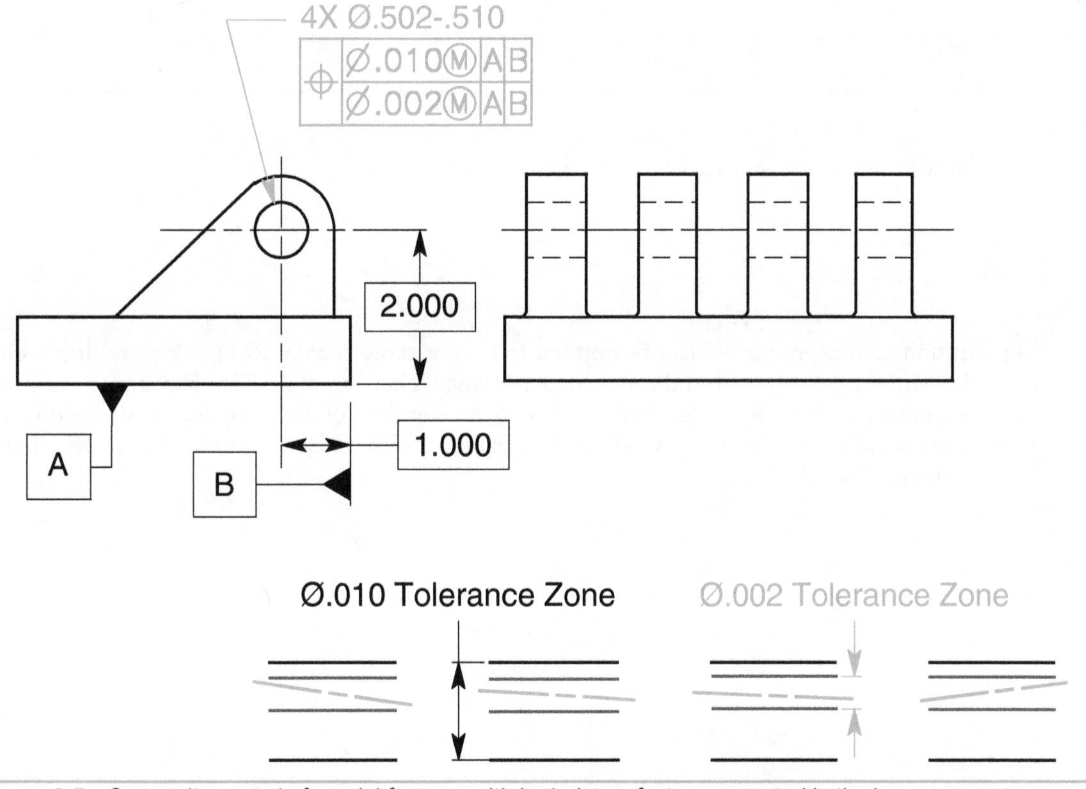

FIGURE 9-5 Composite control of coaxial features with both datum features repeated in the lower segment.

Positional Tolerancing for Coaxial Holes of Different Sizes

Where coaxial holes are of different sizes and the same requirements apply to each hole, as shown in Fig. 9-6, a single composite feature control frame is specified. The number of holes followed by an X precedes the feature control frame. The feature control frame controls the holes similar to the feature control frame in Fig. 9-4.

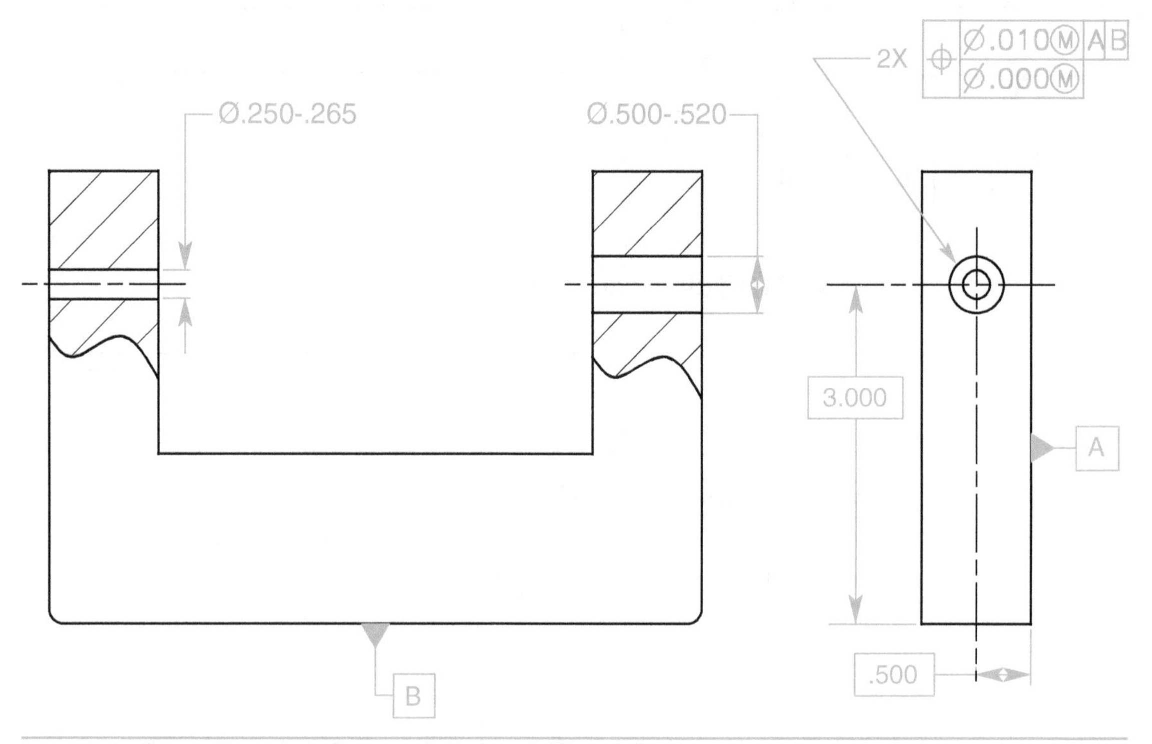

FIGURE 9-6 Composite control of two coaxial holes of different sizes.

Coaxial Features Controlled without Datum References

The position control in Fig. 9-7 can be applied to two or more coaxial features controlling their coaxiality simultaneously within the specified tolerance. There are no datum features in the feature control frame. This control is similar to the lower segment of the composite feature control frame shown in Fig. 9-4. The primary job of the position tolerance is to control location, which in this case is coaxiality.

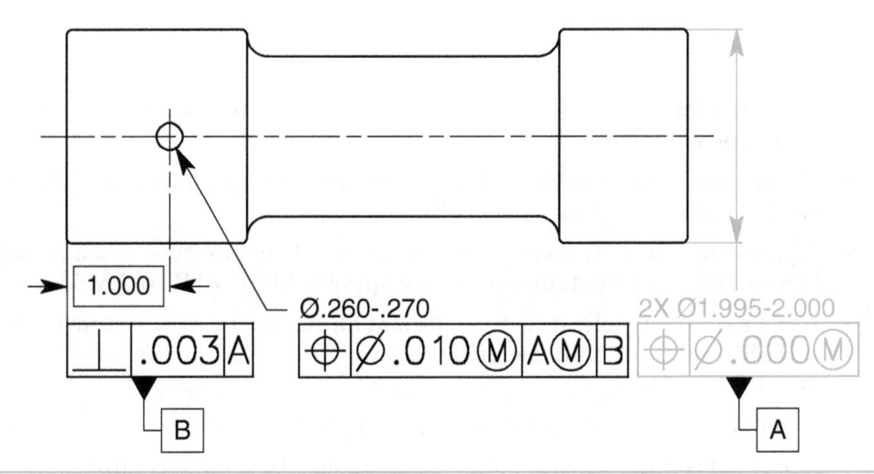

FIGURE 9-7 The position control of two coaxial features.

Tolerancing a Plug and Socket

Where an external, cylindrical, stepped part is required to assemble inside an internal mating part, diameters, such as the datum features, are dimensioned to mate. Some designers feel strongly that internal and external features should not have the same MMCs. They are concerned that a line fit may result. However, it is extremely unlikely that both parts would be manufactured at MMC. If additional clearance is required, tolerance accordingly. Once the datum features have been dimensioned, assign a coaxiality tolerance and tolerance the step features to their virtual conditions.

	Plug	Socket
MMC	.500	.505
Geometric Tolerance	+.000	−.005
Virtual Condition	.500	.500

A mating plug and socket will assemble every time if they are designed to their virtual conditions, as shown in Fig. 9-8.

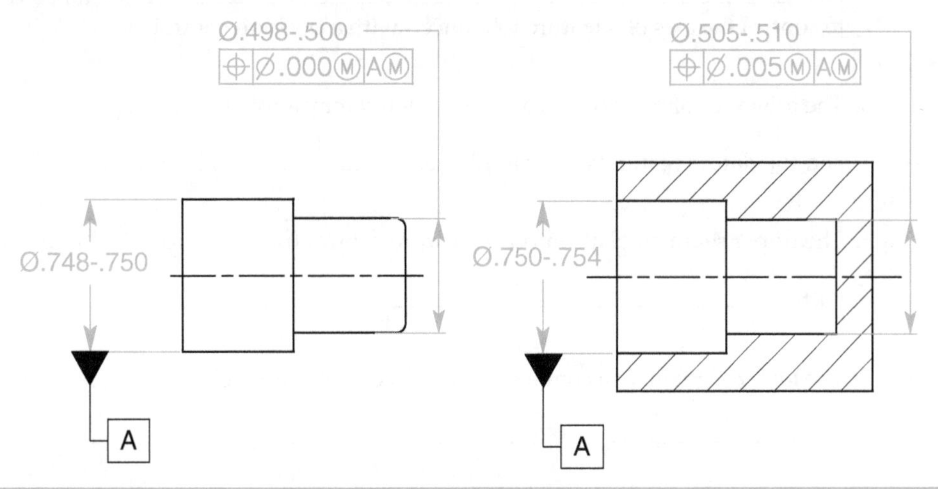

FIGURE 9-8 Plug and socket toleranced to their virtual conditions.

Summary

- The definition of coaxiality is that condition where the axes of two or more surfaces of revolution are coincident.
- The amount of permissible variation from coaxiality may be expressed by a tolerance of position, runout, or profile of a surface.
- A positional coaxiality control has a cylindrical tolerance zone and may apply at MMC, RFS, or LMC, and a datum feature may apply at MMB, RMB, or LMB.
- When a coaxiality tolerance is specified at MMC and a datum feature of size is specified at MMB, bonus and shift tolerances are available in the exact amount of such departures from MMC/MMB.
- A composite positional tolerance may be applied to a pattern of coaxial features.
- A position control, without datum features, may be used to control two or more coaxial features simultaneously within the specified tolerance.
- A mating plug and socket will assemble every time if they are designed to their virtual conditions.

Chapter Review

1. Coaxiality is that condition where the axes of two or more surfaces of revolution are ____.

2. There is a misconception that centerlines or the tolerance block control the ____________ ____ between two cylinders.

3. The ______________________________ control is the appropriate tolerance for coaxial surfaces of revolution that are cylindrical and require an MMC or an LMC.

4. A ______________________________ tolerance zone is used to control the axis of a feature toleranced with a position control.

5. The tolerance of position to control coaxiality may apply at ____________________, and the datum feature(s) may apply at ______________________________.

6. The upper segment of a composite feature control frame controls the location of the hole pattern to ____.

7. The lower segment of a composite feature control frame controls the coaxiality of holes to ____.

8. The smaller tolerance zone framework of a composite feature control frame with no datum features may float ______________________________.

9. The position control, with no datum features, can be applied to two or more coaxial features controlling their ______________________ simultaneously within the specified tolerance.

10. A mating plug and socket will assemble every time if they are designed to their ______________________________.

Problems

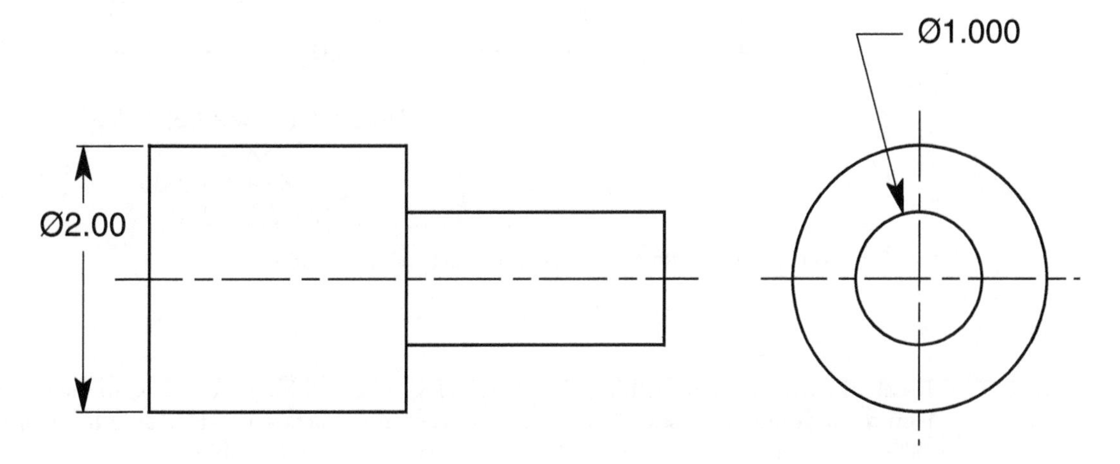

FIGURE 9-9 Controlling coaxiality with positional tolerancing: Probs. 1 through 3.

1. What controls the coaxiality of the two cylinders on the drawing in Fig. 9-9?

______________________________.

2. On the drawing in Fig. 9-9, specify a coaxiality tolerance to control the 1.000-diameter feature within a cylindrical tolerance zone of .004 to the 2.00-diameter feature. Use MMC and MMB wherever possible.

3. Now that the feature control frame has been added to the drawing in Fig. 9-9, if the larger diameter is produced at 2.00 inches and the smaller diameter is produced at 1.000 inch, how much total coaxiality tolerance applies?

______________________________.

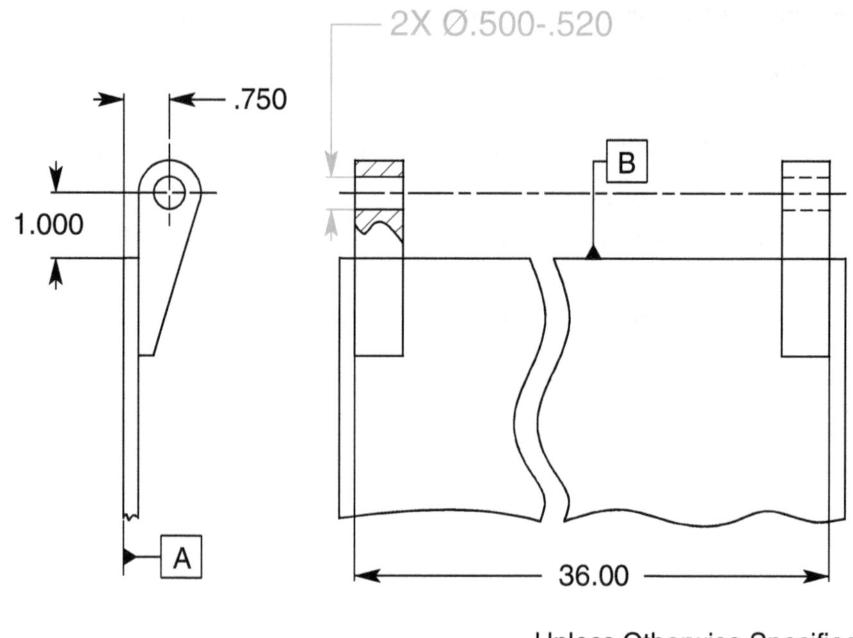

FIGURE 9-10 Controlling coaxiality with composite positional tolerancing: Prob. 4.

4. Locate the two holes in the hinge brackets in Fig. 9-10 within .030 at MMC to the datum features indicated and specify coaxiality to each other. They must be able to accept a .500-diameter hinge pin. Specify MMC and MMB wherever possible.

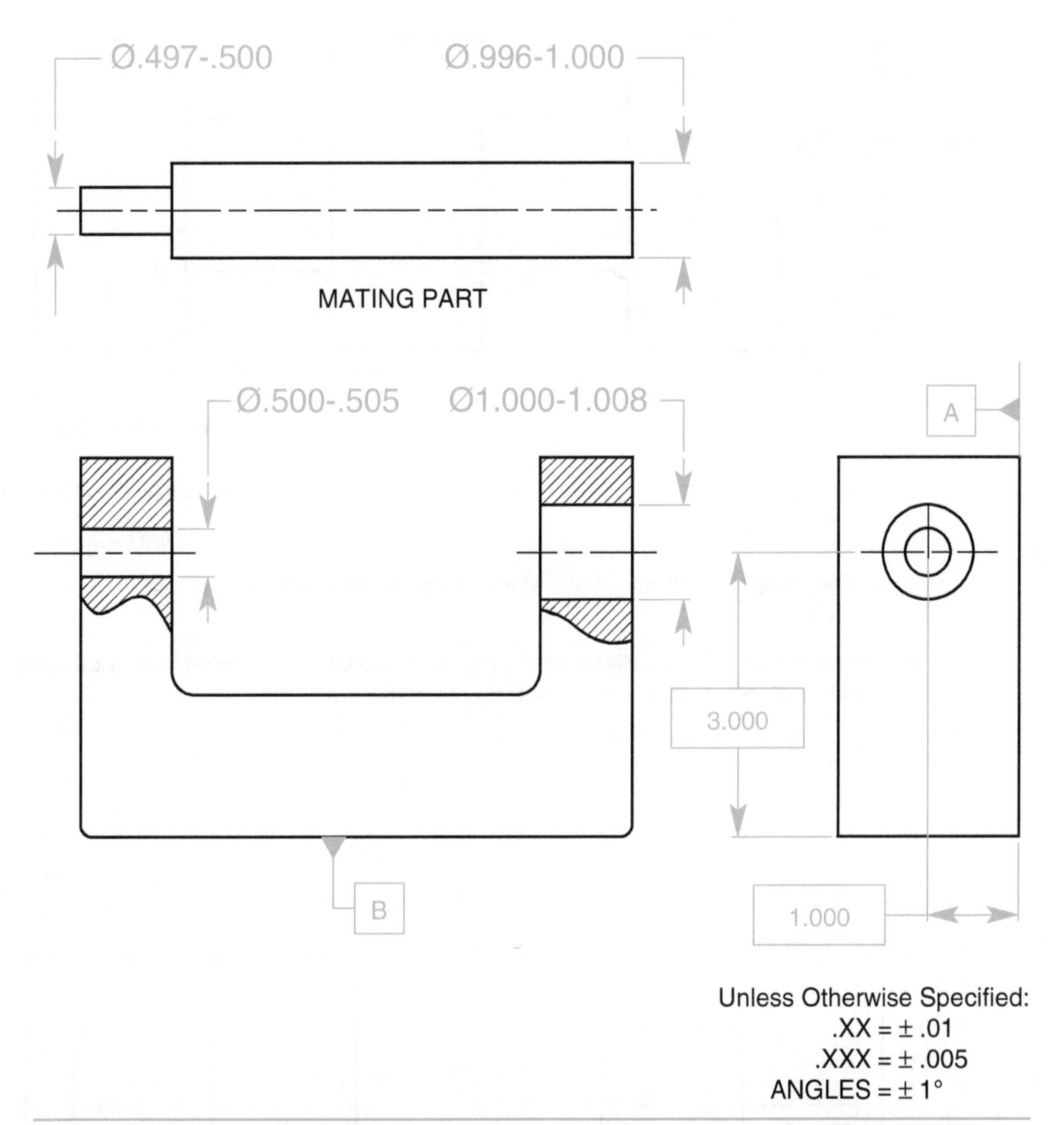

FIGURE 9-11 Controlling two coaxial holes and mating pin of different diameters: Prob. 5.

5. Locate the two coaxial holes in Fig. 9-11 parallel to the back and bottom surfaces of the part within a tolerance of .040 at MMC. Use the appropriate tolerance to control the coaxiality for the two mating parts. Specify MMC and MMB wherever possible.

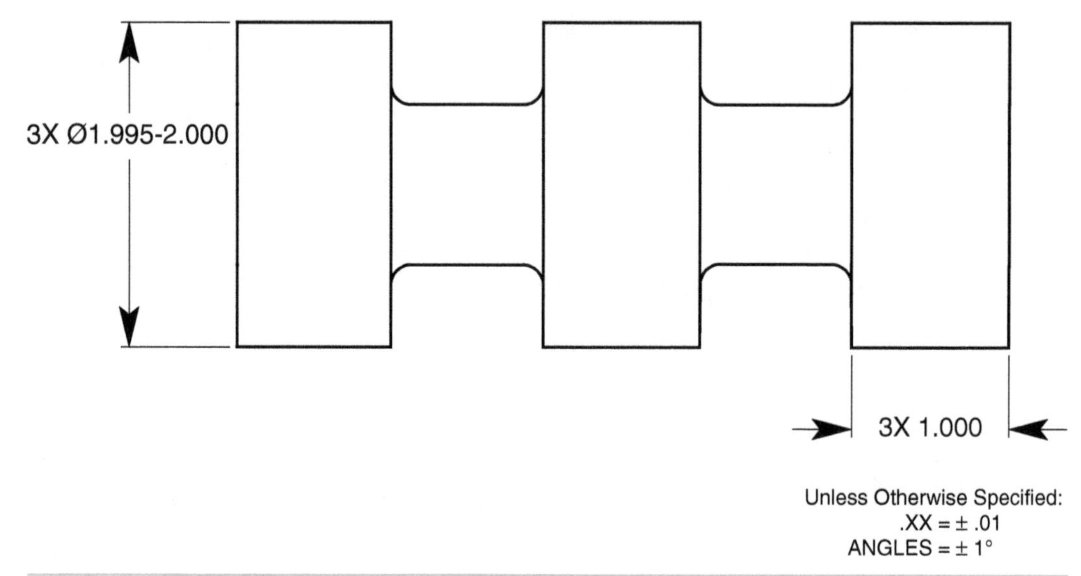

FIGURE 9-12 Controlling coaxiality with positional tolerancing without datum features: Prob. 6.

6. Control the three 3-inch diameters in Fig. 9-12 coaxial to each other with a tolerance of .010 at MMC using the position control without datum features.

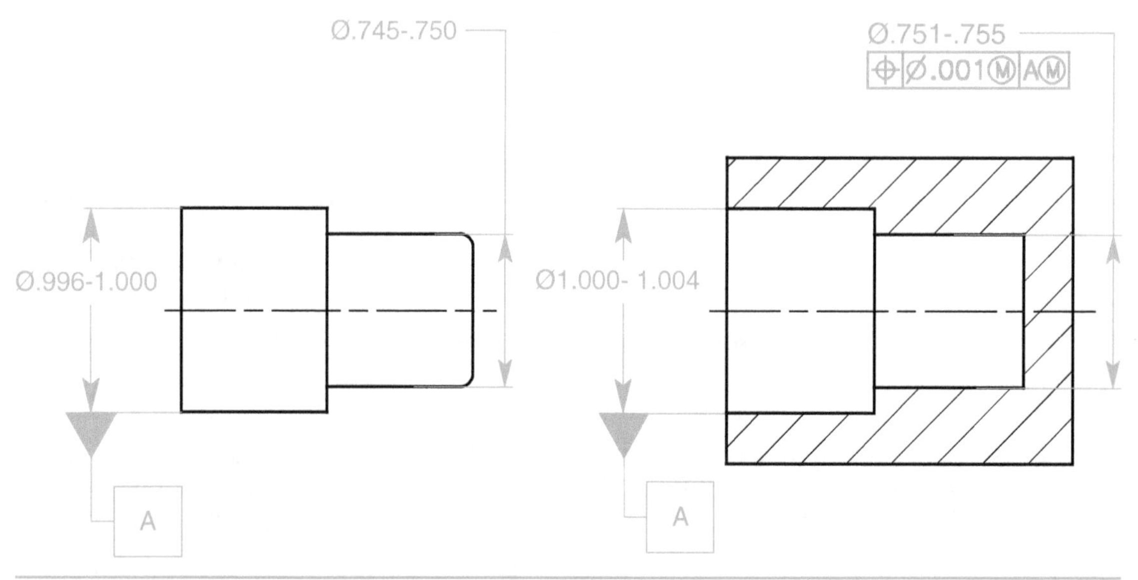

FIGURE 9-13 Controlling coaxiality for a plug and socket: Prob. 7.

7. Control the coaxiality of both parts in Fig. 9-13 so that they will always assemble. Specify MMC and MMB wherever possible.

CHAPTER 10
Runout

Runout is a surface control. It controls surfaces constructed around a datum axis and surfaces constructed perpendicular to a datum axis. Runout controls several characteristics of surfaces of revolution at the same time as the surface is rotated about its datum axis.

Chapter Objectives

After completing this chapter, the learner will be able to:

- *Explain* the difference between circular and total runout
- *Specify* runout and partial runout
- *Explain* the application of common datum features
- *Explain* the meaning of planar and cylindrical datum features
- *Specify* geometric control of individual datum feature surfaces
- *Explain* the relationship between feature surfaces
- *Inspect* runout

Circular and total runout tolerances may be greater than, equal to, or less than the size tolerance of the considered feature. Where a surface is controlled by a runout tolerance, intended interruptions of a surface, such as keyways or holes, do not affect the tolerance zone boundaries. The extent of the boundaries is limited to where there is material.

Definition

Runout is a composite tolerance used to control the functional relationship of one or more features of a part to a datum axis established from a datum feature specified at regardless of material boundary (RMB).

Circular Runout

Circular runout applies independently to each circular element on the surface of a feature either constructed around its datum axis or perpendicular to its datum axis as the part is rotated 360° about its datum axis. The circular runout tolerance zone for surfaces constructed around its datum axis is bounded by two concentric circles centered on the datum axis, in a

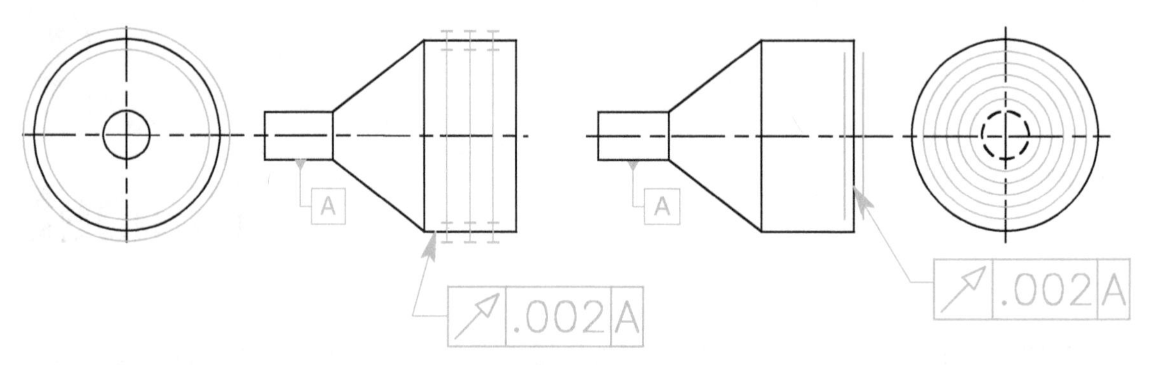

FIGURE 10-1 Circular runout applied to surfaces constructed around a datum axis and perpendicular to a datum axis.

plane that is 90° to the datum axis, and separated by the runout tolerance value, as shown in Fig. 10-1. Because circular runout tolerance applies independently to each circular line element at each measuring position, it may be easily applied to cones and curved profiles constructed around a datum axis. Where applied to surfaces of revolution, circular runout controls a combination of variations in circularity and coaxiality. Where applied to surfaces at a 90° angle to the datum axis, the tolerance zone boundary consists of two parallel circles of equal diameter, centered on the datum axis, and separated by the runout tolerance value, as shown in Fig. 10-1. Circular runout controls variations in perpendicularity of circular elements to its datum axis. In other words, where applied to surfaces perpendicular to a datum axis, circular runout controls wobble.

Total Runout

Total runout is a compound tolerance that provides control of all surface elements of a feature. The total runout tolerance zone about a datum axis is bounded by two coaxial cylinders with a radial separation equal to the tolerance value, as shown in Fig. 10-2. Total runout tolerance is applied simultaneously to all circular and profile measuring positions either around its datum axis or perpendicular to its datum axis as the part is rotated 360° about that datum axis. Where applied to surfaces constructed around a datum axis, total runout controls a combination of surface variations, such as circularity, straightness, coaxiality, angularity, taper, and profile.

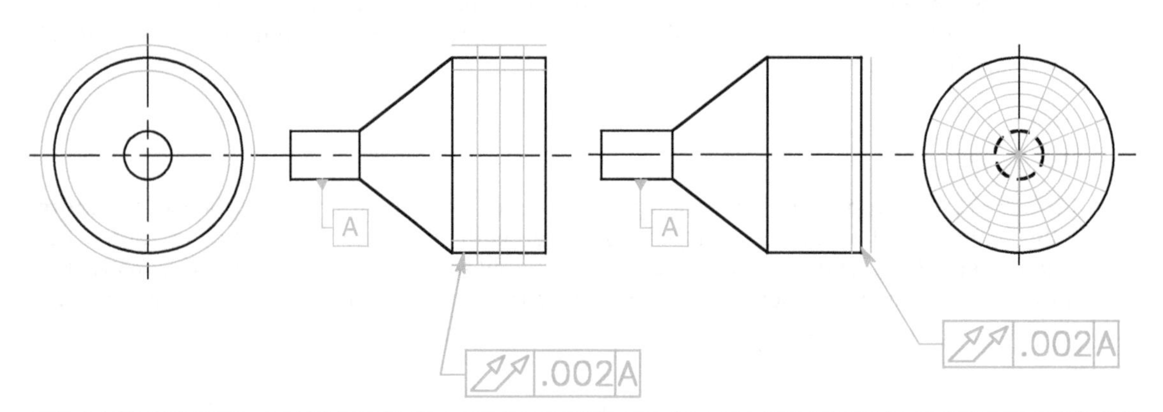

FIGURE 10-2 Total runout applied to surfaces constructed around a datum axis and perpendicular to a datum axis.

Where applied to surfaces at a 90° angle to the datum axis, the tolerance zone boundary consists of two parallel planes perpendicular to the datum axis and separated by the total runout tolerance value. Total runout controls a combination of variations of perpendicularity to the datum axis and flatness. In other words, where applied to surfaces perpendicular to a datum axis, total runout controls wobble and concavity or convexity.

Specifying Runout and Partial Runout

Where specifying runout, the feature control frame is connected to the controlled surface with a leader. In some infrequent instances, the feature control frame may be attached to the extension of a dimension line if the surface to be controlled is small or inaccessible. The feature control frame consists of a runout symbol, the numerical tolerance, and at least one datum feature that applies at RMB. With the possible exception of the free state modifier, circle F, no other symbols are appropriate in the feature control frame. Since runout is a surface control, no material condition applies; consequently, in effect, runout applies at regardless of feature size. Where runout is required for only a portion of a surface, a thick chain line is drawn on one side adjacent to the profile of the surface and dimensioned with a basic dimension, as shown in Fig. 10-3.

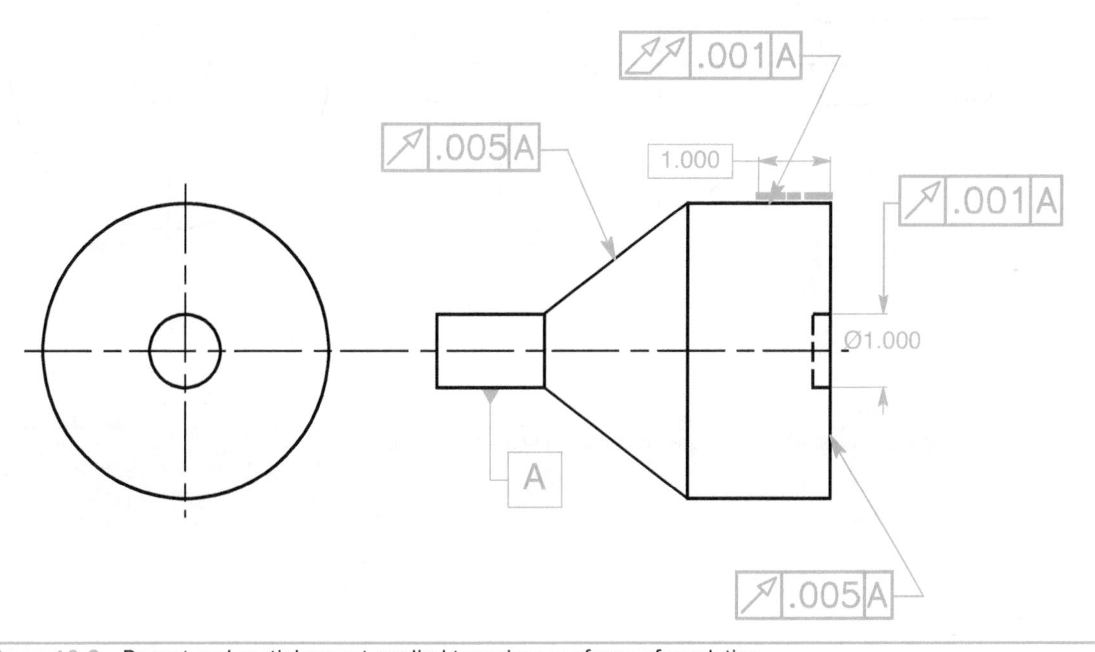

FIGURE 10-3 Runout and partial runout applied to various surfaces of revolution.

Common Datum Features

At least one datum feature must be specified for a runout control. In many cases, two functional datum features are used to support a rotating part, such as the one shown in Fig. 10-4. Datum feature A is no more important than datum feature B and vice versa. Figure 10-4 shows datum features A and B at a slight angle to the true axis, but, taken together, the two datum features tend to equalize their axes, producing a single datum axis truer than the axis that either one of the datum features would have produced independently.

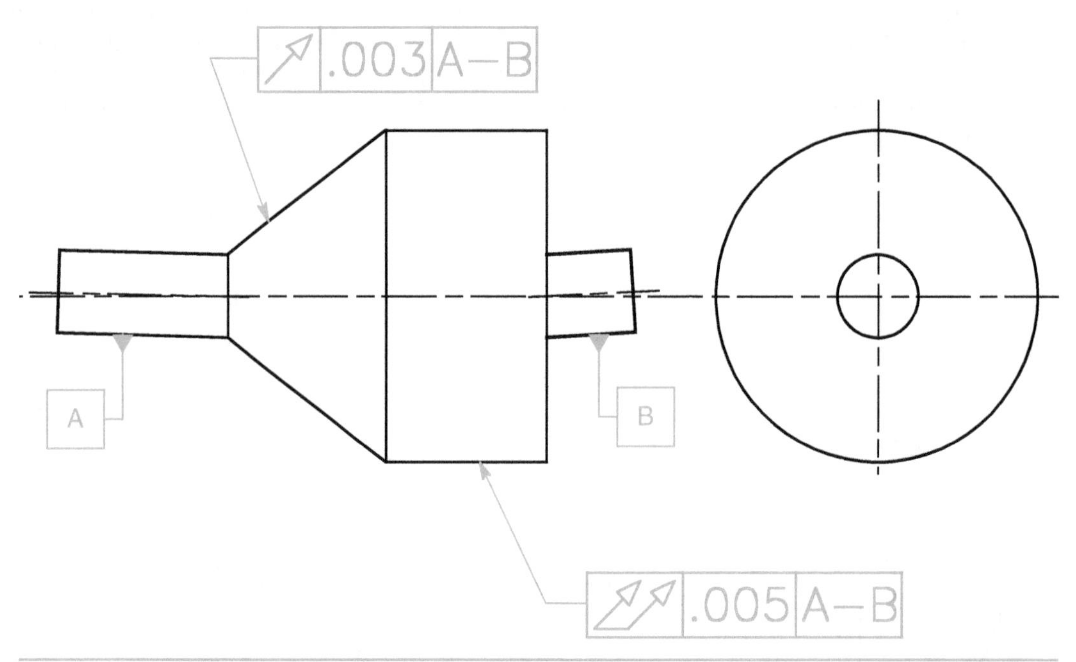

FIGURE 10-4 Common datum features creating a single datum axis.

Planar and Cylindrical Datum Features

The planar and cylindrical datum reference frame specified in Fig. 10-5 is quite a different requirement than the common datum reference frame shown in Fig. 10-4. Datum features A and B are specified in two separate compartments in this feature control frame. Therefore, datum feature A is more important than datum feature B. When inspecting the part, the primary datum feature, datum feature A, must first rest against the inspection equipment, then the part is rotated about the axis of the secondary datum feature, datum feature B, even if datum feature B is not exactly perpendicular to datum feature A.

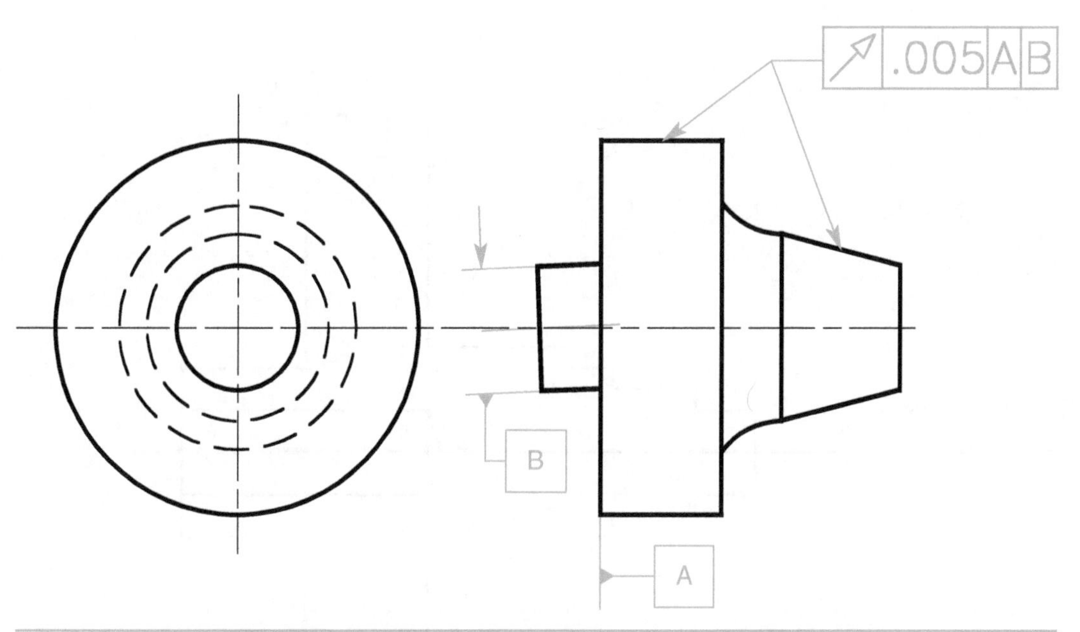

FIGURE 10-5 Planar and cylindrical datum features specified in order of precedence.

Geometric Control of Individual Datum Feature Surfaces

It may be particularly important for datum features to have a form control refinement. Datum features A and B in Fig. 10-6 have a cylindricity refinement of .0005. Design requirements may make it necessary to restrict datum surface variations with respect to straightness, flatness, circularity, cylindricity, and parallelism. Also, it may be necessary to include a runout control for individual datum features on a common datum feature reference, as shown in Fig. 10-6. Datum features A and B are independently controlled with a circular runout tolerance to datum feature A–B. This tolerance is not the same as controlling a feature to itself. In fact, it is expected that datum axis A and datum axis B are coaxial with datum axis A–B, but in the event that datum feature A or datum feature B is out of tolerance with respect to datum feature A–B, the part does not meet design requirements.

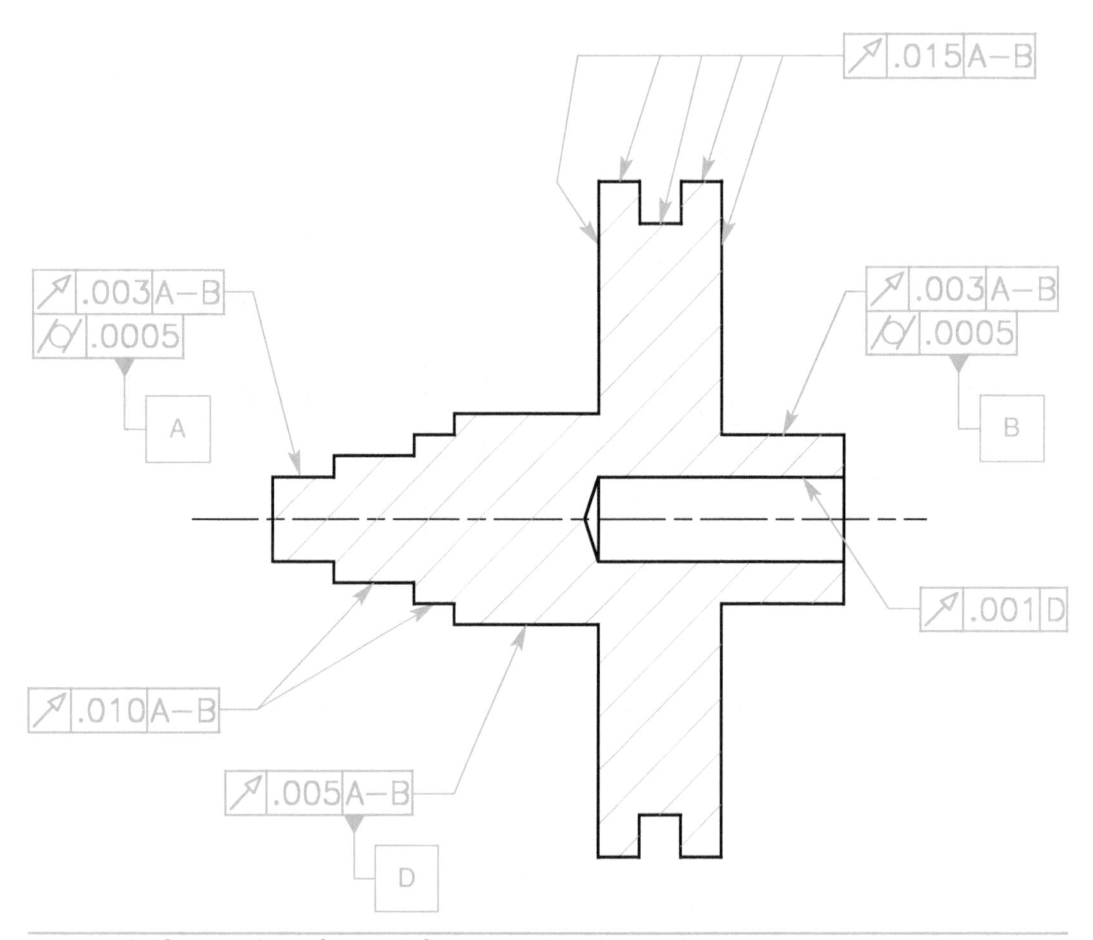

FIGURE 10-6 Common datum features refined with geometric controls.

The Relationship between Feature Surfaces

If two or more surfaces are controlled with a runout tolerance to a common datum reference, the worst-case runout between the two surfaces is the sum of the two tolerances. For example, in Fig. 10-6, the worst-case runout between the largest diameter and datum feature D is the sum of .015 plus .005, or a total of .020. The two axes may, in fact, be coaxial; however, in the worst case, the surface of the larger diameter could be translated .015 in one direction, and the surface of datum feature D could be translated .005 but in the opposite direction, producing a difference of .020 between the two surfaces.

If two features have a specific relationship between them, one should be toleranced directly to the other and not through a common datum axis. Figure 10-6 shows a .500 internal diameter controlled directly with a runout tolerance of .001 to datum feature D rather than controlling each feature to datum feature A–B. If the .500-diameter hole had been toleranced to datum features A–B, the runout tolerance between the hole and datum feature D would be a total of .006.

In Fig. 10-6, the multiple leaders directed from the .015 circular runout feature control frame to the five surfaces of the part may be specified without affecting the runout tolerance. It makes no difference whether one or several leaders are used with a feature control frame as long as the runout tolerance and datum feature(s) are the same.

Inspecting Runout

When inspecting circular runout, the feature first must fall within the specified size limits. It must also satisfy Rule #1. That is, it may not exceed the boundary of perfect form at maximum material condition. The datum feature is then mounted in a chuck or collet. With a dial indicator perpendicular to and contacting the surface to be inspected, the part is rotated 360° about its simulated datum axis, as shown in Fig. 10-7. Several measuring positions are inspected. If full indicator movement (FIM) does not exceed the specified runout tolerance, the feature is acceptable. Runout tolerance may be larger than the size tolerance. If the runout tolerance is larger than the size tolerance and no other geometric tolerance is applied, the size tolerance controls the size and form. The runout tolerance controls orientation and coaxiality. If the size tolerance is larger than the runout tolerance, circular runout refines circularity, orientation, and coaxiality.

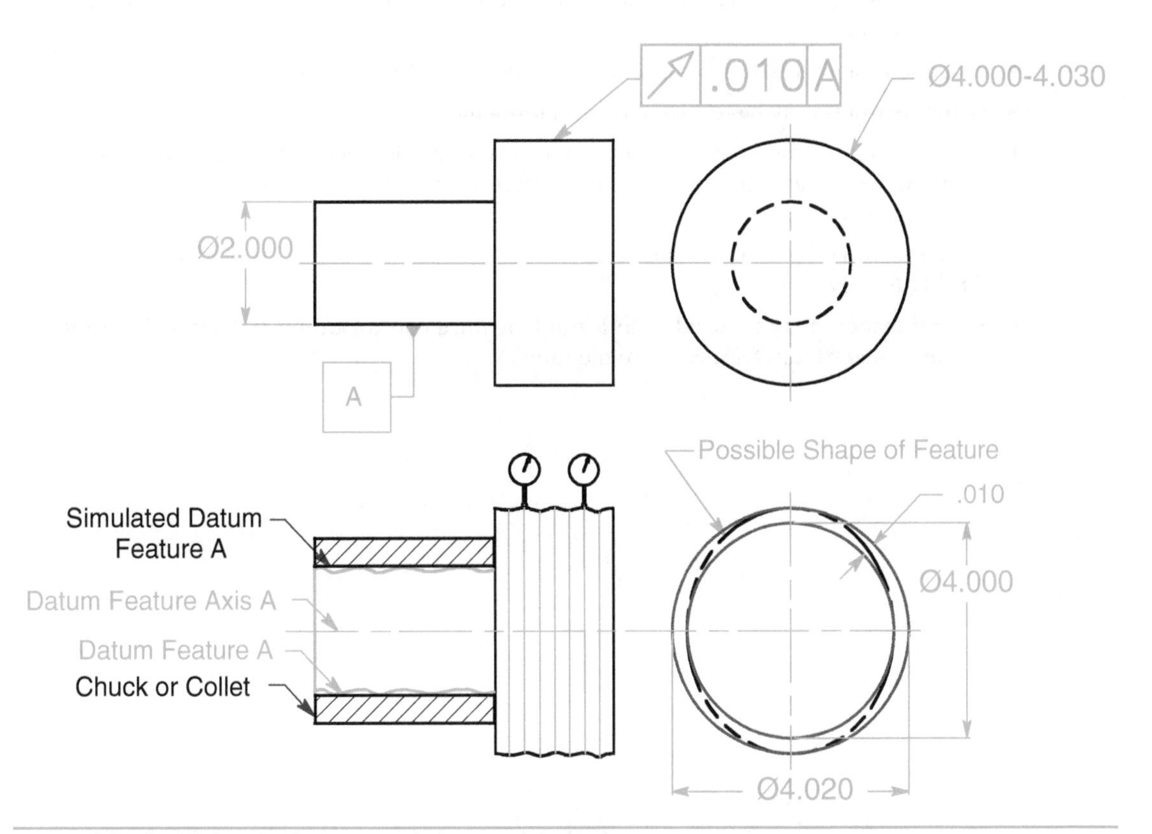

FIGURE 10-7 Inspecting circular runout relative to a datum axis.

The same preliminary inspection procedures required for circular runout are also required for total runout. Just as when inspecting circular runout, a dial indicator is held perpendicular to and contacting the surface being inspected, but the dial indicator is moved along the full length of the feature's profile as the part is rotated 360° about its simulated datum axis. If FIM does not exceed the specified runout tolerance, the feature is acceptable.

Summary

- Runout is a composite tolerance used to control the functional relationship of one or more features of a part to a datum axis established from a datum feature specified at RMB.
- Circular runout applies independently to each circular element on the surface of a part either constructed around its datum axis or perpendicular to its datum axis as the part is rotated 360° about its datum axis.
- Total runout is a compound tolerance that provides control of all surface elements of a feature. Total runout tolerance is applied simultaneously to all circular and profile measuring positions either around its datum axis or perpendicular to its datum axis as the part is rotated 360° about its datum axis.
- Where specifying runout, the feature control frame is connected to the controlled surface with a leader. The feature control frame consists of a runout symbol, the numerical tolerance, and at least one datum feature that applies at RMB. With the possible exception of a free state modifier, no other symbols are appropriate in the feature control frame.
- Two functional datum features, called common datum features, may be used to support a rotating part.
- Face and diameter datum features are specified in order of precedence.
- Datum features may have a form control refinement.
- If two or more surfaces are controlled with a runout tolerance to a common datum reference, the worst-case runout between the two surfaces is the sum of the two individual runout tolerances.
- If two features have a specific relationship between them, one should be toleranced directly to the other.
- Several leaders may be used with a single feature control frame as long as the runout tolerance and datum feature(s) are the same.

Chapter Review

1. Circular runout applies independently to each ______________________ on the surface of a feature either constructed around a datum axis or perpendicular to a datum axis as the part is rotated ______________ about its datum axis.

2. Where applied to surfaces of revolution, circular runout controls a combination of variations in ______________________________________.

3. Total runout is a compound tolerance that provides control of all ______________ of a feature.

4. Total runout tolerance is applied simultaneously to all circular and profile measuring positions either ______________________ or ______________________ to its datum axis as the part is rotated ______________ about that datum axis.

5. Where applied to surfaces constructed around a datum axis, total runout controls a combination of surface variations, such as ______________________________________ ______________________________________.

6. Where applied to surfaces at a 90° angle to the datum axis, total runout controls a combination of variations of ______________________________________.

7. The runout feature control frame consists of ______________________________________ ______________________________________.

8. In many cases, two functional ______________ are used to support a rotating part.

9. The planar and cylindrical datum reference frame is quite a different requirement than the ______________________________________ reference frame.

10. Design requirements may make it necessary to restrict datum surface variations with respect to (other geometric controls) ______________________________________.

11. It may be necessary to include a runout control for individual datum features on a

    ____.

12. If two or more surfaces are controlled with a runout tolerance to a common datum reference, the worst-case runout between two surfaces is the ______________________

    ____.

13. If two features have a specific relationship between them, one should be __________

    ____.

14. Multiple leaders directed from a runout feature control frame may be specified without

    ____.

Problems

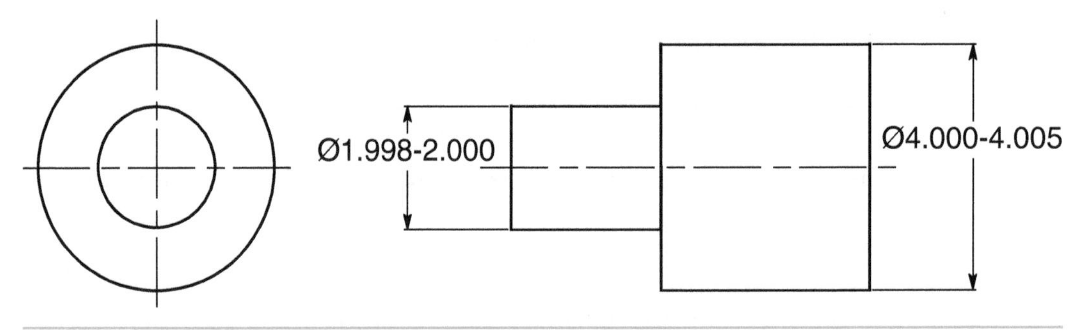

FIGURE 10-8 Controlling a coaxial feature with a runout control: Prob. 1.

1. On the part in Fig. 10-8, control the 4-inch diameter with a total runout tolerance of .002 to the 2-inch diameter.

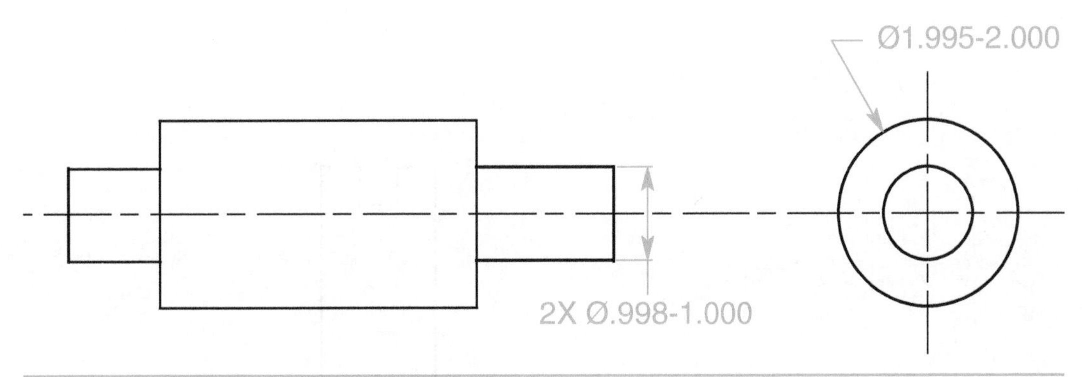

FIGURE 10-9 Controlling a partial runout: Prob. 2.

2. On the drawing in Fig. 10-9, specify a circular runout tolerance of .002 controlling the 2-inch diameter to both of the 1-inch diameters. This control is a partial runout tolerance 1 inch long, starting from the left end of the 2-inch cylinder. Also, specify a circular runout tolerance of .001 for each of the 1-inch diameters.

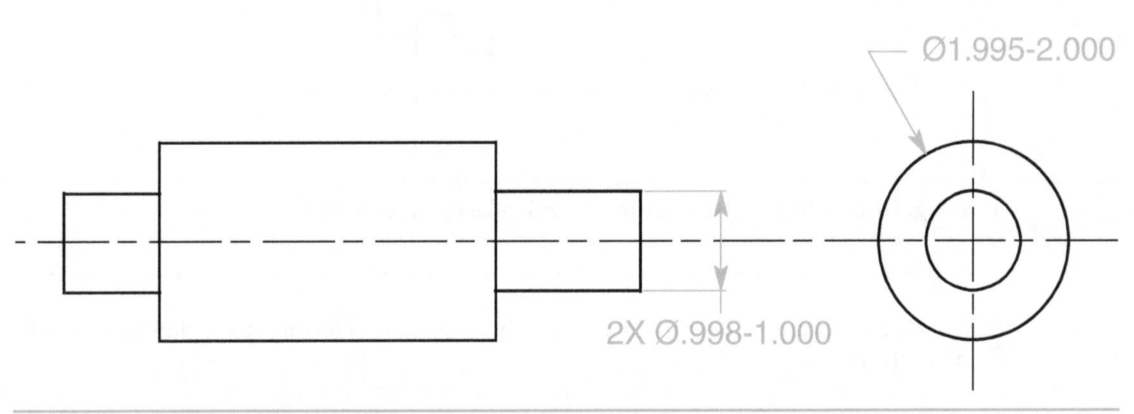

FIGURE 10-10 Datum features toleranced with a cylindricity tolerance: Prob. 3.

3. Tolerance the 2-inch diameter in Fig. 10-10 with a total runout tolerance of .001 to both of the 1-inch diameter shafts. Tolerance each 1-inch diameter shaft with a circular runout tolerance of .001 and a cylindricity tolerance of .0005.

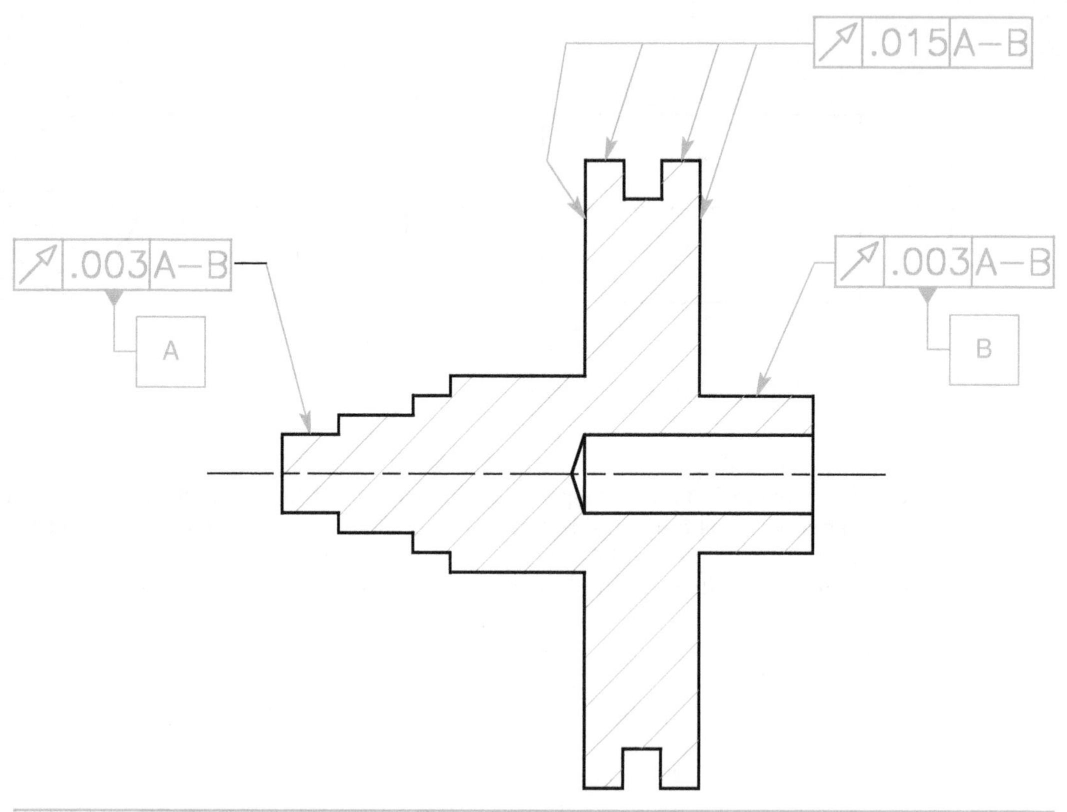

FIGURE 10-11 Multiple features tolerance with one feature control frame: Probs. 4 and 5.

4. In Fig. 10-11, which datum feature, A or B, takes precedence?

____

5. What is the worst possible runout tolerance between datum feature A and datum feature B in Fig. 10-11?

____

CHAPTER 11

Profile

Profile is a powerful and versatile tolerancing tool. It is a surface control that may be used to control just the size and form of a feature or the size, form, orientation, and location of a feature. The profile tolerance controls the orientation and location of features with unusual shapes very much like the position tolerance controls the orientation and location of holes or pins.

Chapter Objectives

After completing this chapter, the learner will be able to:

- *Specify* a profile tolerance
- *Properly apply* datum features for the profile tolerance
- *Explain* the need for a radius control with profile
- *Explain* the combination of a profile tolerance with other geometric controls
- *Specify* coplanarity
- *Properly apply* composite profile tolerancing
- *Properly apply* multiple single-segment profile tolerancing

Definition

Profile of a line is the outline of an object in a plane as the plane passes through the object. Profile of a surface is the result of projecting the profile of an object on a plane or taking cross sections through the object at various intervals.

Specifying a Profile Tolerance

Profile tolerances are used to define a tolerance zone to control size and form or combinations of size, form, orientation, and location of a feature(s) relative to a true profile. The true profile is shown in a profile view or a section view of a part and may be dimensioned with basic size dimensions, basic coordinate dimensions, basic radii, basic angular dimensions, formulas, mathematical data, or undimensioned drawings. Profile of a line and profile of a surface are surface controls. The shape of the tolerance zone is the exact shape of the profile, and it is as wide as the tolerance specified in the feature control frame. The feature control frame is directed to the profile surface with a leader or an extension line. The profile feature control frame contains the profile of a line or profile of a surface symbol and a tolerance. Since the profile controls are surface controls, the diameter symbol and material condition modifiers do not apply in the tolerance section of profile feature control frames.

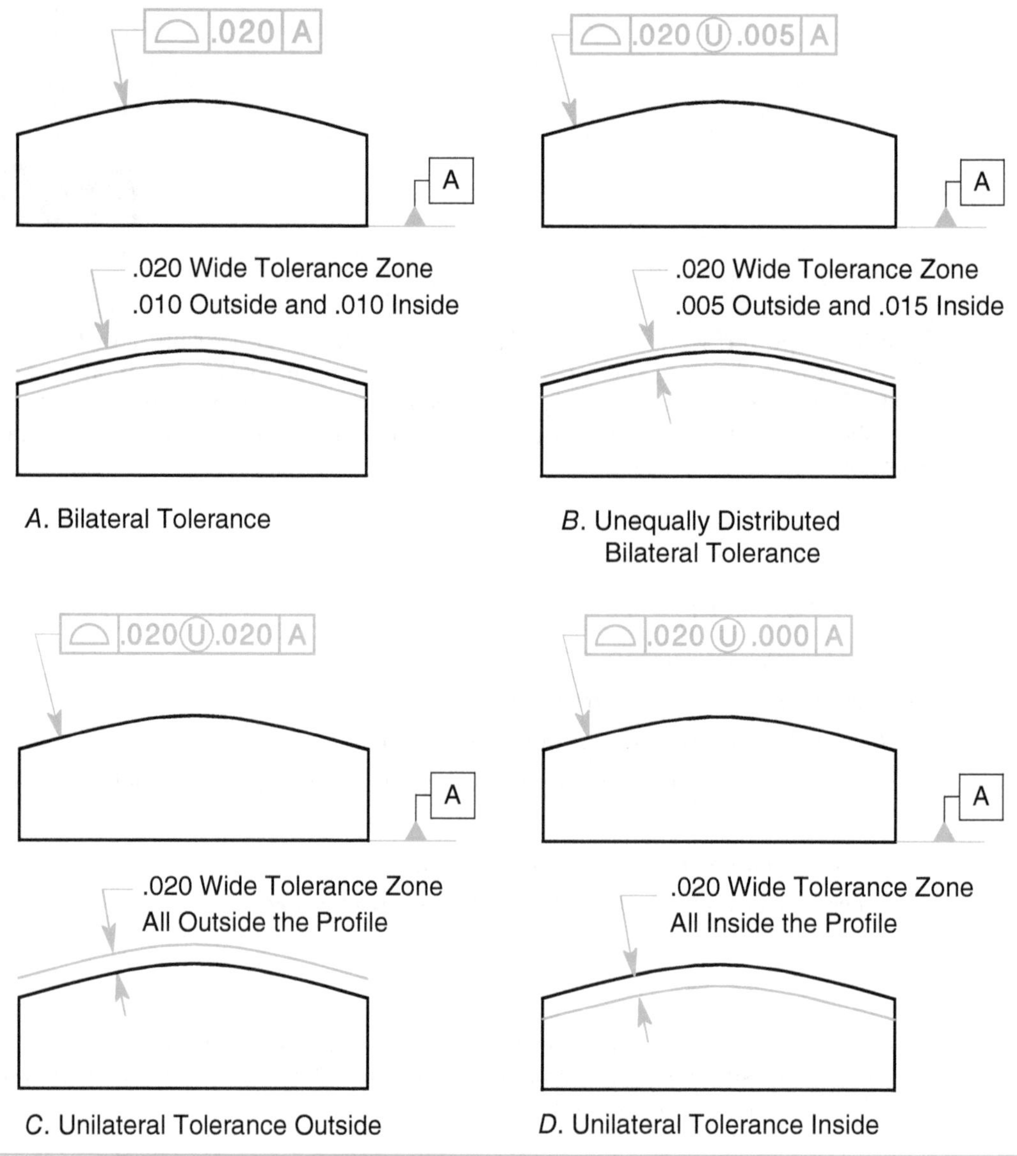

FIGURE 11-1 Specifying profile tolerance.

If the leader from a profile feature control frame points directly to the true profile, the tolerance specified is equally disposed about the true profile. This is the default condition. That is, in Fig. 11-1*A*, the .020 tolerance in the feature control frame is evenly divided, .010 outside and .010 inside the true profile. The unequally disposed profile symbol, the circle U, indicates that the profile of a surface tolerance applies unequally about the true profile. The tolerance following the circle U indicates the amount of the tolerance zone allowing material to be added to the true profile.

If a feature control frame contains all of the tolerance following a circle U symbol, as shown in Fig. 11-1*C*, all of the tolerance is outside the true profile. If a feature control frame contains zero tolerance following a circle U symbol, as shown in Fig. 11-1*D*, all of the tolerance is inside the true profile. If a feature control frame contains a portion of the tolerance following a circle U symbol, as shown in Fig. 11-1*B*, that portion of the tolerance applies outside the true profile (see also Fig. 11-2).

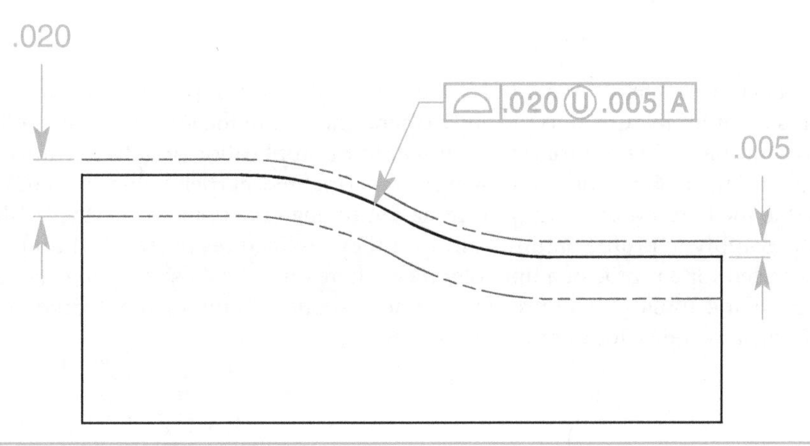

FIGURE 11-2 Unequally disposed profile symbol.

Where a profile tolerance applies all around the profile of a part, the all around symbol is specified, as shown in Fig. 11-3*A*. The all around symbol is indicated by placing a circle around the joint in the leader from the feature control frame to the profile. If the profile is to extend between two points, as shown in Fig. 11-3*B*, the points are labeled, and a note using the between symbol is placed beneath the feature control frame. The profile tolerance applies to the section of the profile between points X and Z where the leader is pointing. If a part such as a casting or forging is to be controlled with a profile tolerance over the entire surface of the part, the all over symbol is specified. The all over symbol is indicated by placing two concentric circles around the joint in the leader from the feature control frame to the profile, shown in Fig. 11-3*C*. Also, the previous method of placing the ALL OVER note beneath the feature control frame is still acceptable. When an unusual profile tolerancing requirement occurs, one not covered by the symbols above, a local note that clearly states the extent and application of the profile tolerance must be included.

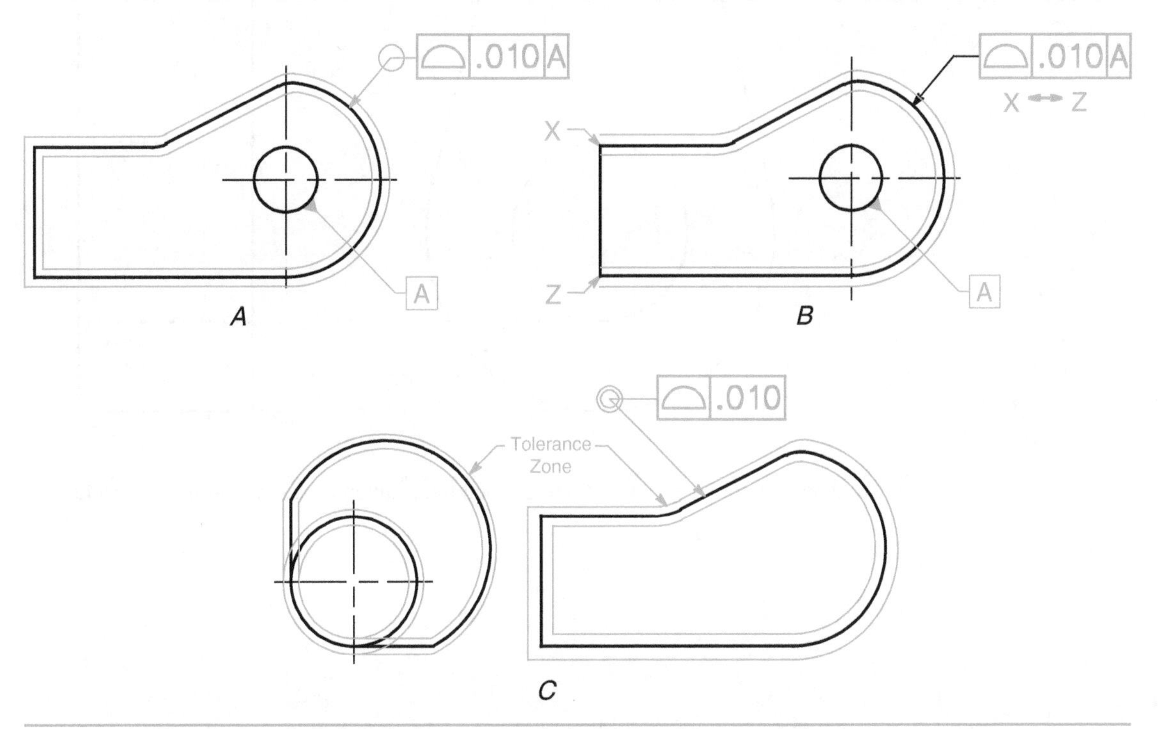

FIGURE 11-3 The all around symbol, the between symbol, and the all over symbol.

The Application of Datum Features

Profile tolerances may or may not have datum features. The profile of a surface control usually requires a datum feature(s) to properly orient and locate the surface. The application of datum features for the profile control is very similar to the application of datum features for the position control. In Fig. 11-4, the profile of a surface is oriented perpendicularly to datum plane A and located to the hole, datum feature B, at maximum material boundary (MMB). Material condition modifiers apply to profile datum features if they are features of size. Datum features are generally not used for a profile of a line tolerance where only the cross section is being controlled. An example of the application of profile of a line without a datum feature would be a profile control specifying a tolerance for a continuous extrusion.

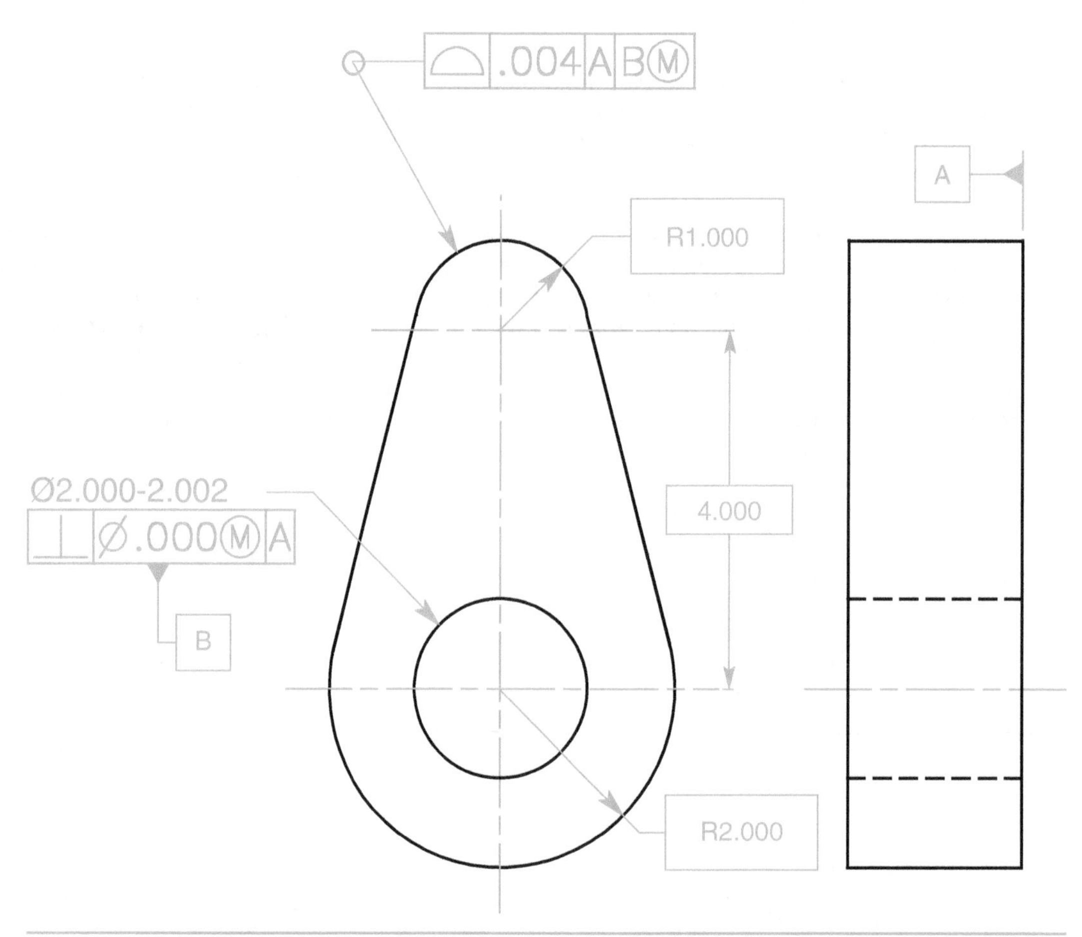

FIGURE 11-4 The orientation of a profile to datum feature A and the location to datum feature B at MMB.

A Radius Refinement with Profile

The profile tolerance around a sharp corner, labeled P in Fig. 11-5, is typically larger than the specified tolerance. Consequently, a sharp corner tolerance will allow a relatively large radius on the part profile. Excessively large radii are shown in Fig. 11-5. If the design requires a smaller radius than the radius allowed by the profile tolerance, a local note, such as ALL CORNERS R.015 MAX or R.015 MAX, is directed to the radius with a leader.

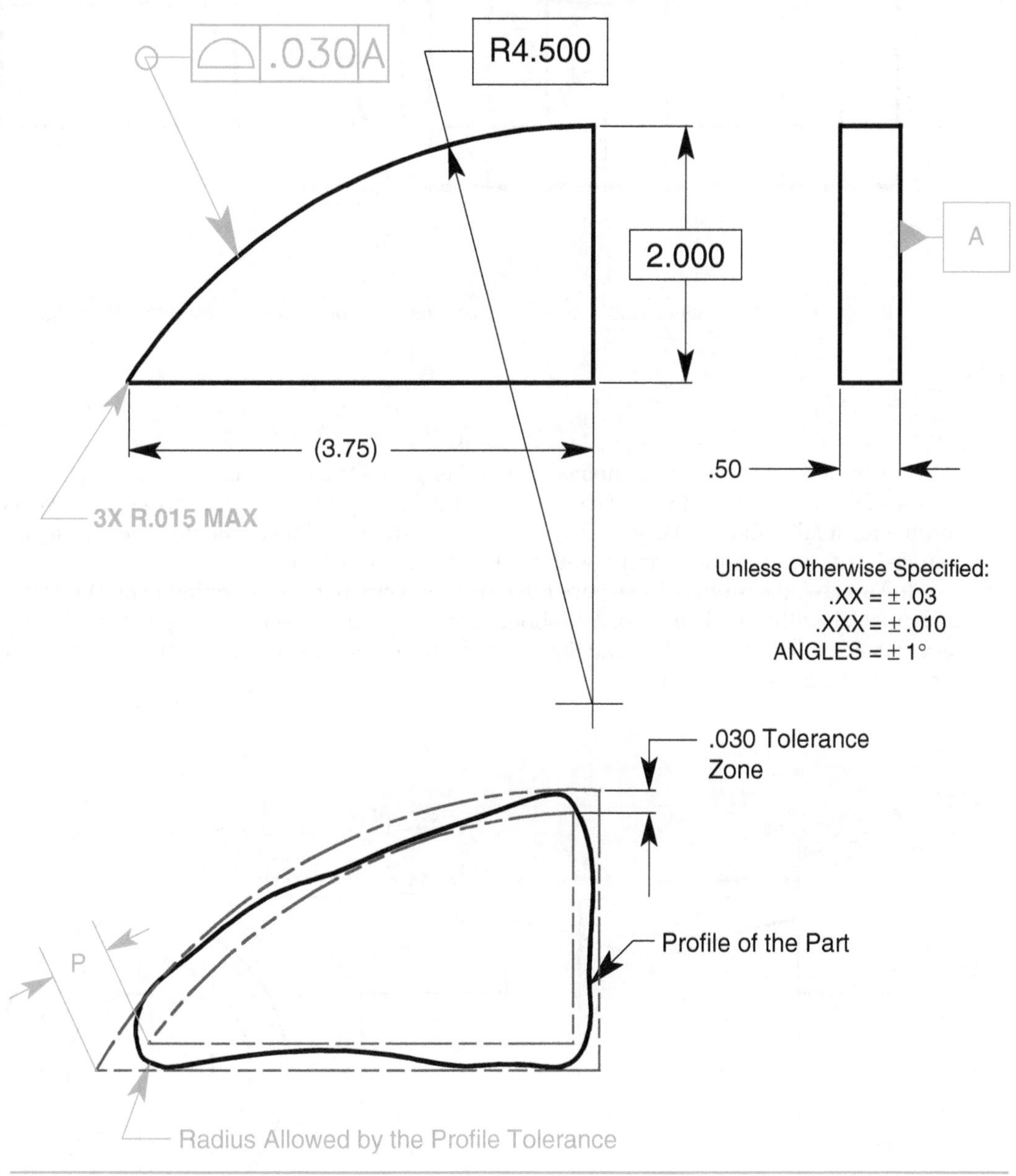

FIGURE 11-5 The profile tolerance allows large radii around sharp points.

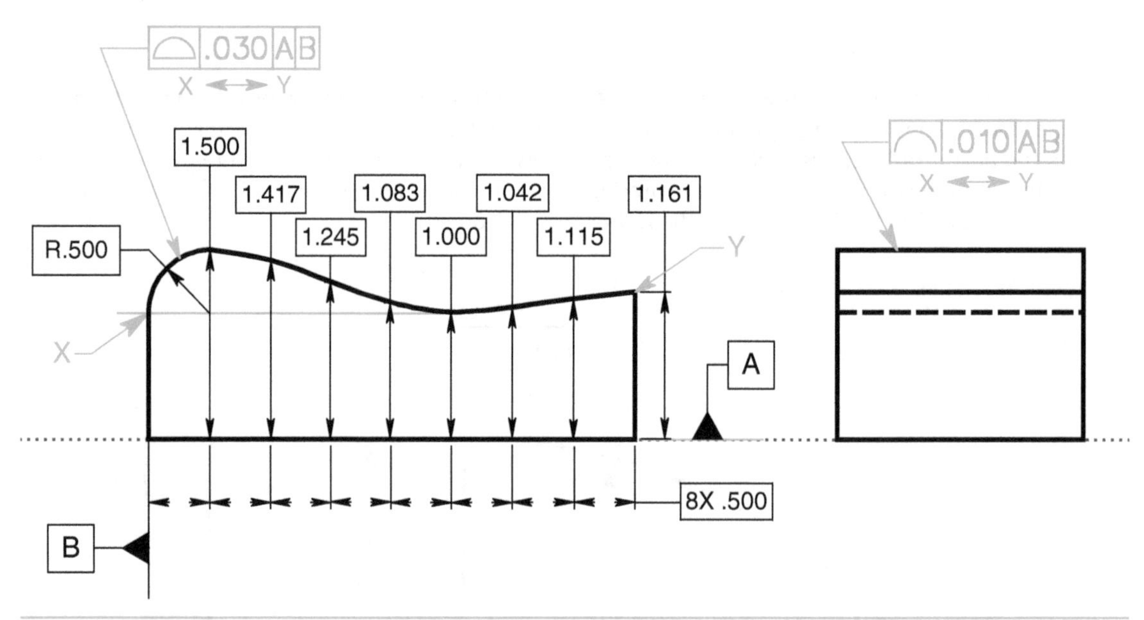

FIGURE 11-6 Profile of a surface refined with a profile of a line tolerance of each line element in the surface.

Combining Profile Tolerances with Other Geometric Controls

The profile tolerance may be combined with other geometric tolerances to refine certain aspects of a surface. In Fig. 11-6, the surface of the profile has a profile of a line refinement. While the profile must fall within a .030 tolerance zone, each individual line element in the profile must be parallel to datum features A and B within a tolerance zone of .010.

In Fig. 11-7, the profile of a surface tolerance has a circular runout refinement. While the profile must fall within a tolerance of .020 about datum feature A–B starting at datum feature C, each circular element in the profile about the datum feature axis must also fall within a circular runout of .004 to datum feature A–B.

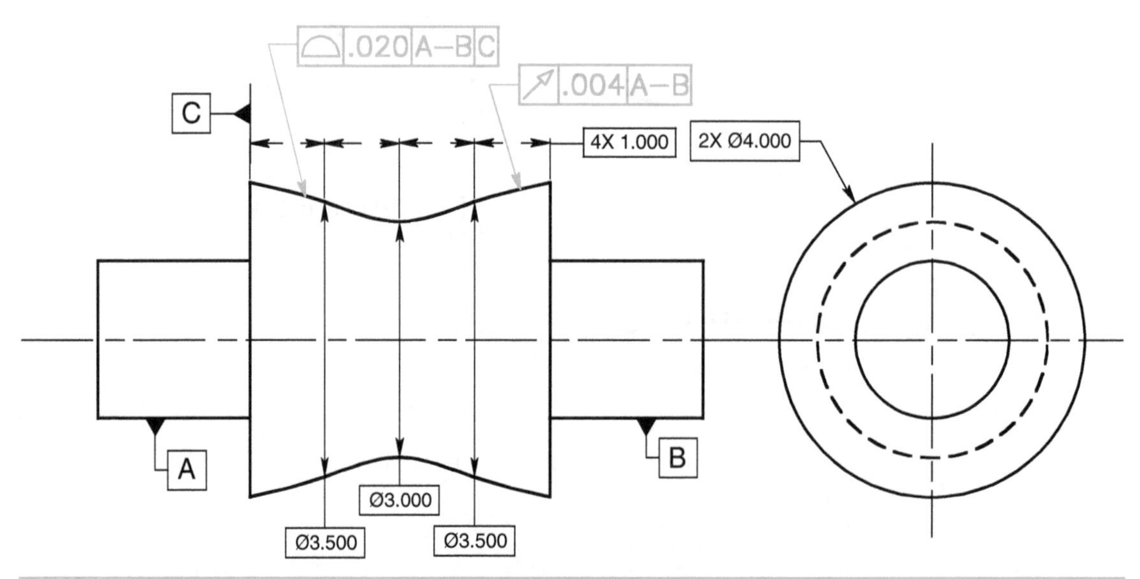

FIGURE 11-7 Profile of a surface tolerance refined with a circular runout control.

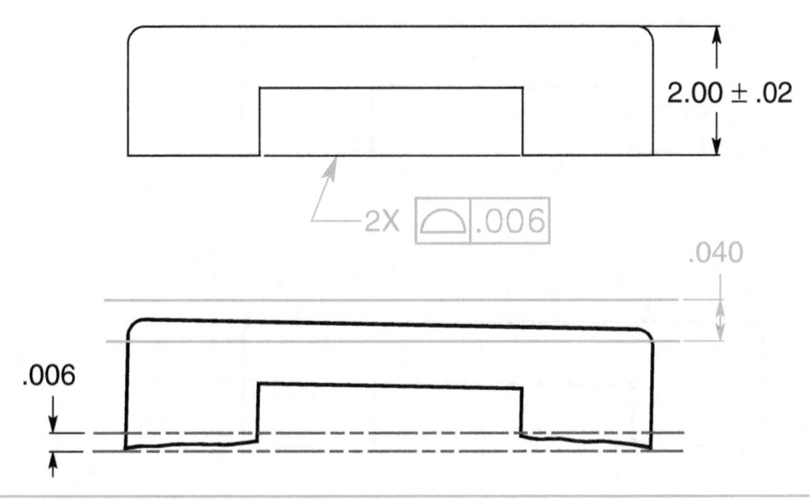

FIGURE 11-8 Specifying coplanarity with a profile control.

Coplanarity

Coplanarity is the condition of two or more surfaces having all elements in one plane. Where coplanarity is required, a profile of a surface tolerance is specified with a leader, directed to an extension line connecting the coplanar surfaces. The number of coplanar surfaces followed by an X precedes the feature control frame. As shown in Fig. 11-8, the two coplanar surfaces are not necessarily parallel to the opposite (top) surface. However, the size of the feature must be within its specified size tolerance. Coplanarity of two or more surfaces specified with a profile tolerance is similar to flatness of a single surface specified with a flatness tolerance.

Where a parallel surface is specified as a datum feature and the datum feature is referenced in the profile feature control frame, as shown in Fig. 11-9, the toleranced surfaces must be coplanar and parallel to the datum feature within the tolerance specified in the feature control frame. If the 2.00-inch dimension is specified with a plus or minus tolerance, as shown in Fig. 11-9, the .006 profile tolerance zone must be parallel to datum feature A and fall within the .040 size tolerance zone. That is, the .006 tolerance zone may float up and down within a .040 size tolerance zone but

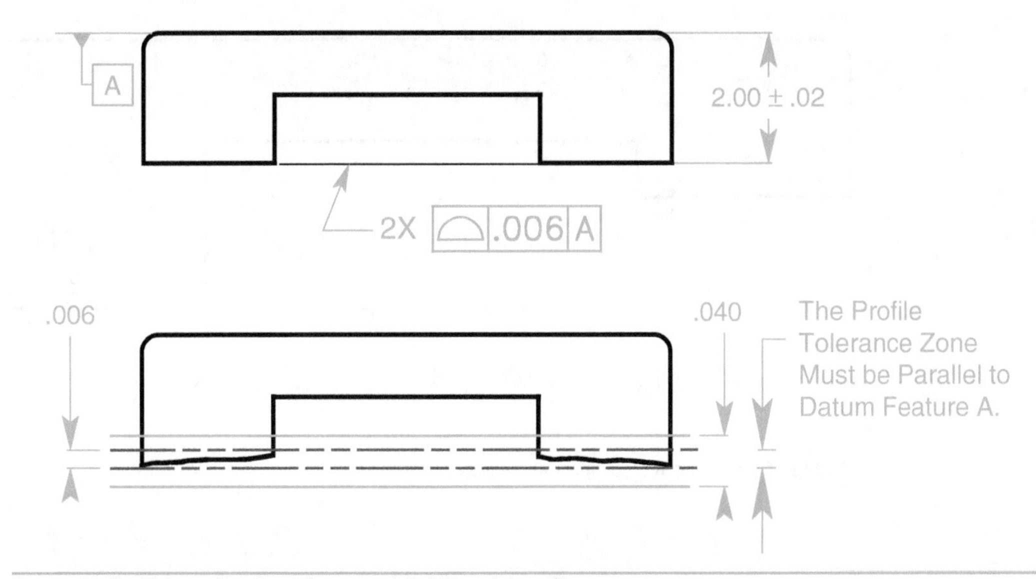

FIGURE 11-9 Two coplanar surfaces parallel to a datum feature.

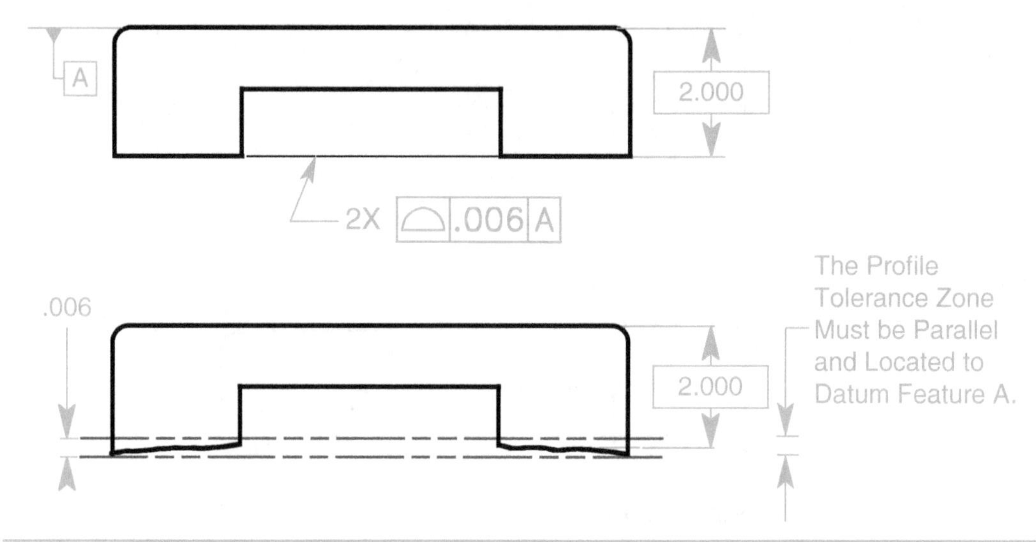

FIGURE 11-10 Two coplanar surfaces parallel and located with a basic dimension to datum feature A.

must remain parallel to datum feature A. Coplanarity of two or more surfaces specified with a profile tolerance, including a datum feature, which identifies a parallel surface, is similar to parallelism of a single surface specified with a parallelism tolerance.

If the 2.000-inch dimension is basic, as shown in Fig. 11-10, the true profile of the two coplanar surfaces is a basic 2.000 inches from datum feature A. The tolerance zone, consisting of two parallel planes .006 apart, is evenly disposed about the true profile. Where the basic dimension is specified, the total tolerance for the feature size, parallelism, and coplanarity is .006. Coplanarity of two or more surfaces specified with a profile tolerance, a basic dimension, and a datum is treated like any other profile control.

A profile tolerance may be used to control two or more stepped surfaces that are required to be flat and parallel within a specific tolerance. The profile tolerance for the part shown in Fig. 11-11 controls the two stepped surfaces. They must be flat, parallel, and stepped 1.000 basic inch apart within a tolerance of .006.

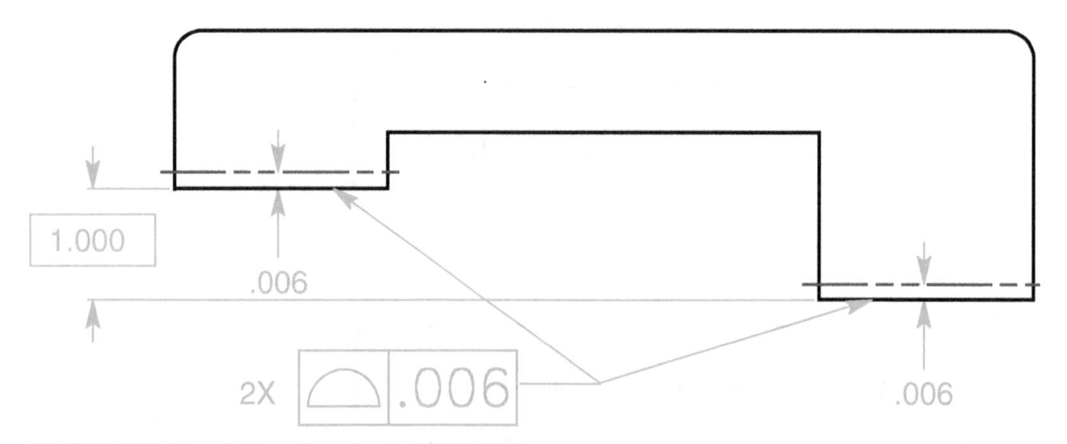

FIGURE 11-11 Profile of a surface for stepped surfaces.

Coplanar surfaces may be used as a datum feature. If this is the case, it is best to attach the datum feature symbol to the coplanar feature control frame, as shown in Fig. 11-12. If there are multiple coplanar surfaces and only a certain few of them are required to be the datum feature, a datum feature symbol is attached to each of those certain few surfaces establishing them as a single common datum feature. The datum feature reference letters, separated by a dash, are entered in one compartment of the feature control frame as a common datum.

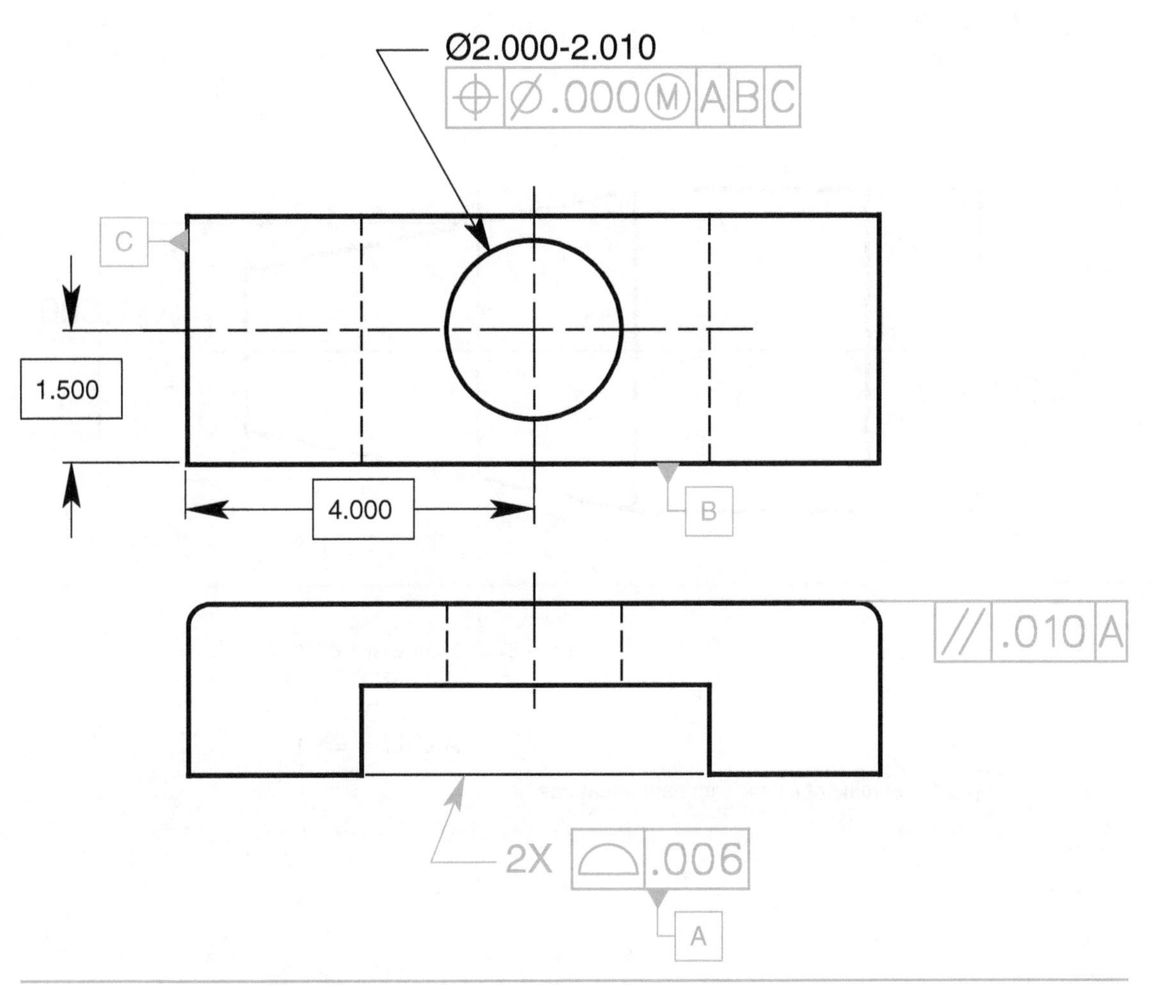

FIGURE 11-12 Two coplanar surfaces specified as a datum feature.

Profile of a Conical Feature

Conicity may be controlled with the profile of a surface tolerance. A conicity tolerance consists of a tolerance zone bounded by two coaxial cones separated by the profile tolerance at a specified basic angle. The conical feature must fall inside the profile tolerance zone, and it must also satisfy the size tolerance requirements. The size tolerance is specified by locating a diameter with a basic dimension and tolerancing that diameter with a plus or minus tolerance. If just the form of a cone is to be toleranced, no datum features are required. Figure 11-13 shows datum features controlling both form and orientation of the cone.

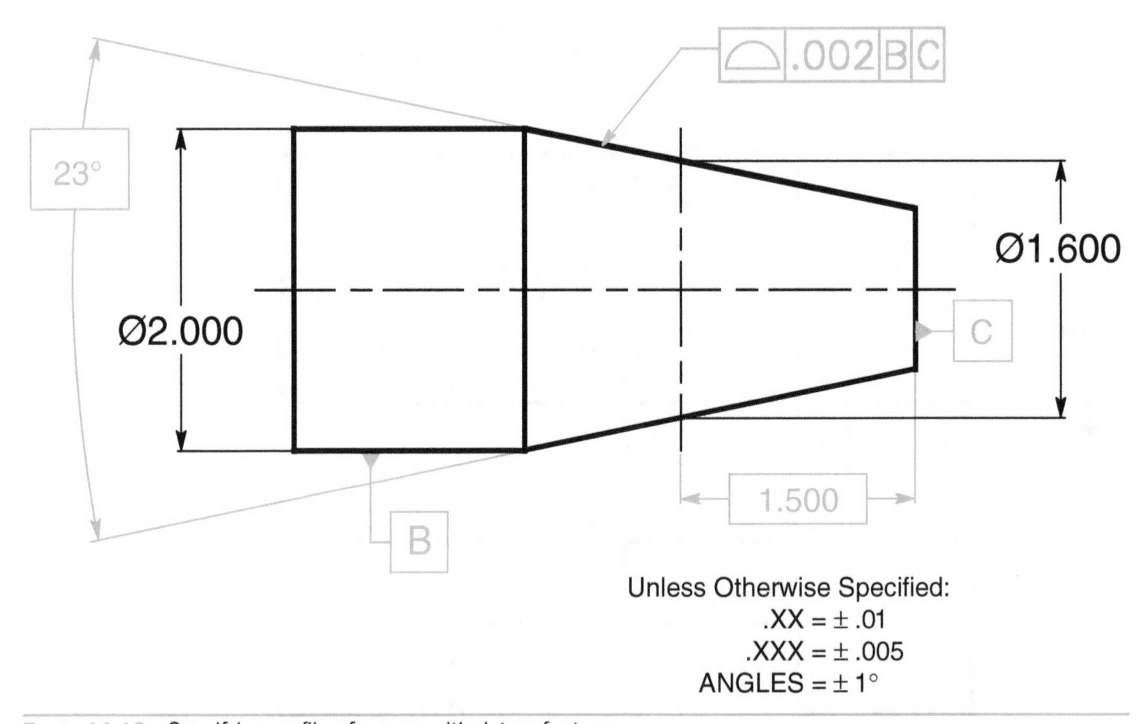

FIGURE 11-13 Specifying profile of a cone with datum features.

Composite Profile Tolerancing

Composite profile tolerancing is very similar to composite positional tolerancing discussed in Chap. 8. A composite profile feature control frame has one profile symbol that applies to the two horizontal segments that follow. The upper segment, called the profile locating control, governs the location relationship between the datum features and the profile. It acts like any other profile control. The lower segment, referred to as the profile feature control, is a smaller tolerance than the profile locating control and governs the size, form, and orientation relationships of the profile. The smaller tolerance zone need not fall entirely inside the larger tolerance zone, but any portion of the smaller tolerance zone that lies outside the larger tolerance zone is unusable. The feature profile must fall inside both profile tolerance zones.

Datum features in the lower segment of a composite feature control frame must satisfy two conditions:

- Datum features in the lower segment must repeat the datum features in the upper segment of the feature control frame. If only one datum feature is repeated, it would be the primary datum feature; if two datum features were repeated, they would be the primary and secondary datum features; and so on.
- Datum features in the lower segment only control orientation.

The profile in Fig. 11-14 must fall within the .010 tolerance zone governing form and orientation to datum feature A. The entire profile, however, may float around within the larger tolerance zone of .040 located to datum features B and C.

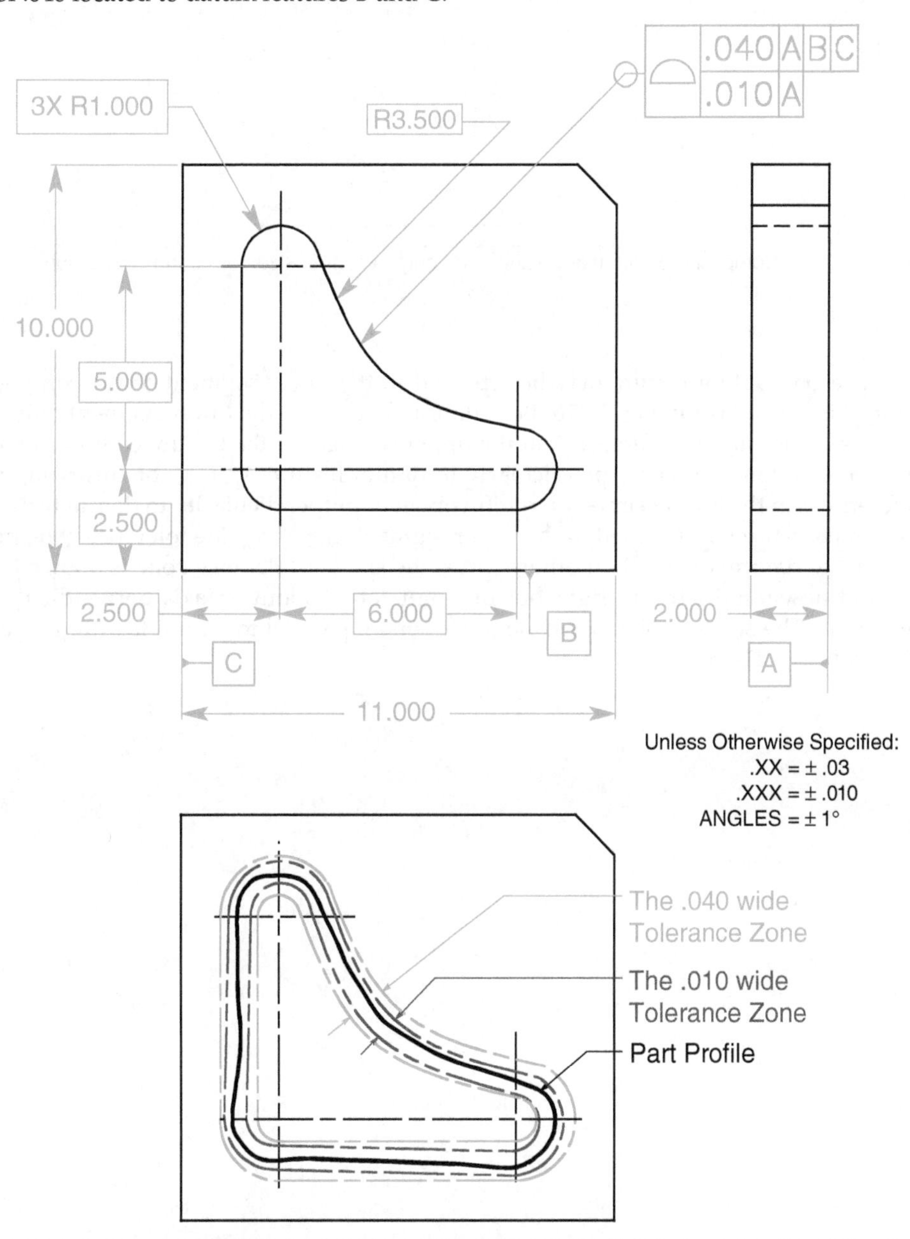

FIGURE 11-14 A feature controlled with composite profile tolerancing.

A composite profile also may be used to control orientation to a larger tolerance with a refinement of size and form to a smaller tolerance in the lower segment of the feature control frame shown in Fig. 11-15. The upper segment governs the orientation relationship between the profile and datum feature A. The lower segment is a smaller tolerance than the profile orienting control and governs the size and form relationship of the profile. The smaller tolerance zone need not fall entirely inside the larger tolerance zone, but any portion of the smaller tolerance zone that lies outside the larger tolerance zone is unusable. The feature profile must fall inside both profile tolerance zones.

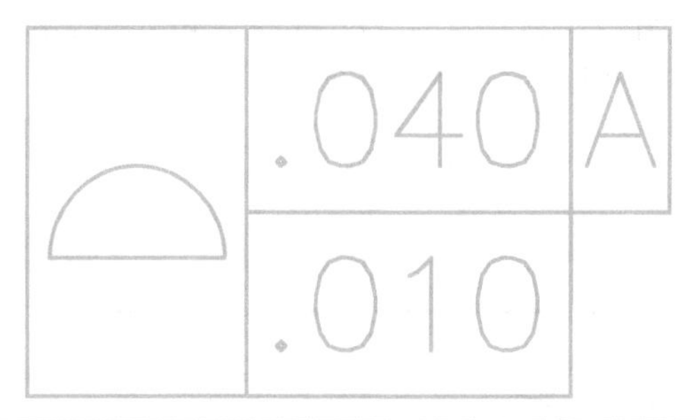

FIGURE 11-15 Composite profile tolerancing used only to control size, form, and orientation.

A second datum feature may be repeated in the lower segment of the composite feature control frame shown in Fig. 11-16. Both datum features in the lower segment only control orientation. Since datum reference A in the upper segment of the feature control frame only controls orientation, that is, perpendicularly to datum feature A, it is not surprising that datum reference A in the lower segment is a refinement of perpendicularity to datum feature A. When datum reference B is included in the lower segment, the .005-wide tolerance zone must remain parallel to datum feature B. In other words, the smaller tolerance zone is allowed to translate up and down and left and right but may not rotate about an axis perpendicular to datum feature A. The smaller tolerance zone must remain parallel to datum feature B at all times, as shown in Fig. 11-16.

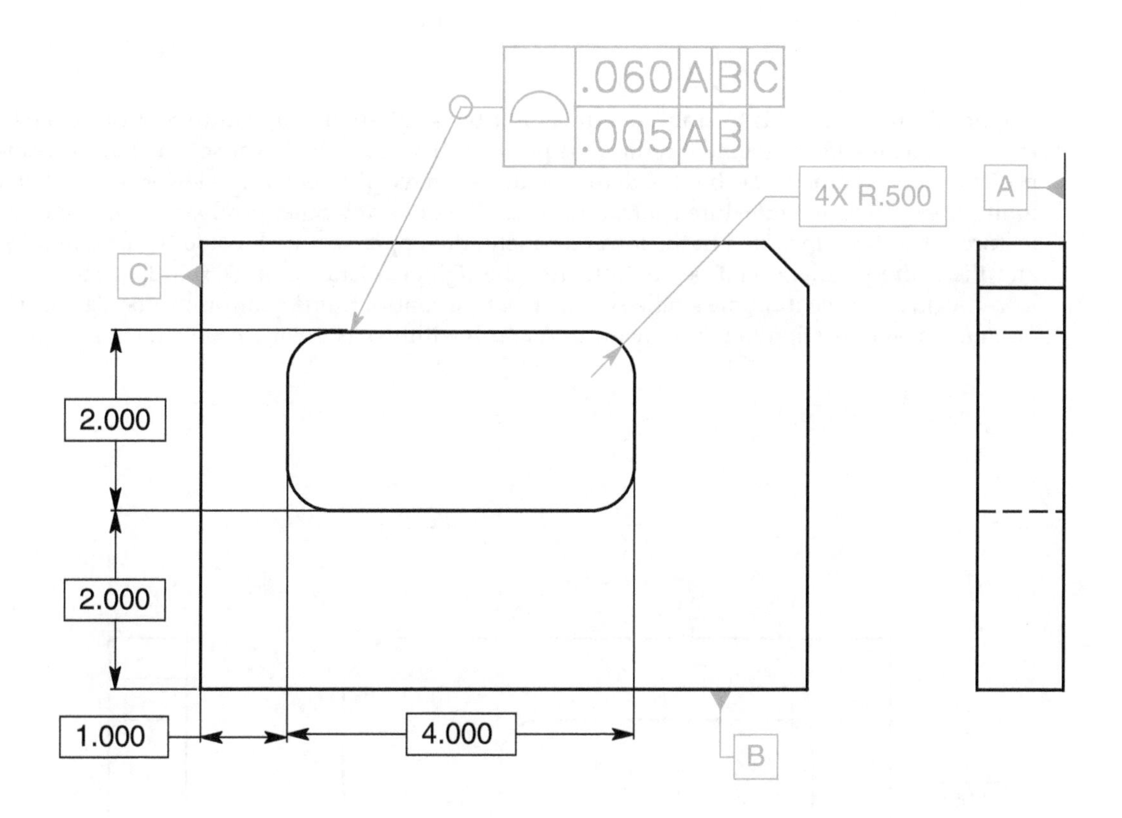

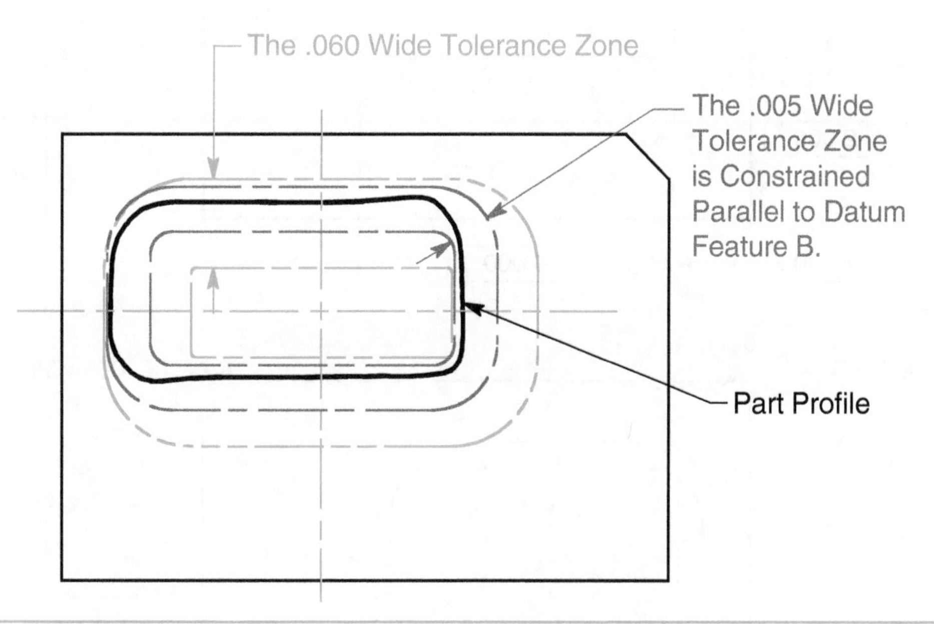

FIGURE 11-16 Composite profile with two datum features repeated in the lower segment.

Multiple Single-Segment Profile Tolerancing

The profile in Fig. 11-17 is toleranced with a multiple single-segment feature control frame. In this example, the lower segment refines the profile just as does the lower segment of the composite feature control frame, but the datum features behave differently. The lower segment of a multiple single-segment feature control frame acts just like any other profile control. If datum feature C had been included in the lower segment, the upper segment would be meaningless, and the entire profile would be controlled to the tighter tolerance of .005. In Fig. 11-17, the lower segment of the multiple single-segment feature control frame controls profile size, form, orientation, and location to datum features A and B within a .005-wide profile tolerance zone.

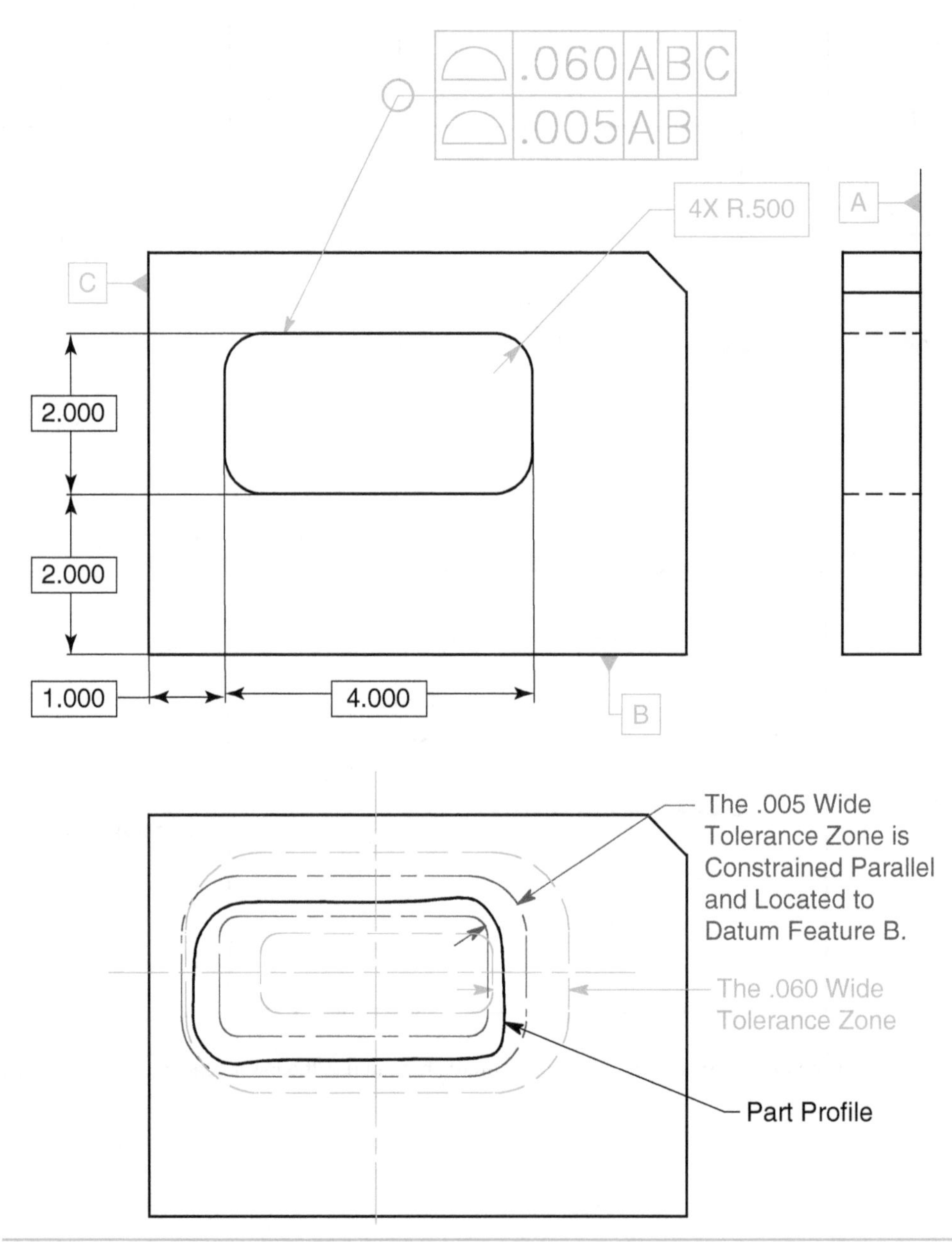

FIGURE 11-17 A multiple single-segment profile tolerance.

In other words, the actual profile must fit inside the profile refinement tolerance, be perpendicular to datum feature A, and be located a basic 2.000 inches from datum feature B within a tolerance of .005. The upper segment, the profile locating control, allows the .005-wide profile refinement tolerance zone to translate back and forth within a profile tolerance of .060. That is, the refinement tolerance zone may translate left and right but may not translate up and down or rotate in any direction.

Inspection

Inspecting a surface that has been controlled with a profile tolerance can be accomplished in a number of ways. The most common methods of inspecting profile are listed below:

- A thickness gage is used to measure variations between a template of the profile and the actual surface.
- An open setup with a dial indicator can be used to inspect some profiles.
- The optical comparator was designed to inspect profiled surfaces. An optical comparator projects a magnified outline of the part onto a screen. The projected outline is then compared to a profile template.
- Coordinate measuring machines designed to inspect profile can be used.
- A gage made to the extreme size and shape of the profile can be used.

Summary

- A profile is the outline of an object.
- The true profile may be dimensioned with basic size dimensions, basic coordinate dimensions, basic radii, basic angular dimensions, formulas, mathematical data, or undimensioned drawings.
- Profile is a surface control.
- Where the leader from a profile feature control frame points directly to the profile, the tolerance specified in the feature control frame is equally disposed about the true profile.
- The all around symbol consists of a circle around the joint in the leader.
- If the profile is to extend between two points, the points are labeled, and a note using the between symbol is placed beneath the feature control frame.
- The all over symbol consists of two concentric circles around the joint in the leader.
- Profile tolerances may or may not have datum features.
- The profile tolerance may be combined with other geometric tolerances to refine certain aspects of a surface.
- Where coplanarity is required, a profile of a surface tolerance is specified.
- Composite profile tolerancing is very similar to composite positional tolerancing.
- Datum features in the lower segment of a composite feature control frame only control orientation.
- A profile may be toleranced with a multiple single-segment feature control frame.

Chapter Review

1. Profile of a line is the ____________________ of an object in a plane as the plane passes through the object.

2. Profile of a surface is the result of ____________________ ____________________ or taking cross sections through the object at various intervals.

3. The true profile may be dimensioned with what kind of dimensions? ____________________ ____________________ ____________________

4. The feature control frame is directed to the profile surface with a ____________________ ____________________.

5. What symbols do not apply in the tolerance section of profile feature control frames? ____________________

6. The unequally disposed profile symbol, the circle U, indicates that the profile of a surface tolerance applies ____________________.

7. The tolerance following the circle U indicates the amount of the tolerance zone ____________________ to the true profile.

8. Where a profile tolerance applies all around the profile of a part, the ____________________ is specified.

9. Draw the all around symbol. ____________________

10. If the profile is to extend between two points, the points are ____________________, and a note using the ____________________ is placed beneath the feature control frame.

11. Draw the between symbol. ____________________

12. If a part is to be controlled with a profile tolerance over the entire surface of the part, the ____________________ symbol is specified.

13. Profile tolerances ______________________ have datum features.

14. The profile of a surface control usually requires a datum feature(s) to properly ______

____.

15. Datum features are generally ______________ for profile of a line tolerance

where only ______________________ is being controlled.

16. If the design requires a smaller radius than the radius allowed by the profile tolerance, a

note, such as ______________________________________,

is directed to the radius with a ______________________________.

17. The profile tolerance may be combined with other ______________________

to ______________________ certain aspects of a surface.

18. Coplanarity is the condition of ______________________________

surfaces having all ______________________________________.

19. Coplanarity is toleranced with the profile of a surface feature control frame, connected with

a ______________, to an ______________ connecting the coplanar surfaces.

20. The number of coplanar surfaces followed by an ______________ precedes the

____.

21. A profile tolerance may be used to control two or more stepped ______________
that are required to be flat and parallel within a specific tolerance.

22. Conicity may be controlled with a ______________________________.

23. Composite profile tolerancing is very similar to ______________________

____.

24. The upper segment of a composite profile feature control frame called, ______________

______________________________________, governs the

____.

25. The lower segment, referred to as the ________________________________,

    is a smaller tolerance than the profile locating control and governs ____________

    ________________________________ of the profile.

26. The feature profile must fall inside ________________________________.

27. Datum features in the lower segment of a composite feature control frame must satisfy two conditions:

    ____

    ________________________________.

28. A second datum feature may be repeated in the lower segment of the composite feature control frame. Both datum features only control ________________________________.

29. The lower segment of a multiple single-segment profile feature control frame acts just like

    ________________________________.

30. The upper segment of a multiple single-segment profile feature control frame allows the smaller tolerance zone to ________________________________
    relative to the datum feature not repeated in the lower segment within the larger tolerance.

Problems

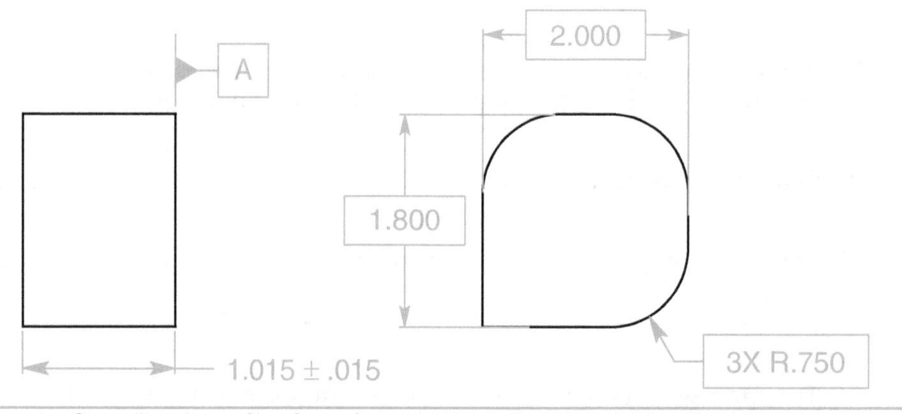

Figure 11-18 Controlling the profile of a surface all around: Prob. 1.

1. Specify a profile of a surface tolerance of .020, perpendicular to datum feature A, and all around the part in Fig. 11-18.

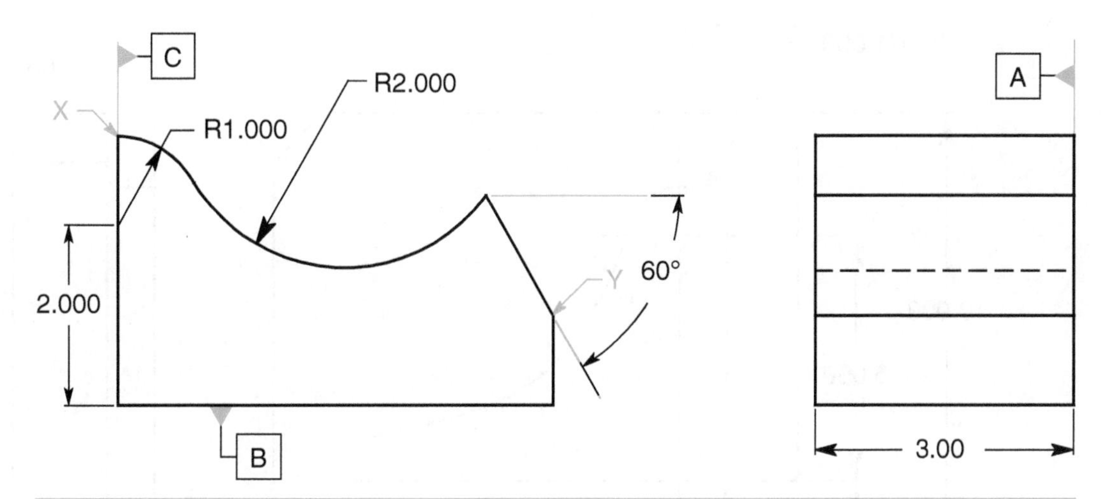

FIGURE 11-19 Controlling the profile of a surface between two points: Prob. 2.

2. Control the top surface between points X and Y in Fig. 11-19 by specifying a profile tolerance of .030, located to datum features A, B, and C.

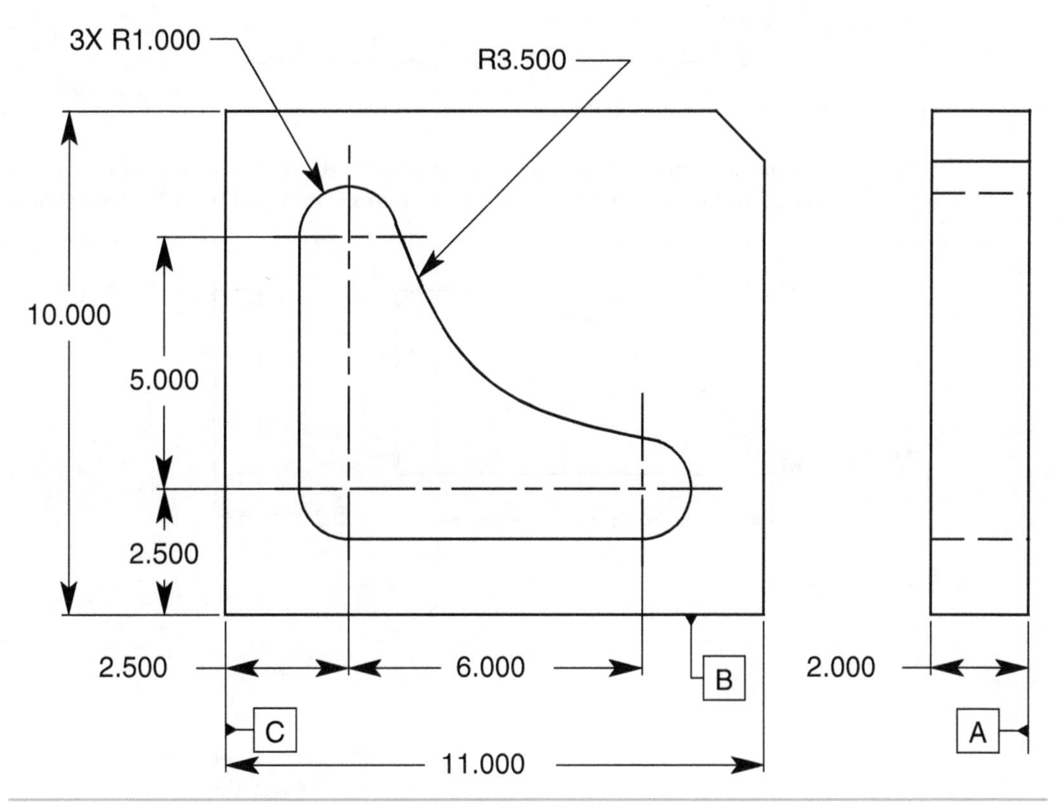

FIGURE 11-20 Control the location of a profile: Prob. 3.

3. Control the entire surface of the die cavity in Fig. 11-20 to the datum features indicated within a tolerance of .015 in the direction that would remove material from the true profile.

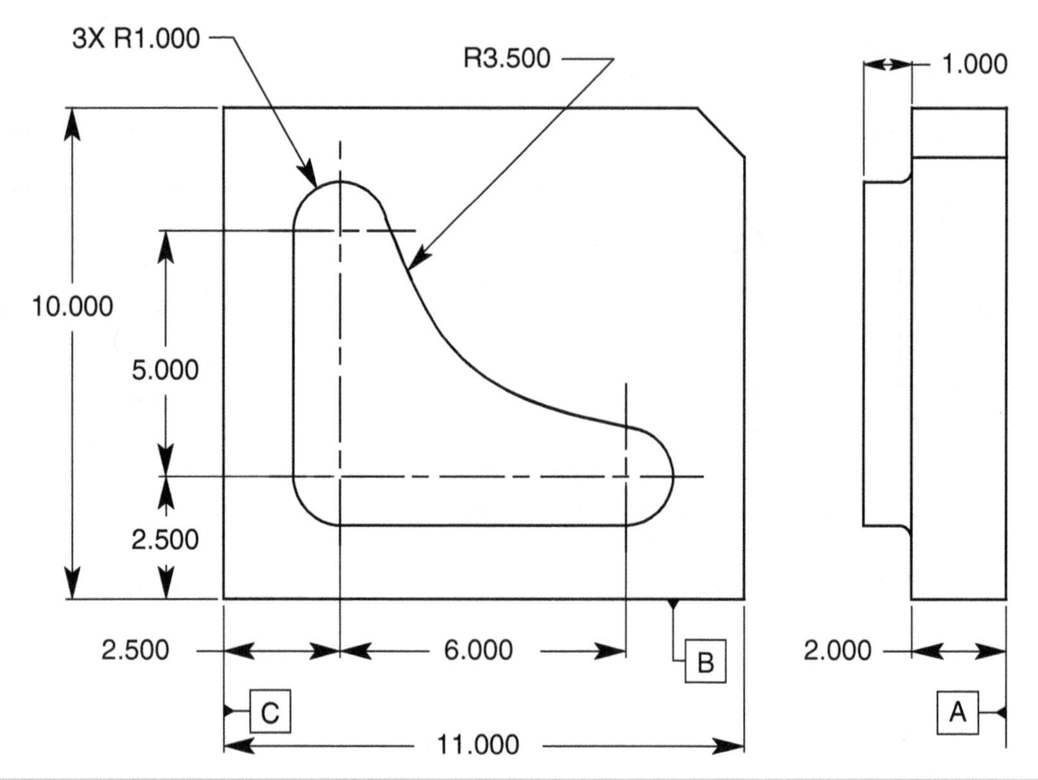

FIGURE 11-21 Locating a mating profile: Prob. 4.

4. Control the entire surface of the punch in Fig. 11-21 to the datum features indicated within a tolerance of .015 in the direction that would remove material from the true profile.

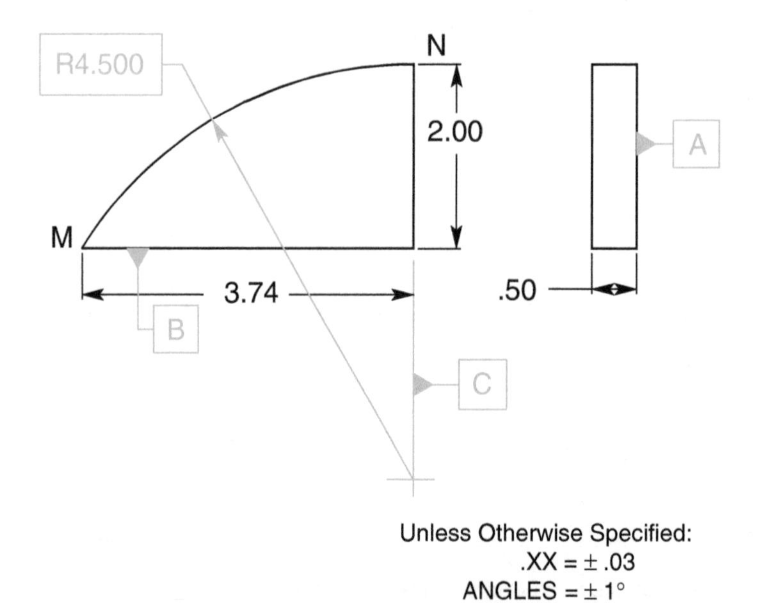

FIGURE 11-22 Control a profile with combined tolerances: Prob. 5.

5. Use the profile of a surface control to tolerance the curve in Fig. 11-22 within .030 between points M and N. Each line element in the profile must be parallel to datum features B and C within .010. The point at M may not exceed a radius of .010.

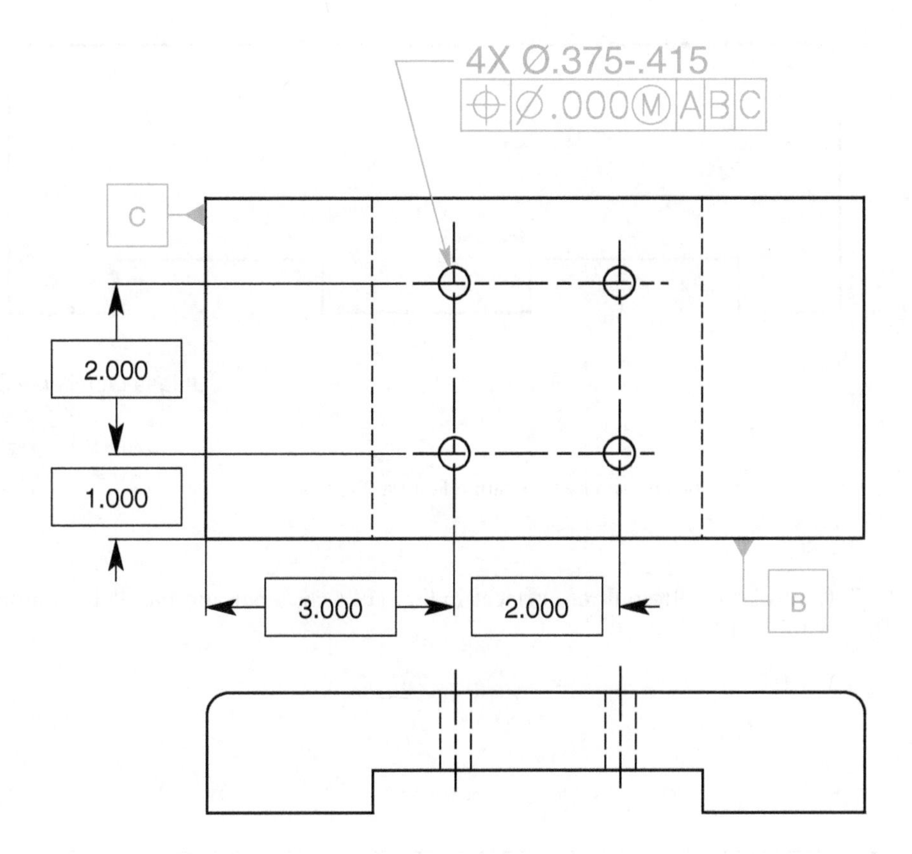

FIGURE 11-23 Controlling coplanarity with the profile of a surface: Prob. 6.

6. The primary datum feature is the two lower coplanar surfaces in Fig. 11-23. Specify the primary datum feature to be coplanar within .004.

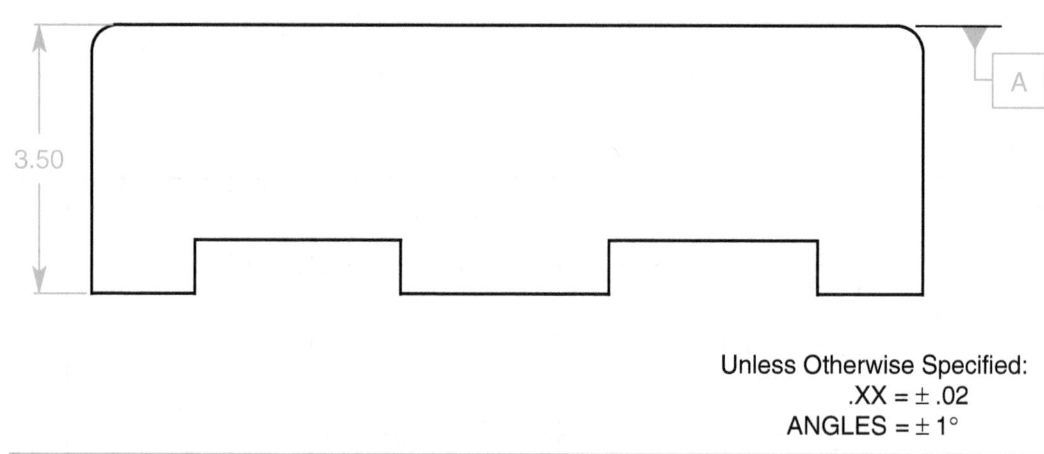

FIGURE 11-24 Controlling coplanarity to a datum feature: Prob. 7.

7. Control the bottom three surfaces in Fig. 11-24 coplanar and parallel to datum feature A within .010.

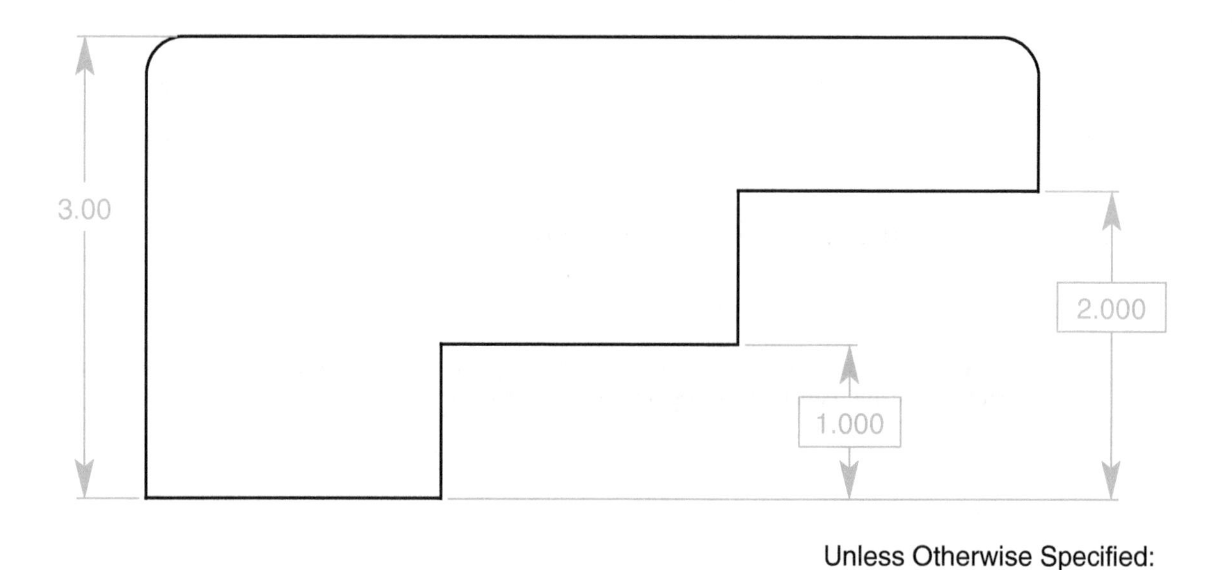

FIGURE 11-25 Controlling a stepped part with a profile of a surface: Prob. 8.

8. Control the three bottom surfaces in Fig. 11-25 flat, parallel, and stepped 1.000 basic inch apart within a tolerance of .005.

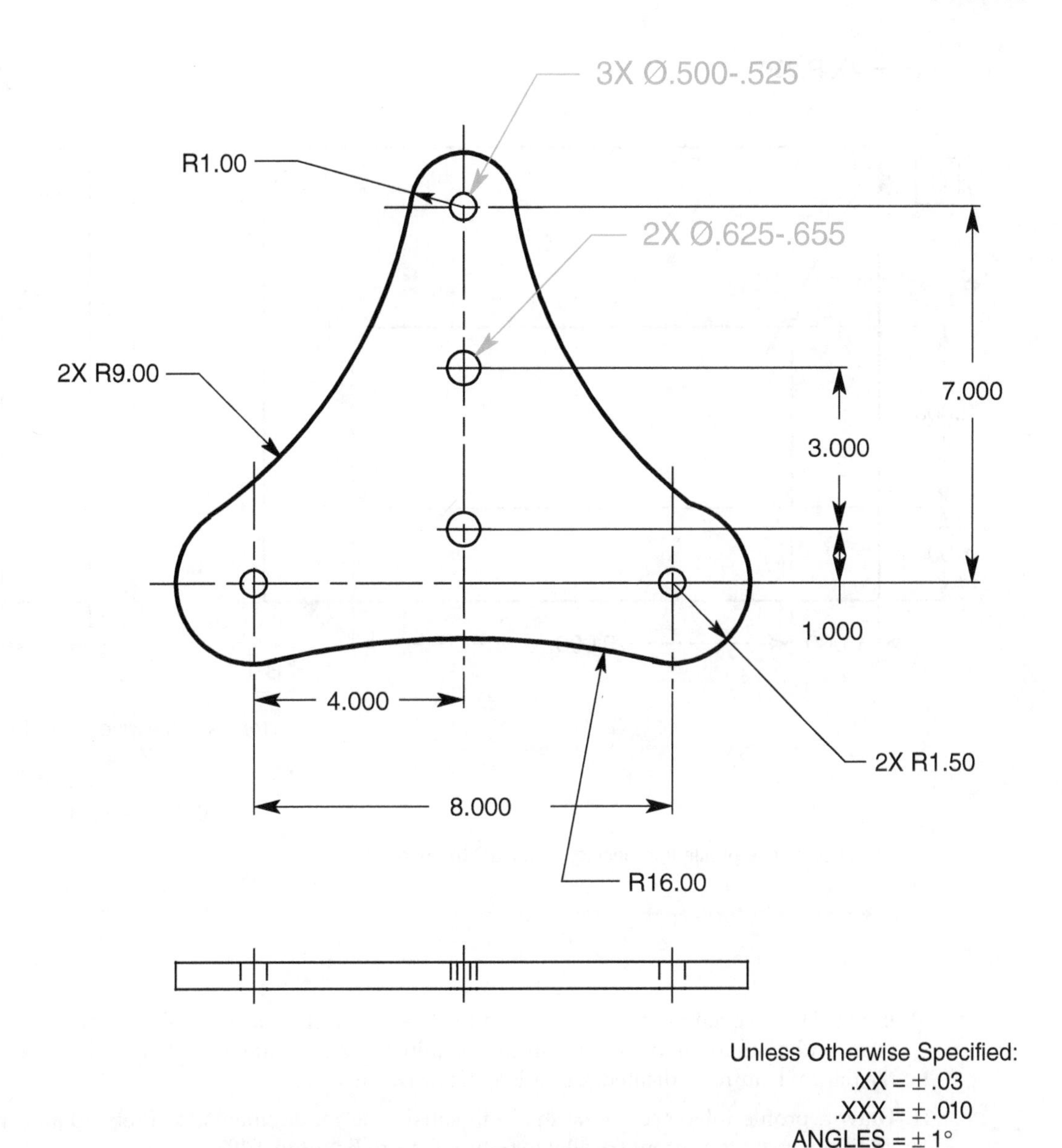

FIGURE 11-26 Controlling a profile to datum features of size: Prob. 9.

9. Tolerance the drawing in Fig. 11-26. Specify controls locating the hole patterns to each other and perpendicular to the back of the part. The holes are for ½-inch and 5/8-inch bolts, respectively. Specify a control locating the outside profile of the part to the hole patterns and perpendicular to the back of the part within a tolerance of .060. Specify MMC and MMB wherever possible.

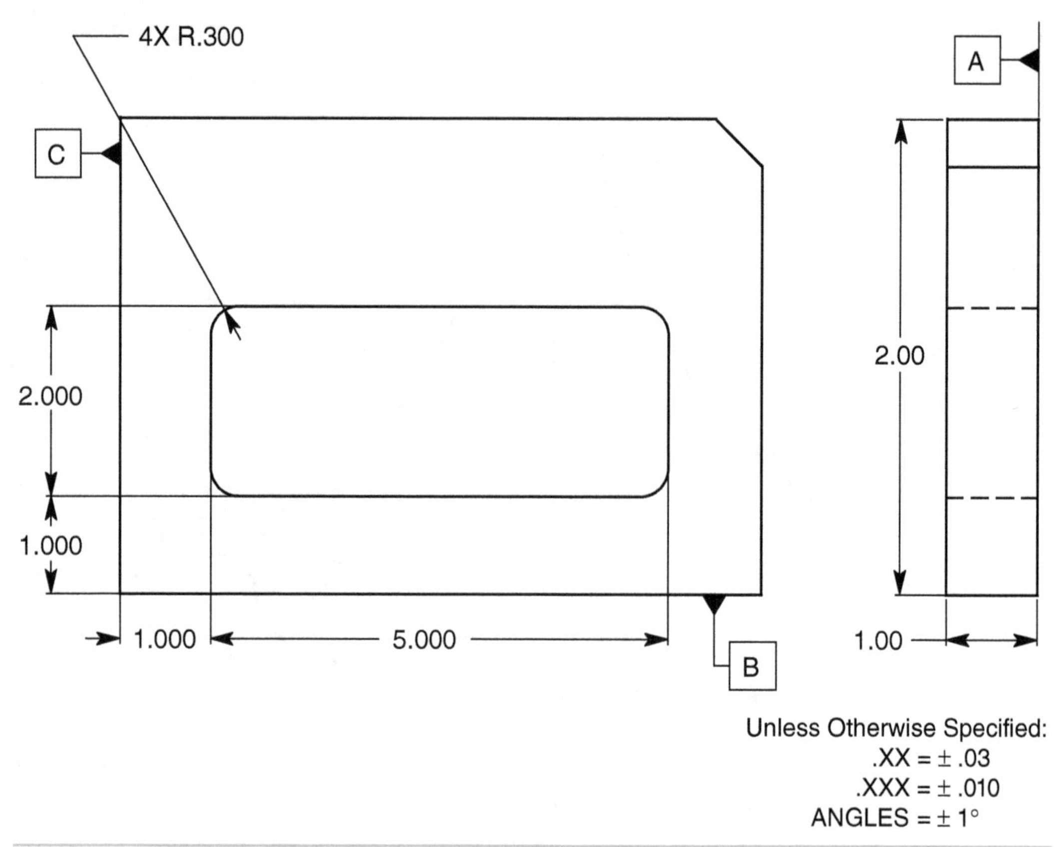

FIGURE 11-27 Composite profile tolerancing: Probs. 10 through 12.

10. In Fig. 11-27, specify a profile tolerance for the center cutout that will control the size, form, and orientation to datum feature A within .010 and locate the cutout within .060 to the datum features indicated. Complete the drawing.
11. Draw a profile tolerance below that will satisfy the requirements for Prob. 10 and, in addition, *orient* the cutout parallel to datum feature B within .010.

12. Draw a profile tolerance below that will satisfy the requirements for Prob. 10 and, in addition, *locate* the cutout to datum feature B within .010.

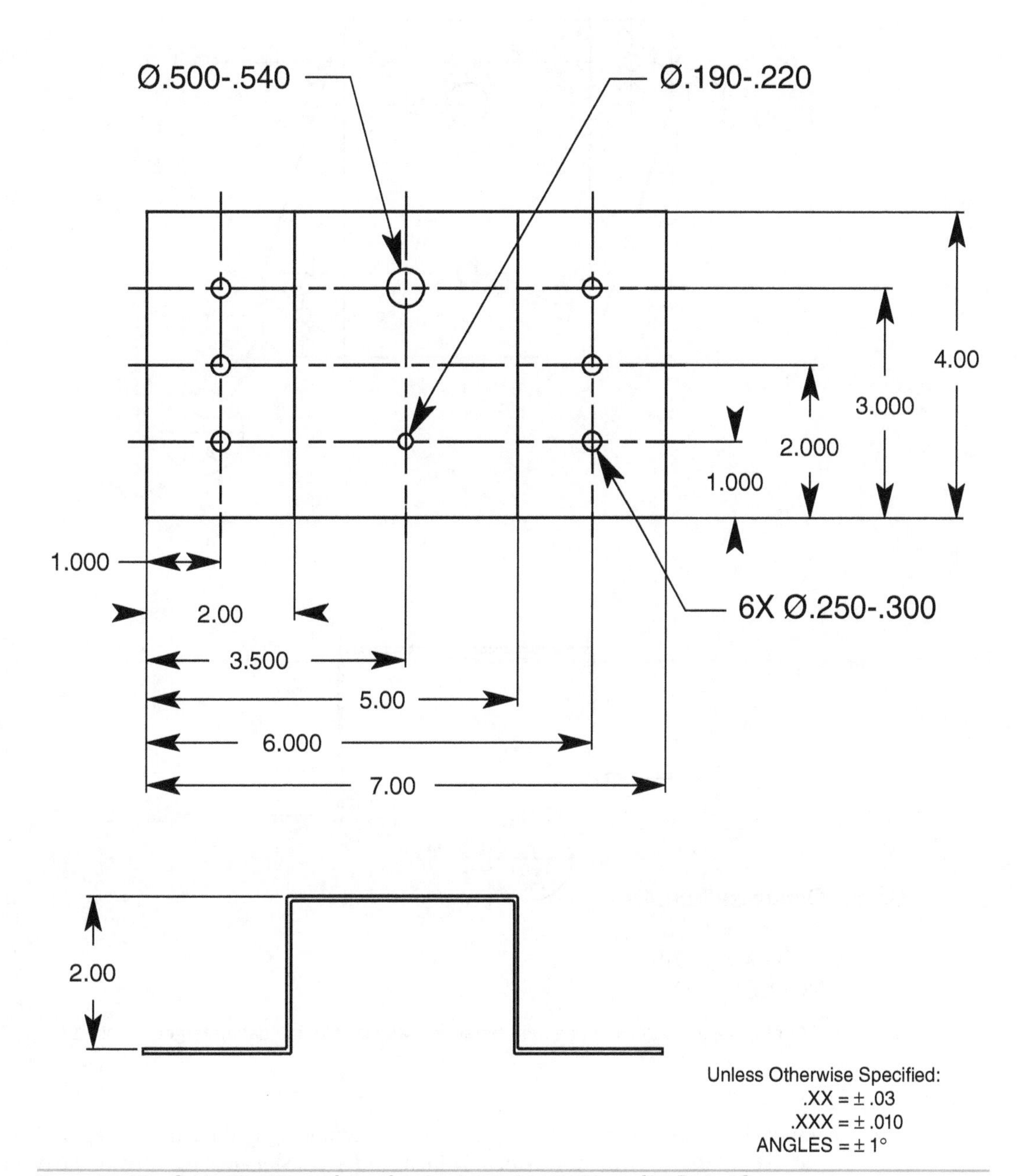

FIGURE 11-28 Controlling the surface of a sheet metal part with profile: Prob. 13.

13. Specify the lower two surfaces of the bottom of the sheet metal part in Fig. 11-28 coplanar within .020. Tolerance the holes with geometric tolerancing. The MMC for each hole is the virtual condition for the mating fastener. Specify the profile of the top surface of the sheet metal part within .040. Use MMC and MMB wherever possible.

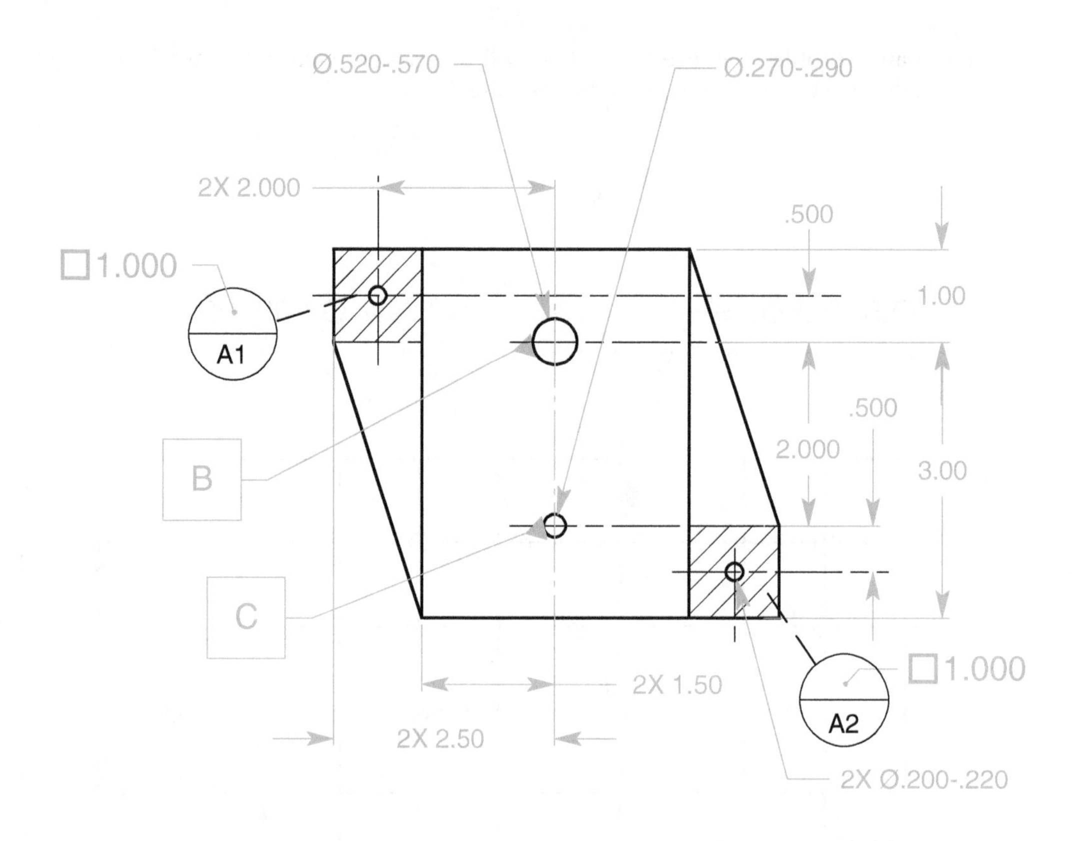

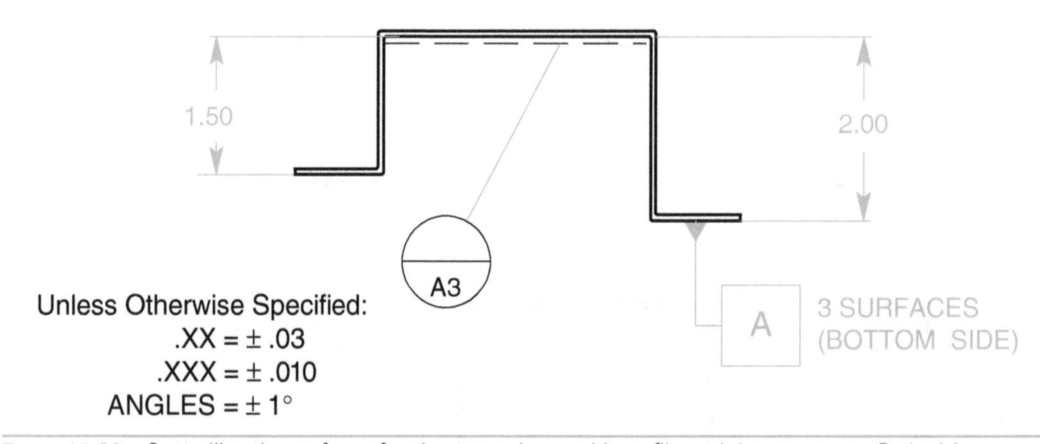

Unless Otherwise Specified:
.XX = ± .03
.XXX = ± .010
ANGLES = ± 1°

FIGURE 11-29 Controlling the surface of a sheet metal part with profile and datum targets: Prob. 14.

14. Tolerance the holes in Fig. 11-29 with geometric tolerancing to datum features A, B, and C. The sizes of the holes are for ½-inch, ¼-inch, and #10 fastener sizes. Specify the profile of the top surface of the part within a tolerance of .030 to datum features A, B, and C. Use MMC and MMB wherever possible.

CHAPTER 12

A Strategy for Tolerancing Parts

When tolerancing a part, the designer must determine the attributes of each feature or pattern of features and the relationship of these features to their datum features and to each other. In other words, the designer must ask the following questions of each feature:

1. What features are the most appropriate datum features?
2. What are the size and the size tolerance?
3. What is the form tolerance?
4. What are the location and orientation dimensions, and tolerances?
5. What material conditions apply to features of sizes?

All of these questions must be answered in order to properly tolerance a part. Some designers believe that parts designed with a solid modeling CAD program do not require tolerancing; nothing could be further from the truth. The dimensioning and tolerancing standard clearly indicates that, except reference, maximum, minimum, or stock, each dimension shall have a tolerance.

Chapter Objectives

After completing this chapter, the learner will be able to:

- *Tolerance* a feature(s) of sizes located to plane surface datum features
- *Tolerance* a feature(s) of sizes located to a datum feature(s) of sizes
- *Tolerance* a pattern of features located to a second pattern of features as a datum

Locating Features of Size to Plane Surface Datum Features

Once the drawing has been lined out and dimensioned, as shown in Fig. 12-1, it is ready to be toleranced. The first step in tolerancing a drawing is to specify the datum features. The 2-inch-diameter hole in Fig. 12-1 is dimensioned up from the bottom edge and over from the left edge of the part.

Consequently, the bottom and left edges are implied datum features. Where geometric dimensioning and tolerancing is applied, datum features must be specified. If the designer has decided that the bottom edge is more important to the part design than the left edge, the datum feature for the bottom edge, datum feature B, will precede the datum feature for the left edge, datum feature C, in the feature control frame shown in Fig. 12-3.

Datum features B and C control not only location but also orientation. If the hole in Fig. 12-2 is controlled with the feature control frame in Fig. 12-3, the hole is to be parallel to datum surface B and parallel to datum surface C within the tolerance specified in the feature control frame. The only orientation relationship between the hole and datum features B and C is parallelism. Parallelism can be controlled with the primary datum feature, datum feature B, but in only one direction. The secondary datum feature must contact the datum reference frame with a minimum of two points of contact; only two points of contact are required to control parallelism in the other direction. If the feature control frame in Fig. 12-3 is specified to control the hole in Fig. 12-2, the cylindrical tolerance zone is located from and parallel to datum features B and C, establishing both location and orientation for the hole.

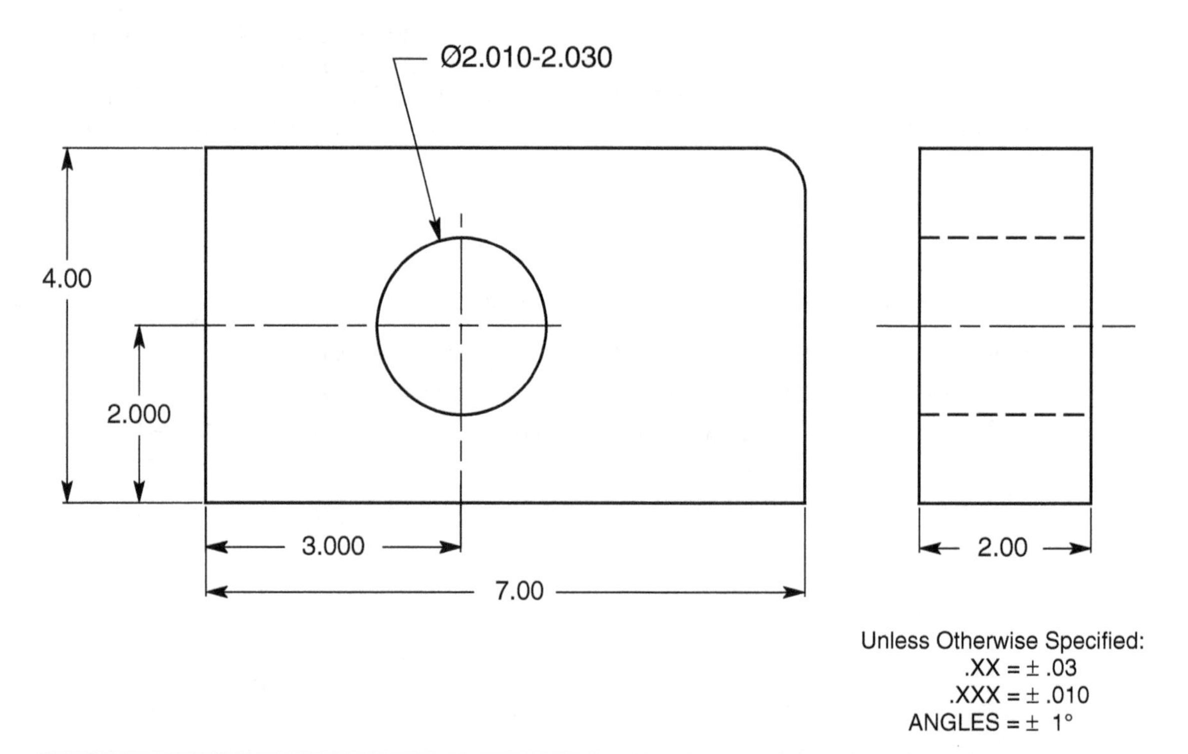

FIGURE 12-1 A feature of size located to plane surface datum features on an untoleranced drawing.

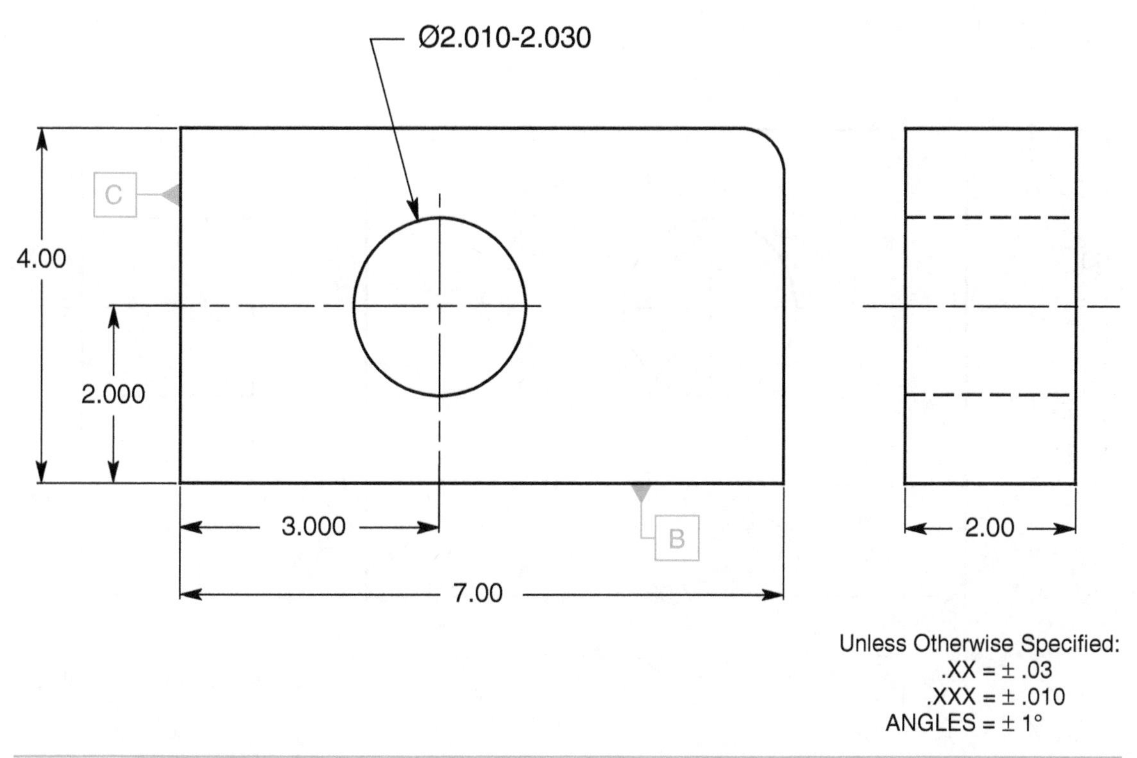

FIGURE 12-2 Feature of size located to specified datum features.

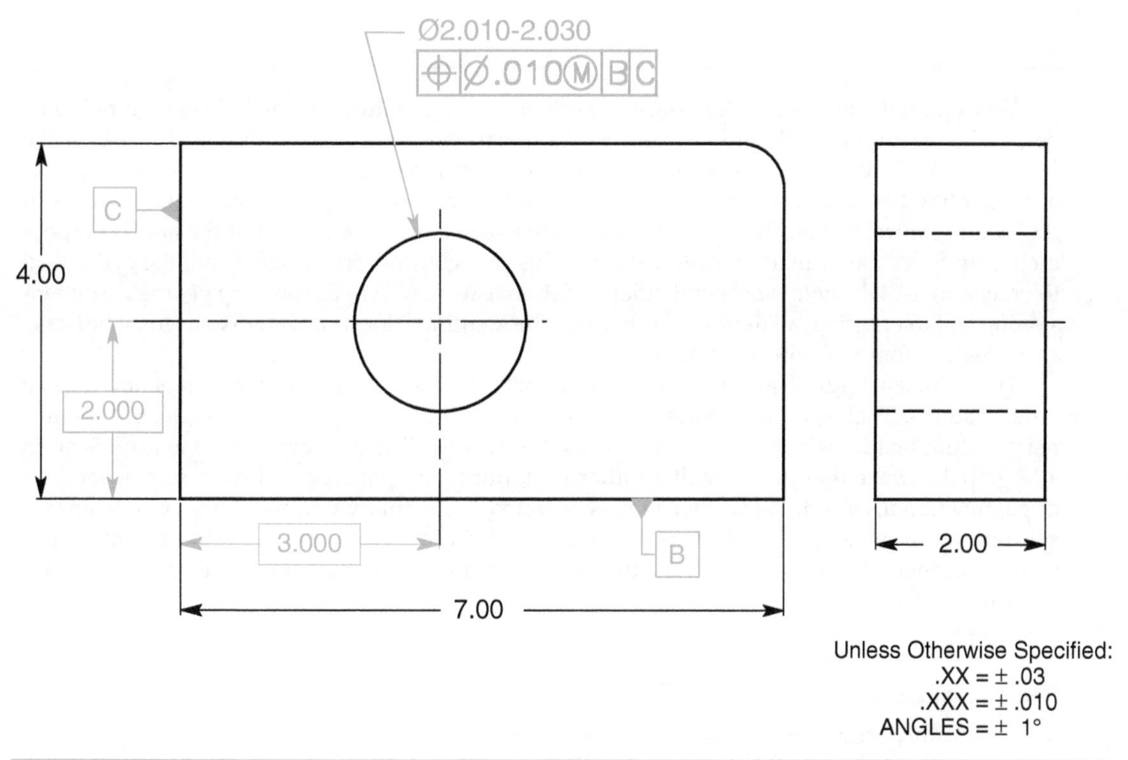

FIGURE 12-3 A position tolerance locating and orienting the hole to datum features B and C.

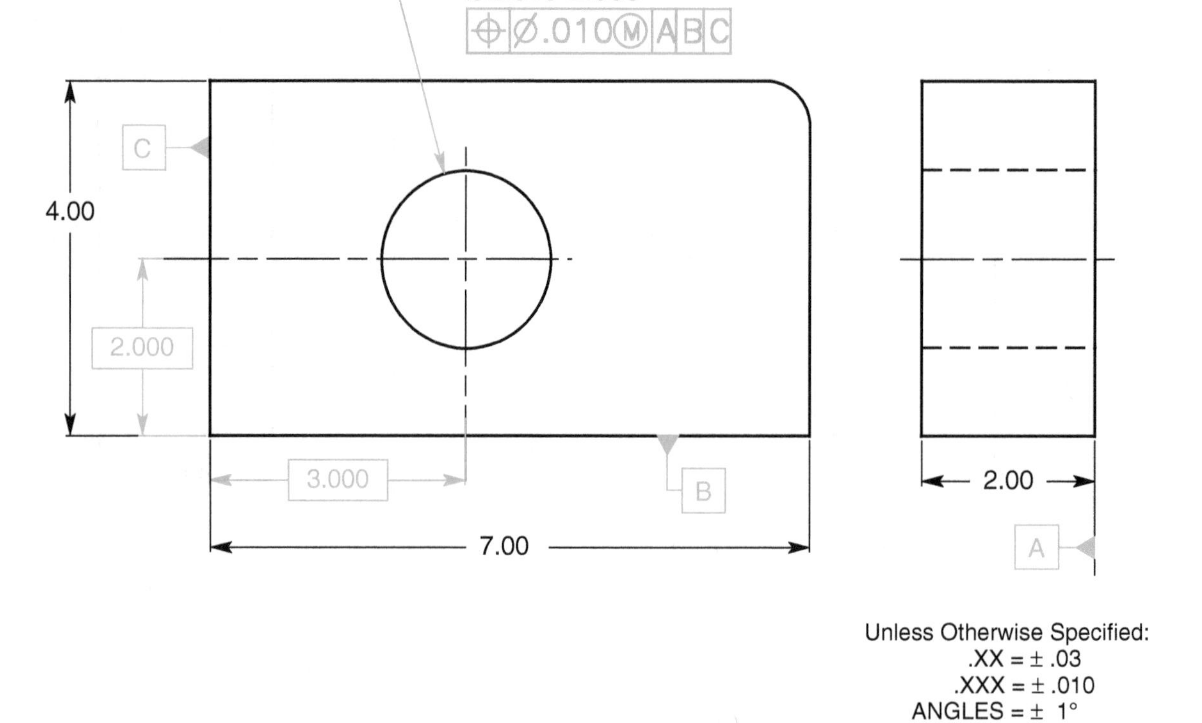

FIGURE 12-4 A position tolerance locating and orienting the 2-inch-diameter hole to a datum reference frame established by datum features A, B, and C.

Very often, the front or back surface or both are mating features. If the hole is required to be perpendicular to one of these features, an additional datum feature symbol is attached to the front or back of the part, the more important of the two surfaces. In Fig. 12-4, the back surface has been identified as datum feature A. Datum feature A is specified as the primary datum feature in the feature control frame. The only relationship between datum feature A and the hole is perpendicularity. Since the primary datum feature in this case controls only orientation, the cylindrical tolerance zone of the hole is perpendicular to datum feature A. Where applying geometric dimensioning and tolerancing, all datum features must be identified, location dimensions must be basic, and a feature control frame must be specified.

The primary datum feature is the most important datum feature and is independent of all other features. That is, other features are controlled to the primary datum feature, but the primary datum feature is not controlled to any other feature. The primary datum feature is often a large flat surface that mates with another part, but many parts do not have flat surfaces. A large, functional, cylindrical surface may be selected as a primary datum feature. Other surface configurations are also selected as primary datum features even if they require datum targets to support them. In the final analysis, the key factors in selecting a primary datum feature are the following:

- Functional surfaces
- Mating surfaces
- Readily accessible surfaces
- Sufficiently large, accessible surfaces that will provide repeatable positioning

The only appropriate geometric tolerance for a primary datum feature is a form control. All other geometric tolerances control features to datum features. On complicated parts, it is possible to have a primary datum feature positioned to another datum reference frame. However, in most cases, it is best to have only one datum reference frame.

Rule #1 controls the flatness of datum feature A in Fig. 12-4 since no other control is specified. The size tolerance, a title block tolerance of ±.030, or a total of .060, controls the form of the size feature of which datum feature A is one side. If Rule #1 doesn't sufficiently control flatness, a flatness tolerance must be specified. If the side opposite datum feature A must be parallel within a smaller tolerance than the tolerance allowed by Rule #1, a parallelism control is specified, as shown in Fig. 12-5. If required, a parallelism control can also be specified for the sides opposite datum features B and C.

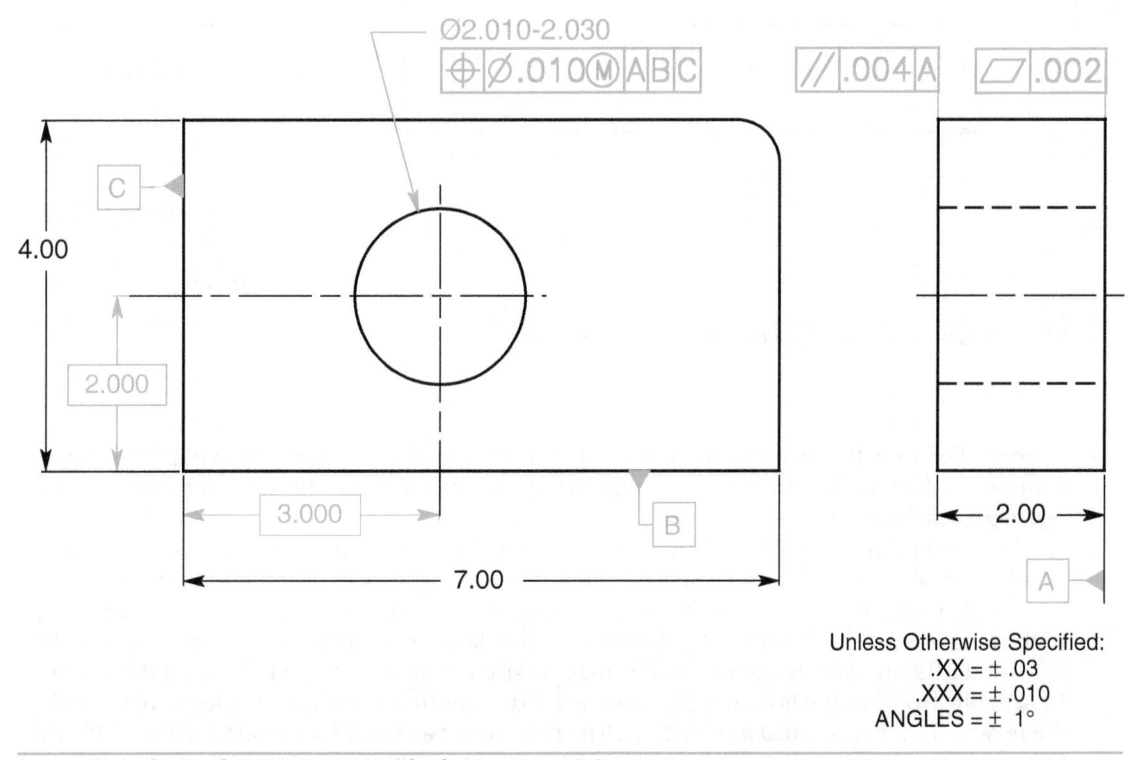

FIGURE 12-5 Datum features controlled for form and parallelism.

In Fig. 12-5, datum feature B is specified as the secondary datum feature, the more important of the two location datum features. It may be more important because it is larger than datum feature C or because it is a mating surface. When producing or inspecting the hole, datum feature B must contact the datum reference frame with a minimum of two points of contact. If not otherwise toleranced, datum features B and C are perpendicular to datum feature A and to each other within the ±1° angularity tolerance in the title block. However, as shown in Fig. 12-6, datum feature B is controlled to datum feature A with a perpendicularity tolerance of .004. Datum feature C is specified as the tertiary (third) datum feature; it is the least important datum feature. When producing or inspecting the hole, datum feature C is placed in a datum reference frame with a minimum of one point of contact. The orientation of datum feature C may

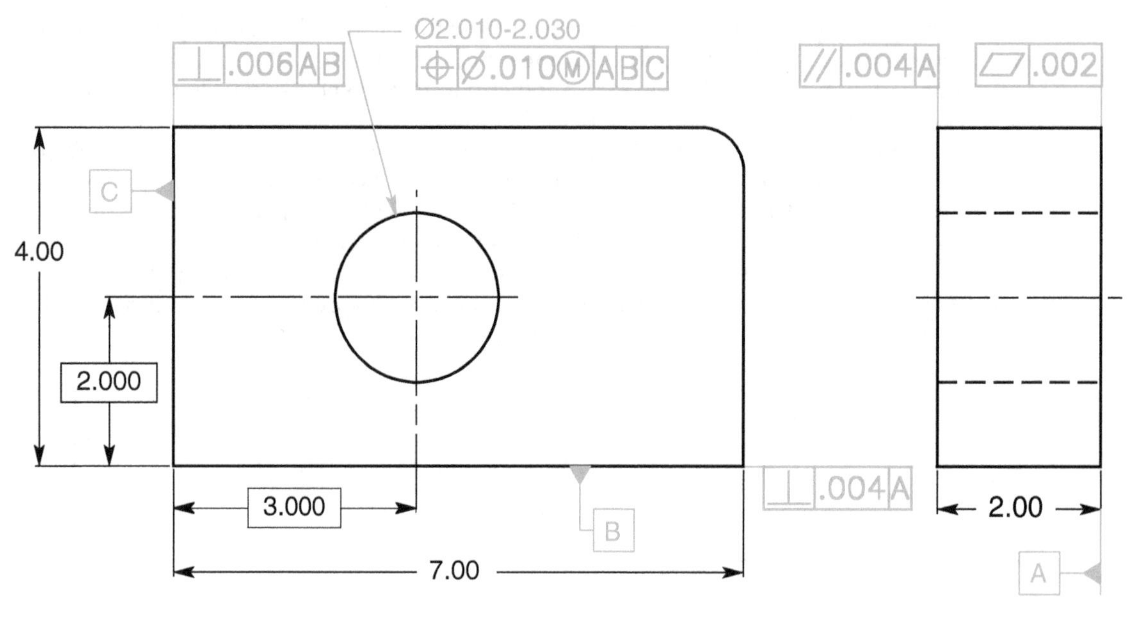

FIGURE 12-6 Datum features controlled for form and orientation.

be controlled to both datum features A and B. For the 2-inch-diameter hole in Fig. 12-6, datum feature A is the reference for orientation (perpendicularity), and datum features B and C are the references for location.

The size and the size tolerance may be determined by using one of the fastener formulas, a standard fit table, manufacturer's specifications of mating parts, or customer's requirements.

The size tolerance of a feature controls not only its size but also its form, Rule #1. According to the drawing in Fig. 12-1, the size of the 2-inch-diameter hole can be made anywhere between 2.010 and 2.030 in diameter, which means that the size tolerance is .020. However, if the machinist actually produces the hole at a diameter of 2.020, according to Rule #1, the form tolerance for the hole is .010, that is, 2.020 minus 2.010. The hole must be straight and round within .010. The hole size can be produced even larger, up to a diameter of 2.030, in which case the form tolerance is even larger. If the straightness and/or circularity tolerance, automatically implied by Rule #1, does not satisfy the design requirements, an appropriate form tolerance must be specified.

The location tolerance between mating features comes from the difference between their size dimensions and tolerances. If a mating shaft with a virtual condition of 2.000 inches in diameter must fit through the hole in Fig. 12-6, the location tolerance can be as large as the difference between the 2.010-diameter hole and the 2.000-inch-diameter, virtual condition shaft for a positional tolerance of .010. A positional tolerance for locating and orienting features of size is always specified with a material condition modifier. The maximum material condition (MMC) modifier (circle M) has been specified for the hole in Fig. 12-6. The MMC modifier is typically specified for features in static assemblies. The regardless of feature size (RFS) modifier is typically used for high-speed, dynamic assemblies. The least material condition (LMC) modifier is used where a specific minimum edge distance must be maintained.

If the location tolerance diameter of .010 at MMC specified for the hole is also acceptable for orientation, the position control included in Fig. 12-6 is adequate. If an orientation refinement of the hole is required, a smaller perpendicularity tolerance, such as the one in Fig. 12-7, may be stipulated.

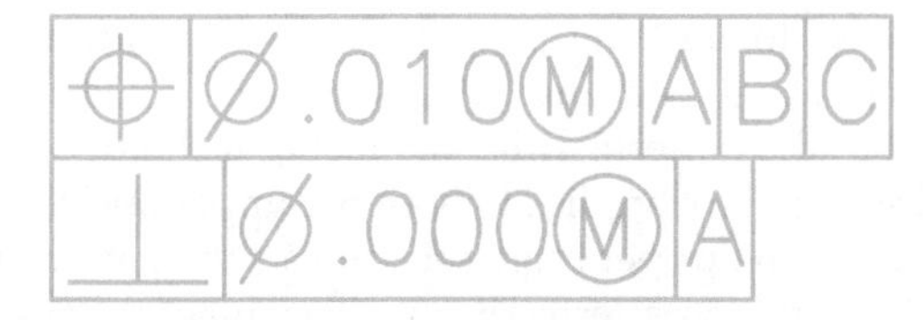

FIGURE 12-7 A location tolerance with a perpendicularity refinement.

If the hole is actually produced at a diameter of 2.020, there is a .010 bonus tolerance that applies to both the location and the orientation tolerances. Consequently, the total positional tolerance is .020 in diameter, and the total perpendicularity tolerance is .010 in diameter.

The same tolerancing techniques specified for the single hole in the drawing above also applies to a pattern of holes shown in Fig. 12-8. The hole pattern is located to a datum reference frame established by datum features A, B, and C and to each other with basic dimensions. The note 4X Ø.510-.540 and the geometric tolerance apply to all four holes. The pattern of four cylindrical tolerance zones, .010 in diameter at MMC, is located and oriented with basic dimensions to datum features A, B, and C. The axis of each hole must fall completely inside its respective tolerance zone.

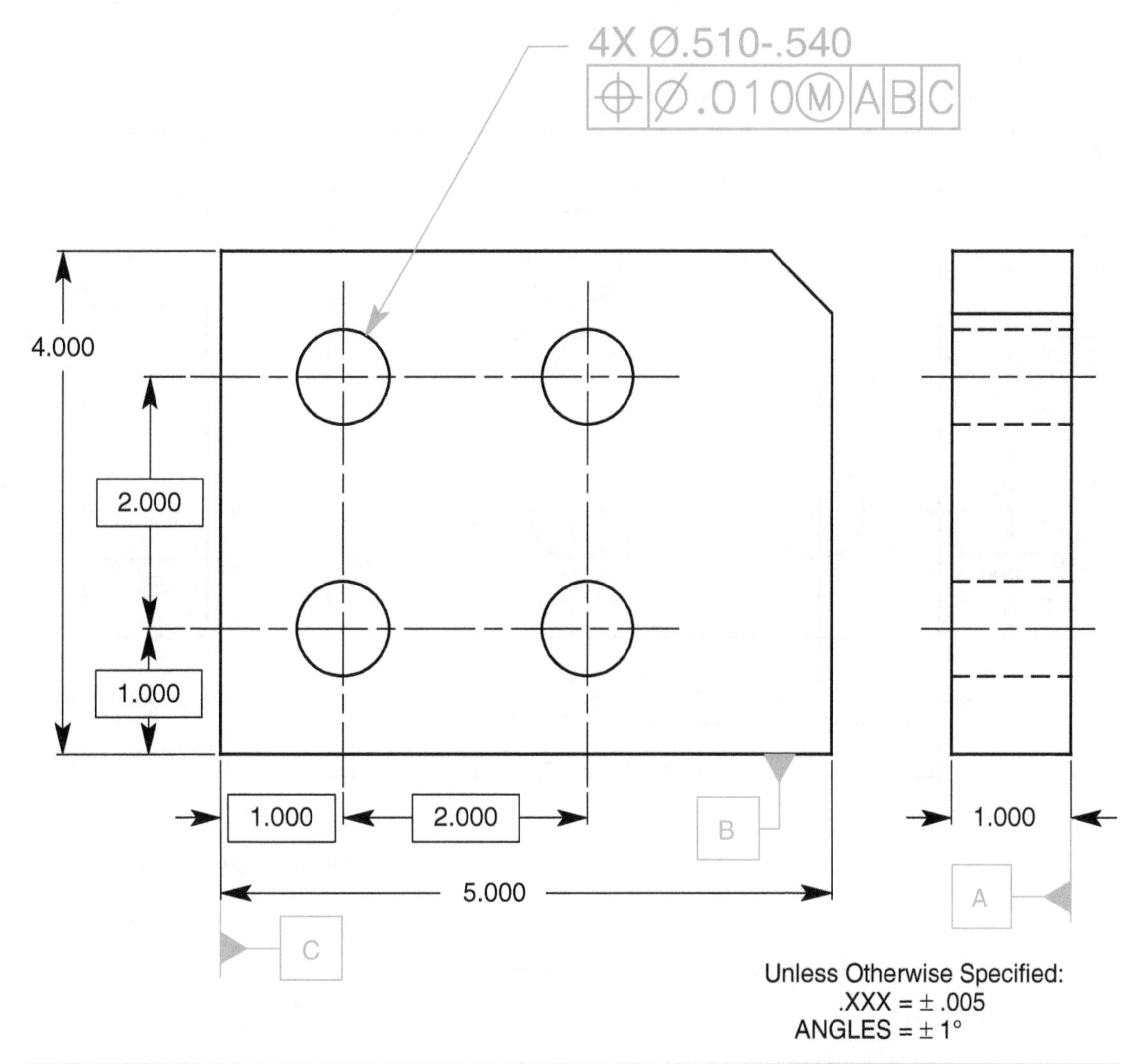

FIGURE 12-8 A geometric tolerance applied to a pattern of features.

Composite geometric tolerancing is applied only to patterns of features and is employed when the tolerance between the datum features and the pattern is not as critical as the tolerance between features within the pattern. This tolerancing technique is often used to reduce the cost of the part. The position symbol applies to both the upper and the lower segments of a composite feature control frame. The upper segment controls the pattern in the same way that a single feature control frame controls a pattern. The lower segment refines the feature-to-feature location relationship; the primary function of the position tolerance is location.

The pattern in Fig. 12-9 is located with basic dimensions to datum features A, B, and C within four cylindrical tolerance zones .040 in diameter at MMC. The relationship between features located to each other with basic dimensions as well as perpendicularity to datum feature A is controlled by four cylindrical tolerance zones .010 in diameter at MMC. The axis of each feature must fall completely inside both of its respective tolerance zones.

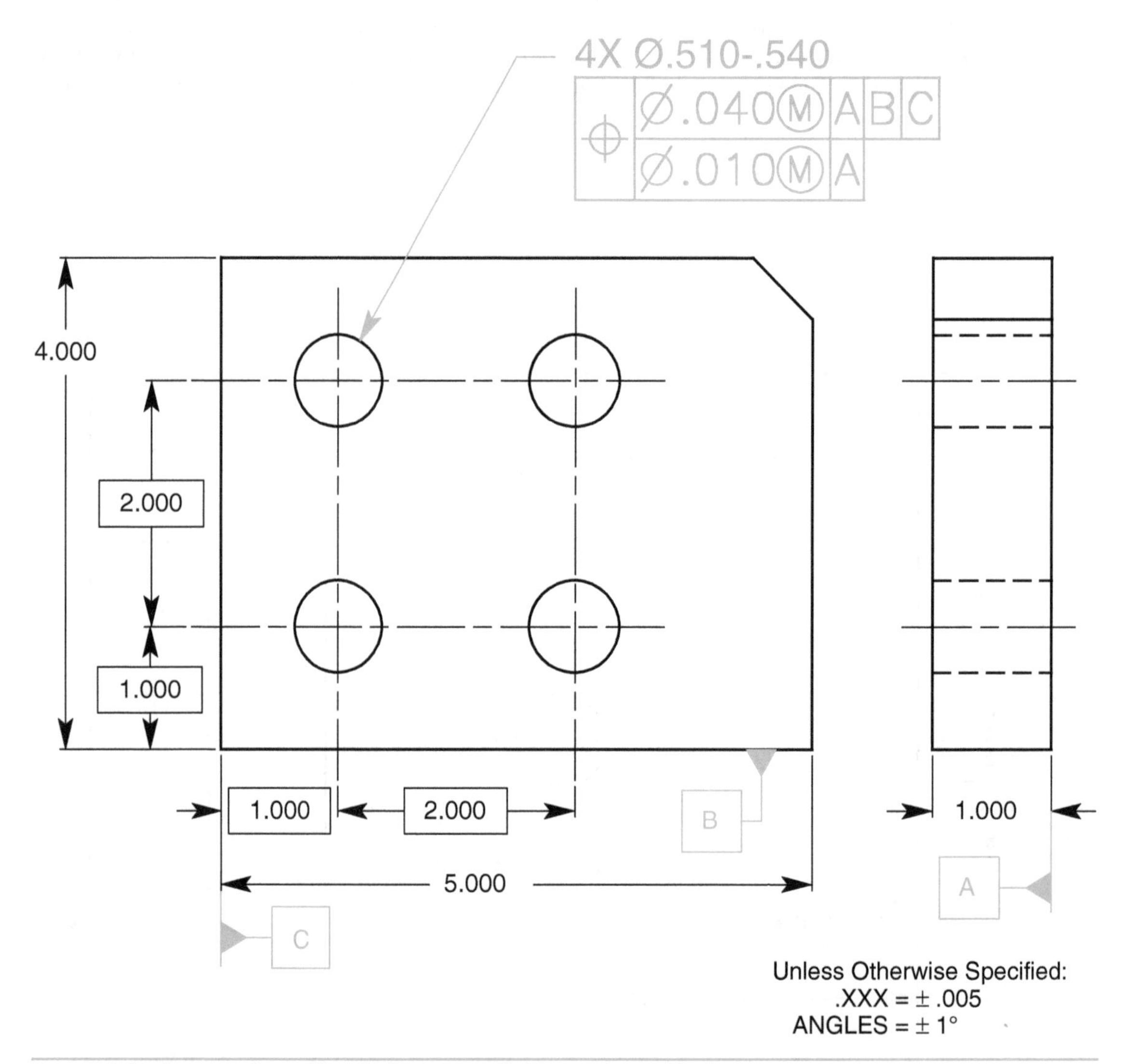

FIGURE 12-9 A composite positional tolerance applied to a pattern of features.

Locating Features of Size to Datum Features of Size

Another common geometry with industrial applications is a pattern of holes located to a datum feature of size, such as an inside or an outside diameter.

The back surface and the inside diameter of the center hole shown in Fig. 12-10 are mating surfaces. The virtual condition of the mating shaft is 1.250 in diameter, and the eight-hole pattern is toleranced for ½-inch bolts.

Since the back surface and the inside diameter of this part are identified as mating surfaces, they are good candidates for datum features. The back of the part is a large flat surface that will provide excellent orientation control. It is identified as datum feature A and is controlled with a flatness tolerance of .002. The inside diameter is also a critical feature; consequently, it is identified as the secondary datum feature, datum feature B, and is controlled to datum feature A with a perpendicularity tolerance.

Because the mating shaft has a virtual condition of 1.250, the corresponding hole is toleranced with the same virtual condition. Datum feature B has been assigned a zero perpendicularity

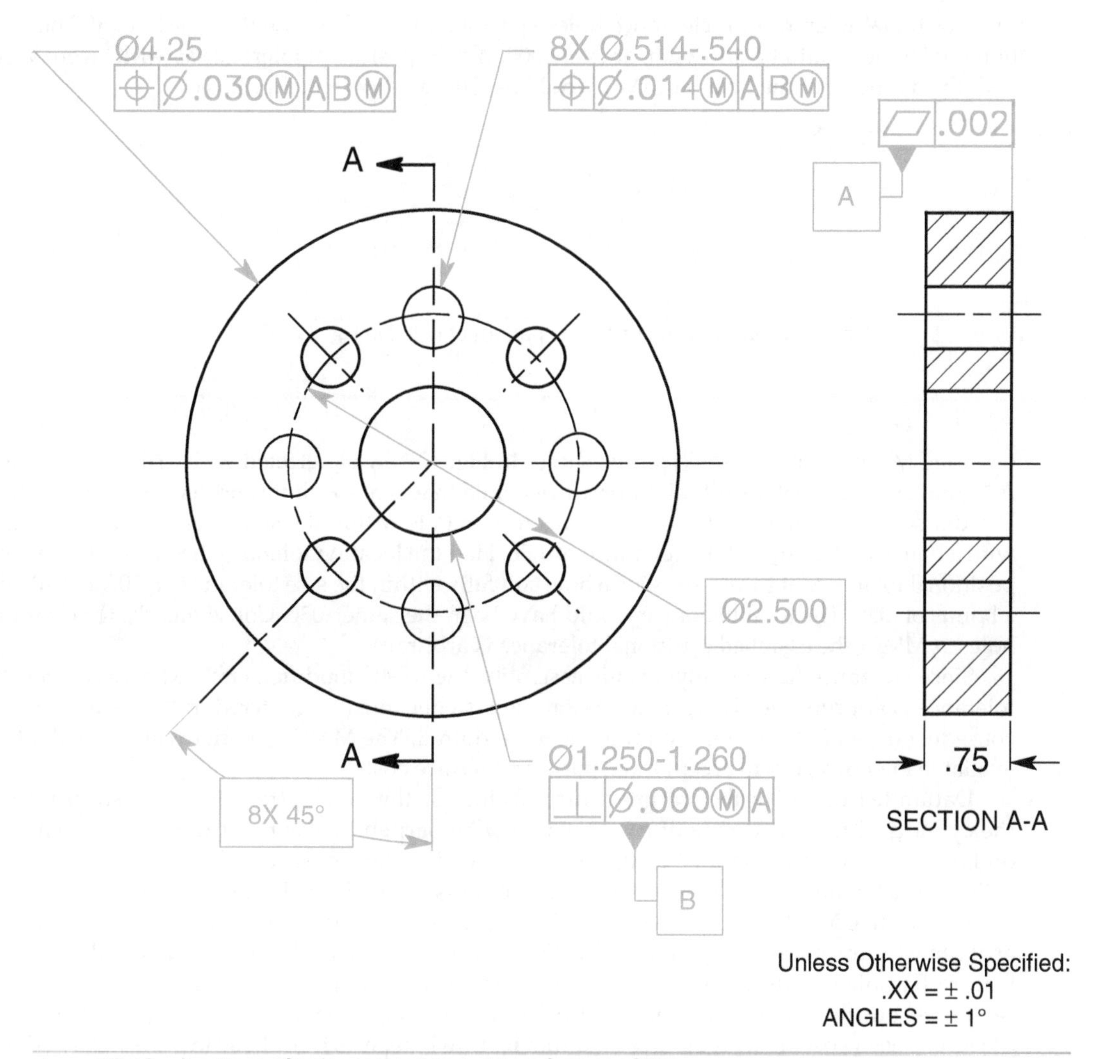

FIGURE 12-10 A pattern of holes located to a datum feature of size.

tolerance at MMC. Where zero tolerance at MMC is employed, all of the tolerance comes from the bonus. Both the virtual condition and MMC are the same diameter. If the machinist produces datum feature B at a diameter of 1.255, the hole must be perpendicular to datum feature A within a cylindrical tolerance zone of .005.

In Fig. 12-10, the eight-hole pattern is placed on a basic 2.500-diameter bolt circle, with a basic 45° angle between each hole. If the back of this part is to mate with another part and these holes are clearance holes used to bolt the parts together, the holes should be perpendicular to the mating surface. Consequently, it is appropriate to control the eight-hole pattern to the primary datum feature, datum feature A. The center of the bolt circle is positioned on the axis of the center hole, datum feature B. As a result, the pattern is perpendicular to datum feature A and located to datum feature B.

Next, the clearance holes are toleranced for ½-inch fasteners with a positional tolerance of .014 at MMC. The fastener formula is

$$\text{Hole diameter at MMC} = \text{Fastener at MMC} + \text{Geometric Tolerance at MMC}$$

$$\text{Hole diameter} = .500 + .014 = .514$$

The positional tolerance for clearance holes specified at MMC is essentially arbitrary. The positional tolerance could be .010, .005, or even .000. If zero positional tolerance at MMC were specified, the diameter of the holes at MMC would be .500, as shown in Fig. 12-11.

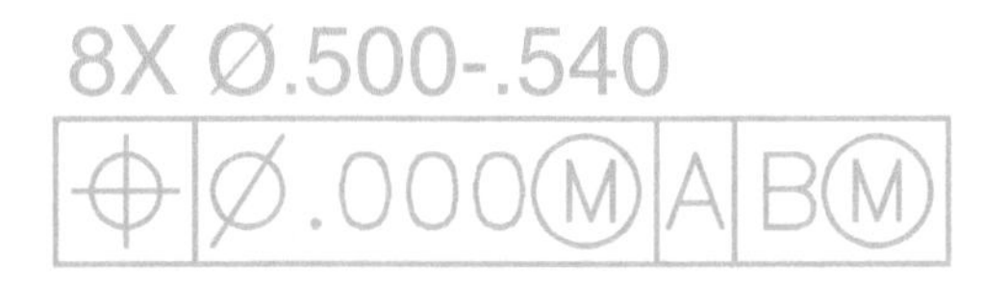

FIGURE 12-11 A zero positional tolerance for the pattern of holes in Fig. 12-10.

The LMC hole size for the clearance holes in Fig. 12-10 was selected with drill sizes in mind. A 17/32 (.531)–diameter drill might produce a hole that is a few thousandths oversize, resulting in a diameter of perhaps .536. A .536-diameter hole falls within the size tolerance of .514 to .540 with a bonus of .022 and a total tolerance of .036. Had the location tolerance been specified at zero positional tolerance at MMC, the Ø.536 hole still falls within the size tolerance of .500 to .540 with a bonus of .036. The total tolerance would have been the same, .036. Consequently, for clearance holes at MMC, the assigned positional tolerance is arbitrary.

Since clearance holes imply a static assembly, the MMC modifier, circle M, placed after the tolerance is appropriate. There is no reason the fastener must be centered in the clearance hole; consequently, an RFS material condition is not required. The MMC modifier will allow all of the available tolerance; it will accept more parts and reduce costs.

Datum feature A, in the feature control frame for the hole pattern location, specifies that the cylindrical tolerance zone of each hole must be perpendicular to datum plane A. The secondary datum feature, datum feature B, is the locating datum feature. Datum feature B is the 1.250-diameter hole. The center of the bolt circle is located on the axis of datum feature B. Because a circle M follows datum feature B in the feature control frame, datum feature B applies at its maximum material boundary (MMB). In this particular situation, the MMB of datum feature B applies at its virtual condition with respect to the preceding datum feature, datum feature A, but the virtual condition and the MMC are the same since zero perpendicularity at MMC has been specified. As the size of datum feature B departs from 1.250 toward 1.260 in diameter, the pattern gains shift tolerance in the exact amount of such departure. If datum feature B is produced at a diameter of 1.255, there is a cylindrical tolerance zone .005 in diameter about the axis of datum feature B within which the axis of the bolt circle may shift. In other words, the axis of the pattern, as a whole, may shift in any direction within a cylindrical tolerance zone of .005 in diameter. Shift tolerance may be determined with graphic analysis techniques discussed in Chap. 13.

One of the most common drawing errors is the failure to control coaxiality. The feature control frame beneath the 4.25-diameter size dimension controls the coaxiality of the outside diameter to the inside diameter. Coaxiality may be toleranced in a variety of ways, but it must be controlled to avoid incomplete drawing requirements. Many designers omit this control, claiming that it is "overkill," but sooner or later, they will buy a batch of parts that will not assemble because coaxial features are out of coaxiality tolerance.

Some designs require patterns of features to be clocked to a third datum feature. That is, where a pattern is not allowed to rotate about its center axis, a third datum feature is assigned and included in the feature control frame to prevent rotation. The hole pattern in Fig. 12-12 is toleranced the same way as the hole pattern in Fig. 12-10 except that it has been clocked to a third datum feature. The flat on the outside diameter has been designated as datum feature C and specified as the tertiary datum feature in the feature control frame, preventing clocking of the hole pattern about datum feature B.

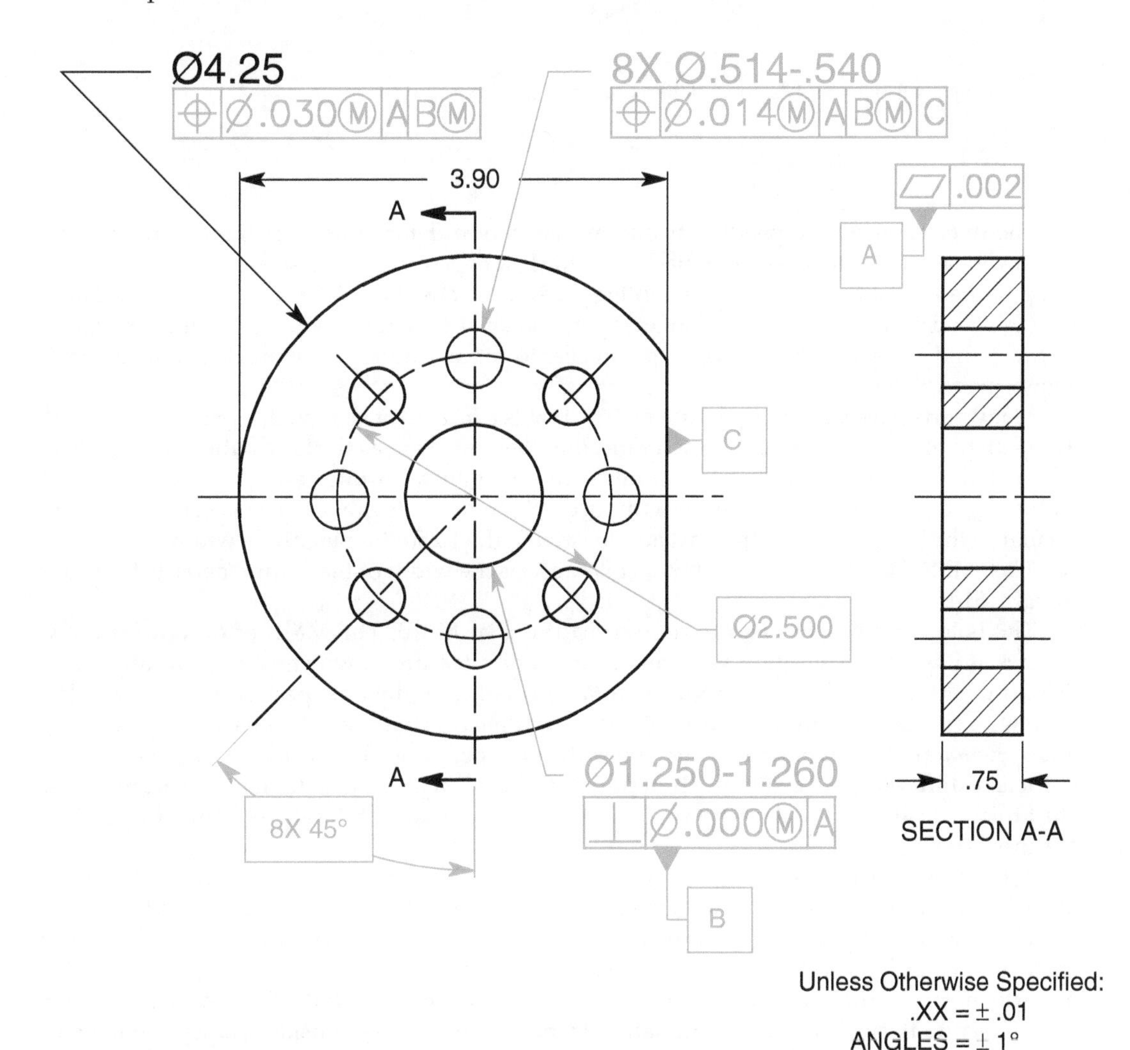

FIGURE 12-12 A pattern of holes located to a datum feature of size and clocked to a flat surface.

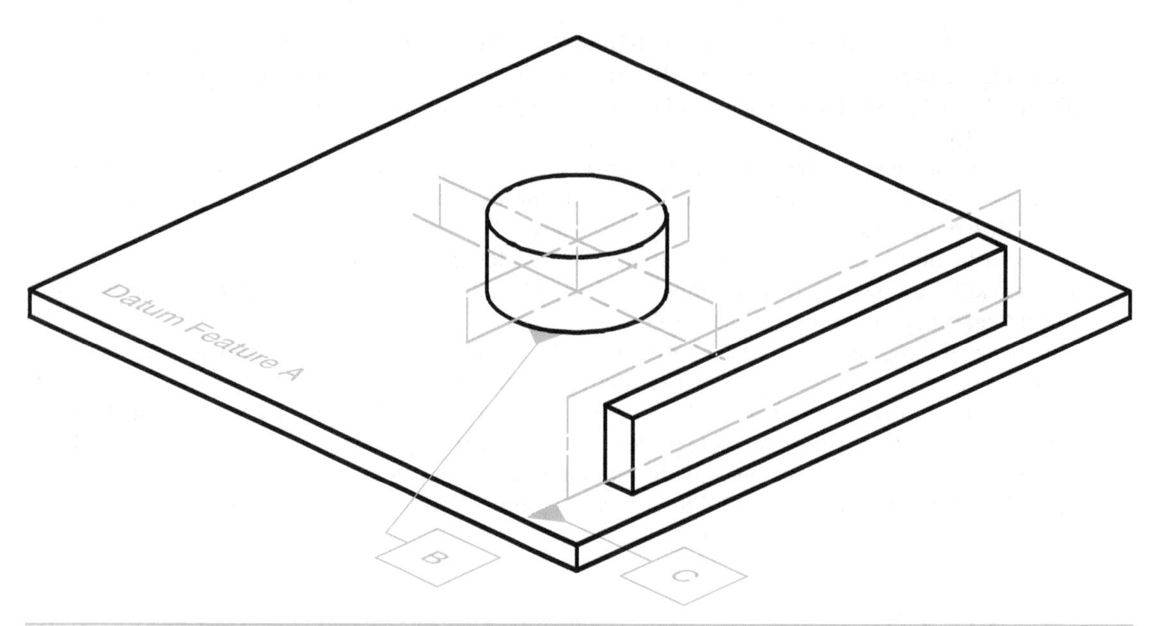

FIGURE 12-13 A special fixture establishing a datum reference frame for the part in Fig. 12-12.

The three mutually perpendicular datum planes of a datum reference frame can be established with a coordinate measuring machine (CMM), an open setup, or a special fixture. The open setup or the special fixture for the part in Fig. 12-12 will consist of a plane flat surface for datum feature A, a precision pin 1.250 in diameter perpendicular to datum feature A to simulate datum feature B, and a movable flat surface perpendicular to datum feature A to simulate datum feature C shown in Fig. 12-13.

Many parts have a clocking datum feature that is a feature of size, such as a hole or keyseat. The pattern of holes in Fig. 12-14 is toleranced in the same way as the hole pattern in Fig. 12-10 except that it has been clocked to datum feature C, which, in this case, is a feature of size. Datum feature C is a ½-inch keyseat with its own geometric tolerance. The keyseat is perpendicular to the back surface of the part and located to the 1.250-diameter hole within a tolerance of .002 at MMC. The keyseat gains bonus tolerance as the width of the feature departs from .502 toward .510.

The hole pattern is clocked to datum feature C at MMB. The MMB of datum feature C applies at its virtual condition with respect to datum feature B, which is .502 minus .002, or .500. If the keyseat is actually produced at a width of .505, the hole pattern has a shift tolerance with respect to datum feature C of .005. That means that the entire pattern can shift up and down and can rotate within the .005 shift tolerance zone. This is assuming that there is sufficient shift tolerance available from datum feature B with respect to datum feature A. The shift tolerance allowed will be the smaller of the available shift tolerances from either datum feature B or C.

Tolerances on parts like the one in Fig. 12-14 are complicated and sometimes difficult to visualize. It is helpful to draw the gage that could be used to inspect the part. On a print, a simple sketch of the gage around the part is sufficient. However, it is even better to sketch the gage on a piece of tracing paper so that the part can move inside the gage. This sketch is sometimes called a "cartoon gage." The sketch illustrates how the part can shift about the 1.250 center diameter, datum feature B, and the ½-inch key, datum feature C. Finally, the outside diameter of the part must be sufficiently coaxial to fit inside the 4.290 diameter. Visualization of shift tolerances can be greatly enhanced with the use of a gage sketch like the one shown in Fig. 12-15.

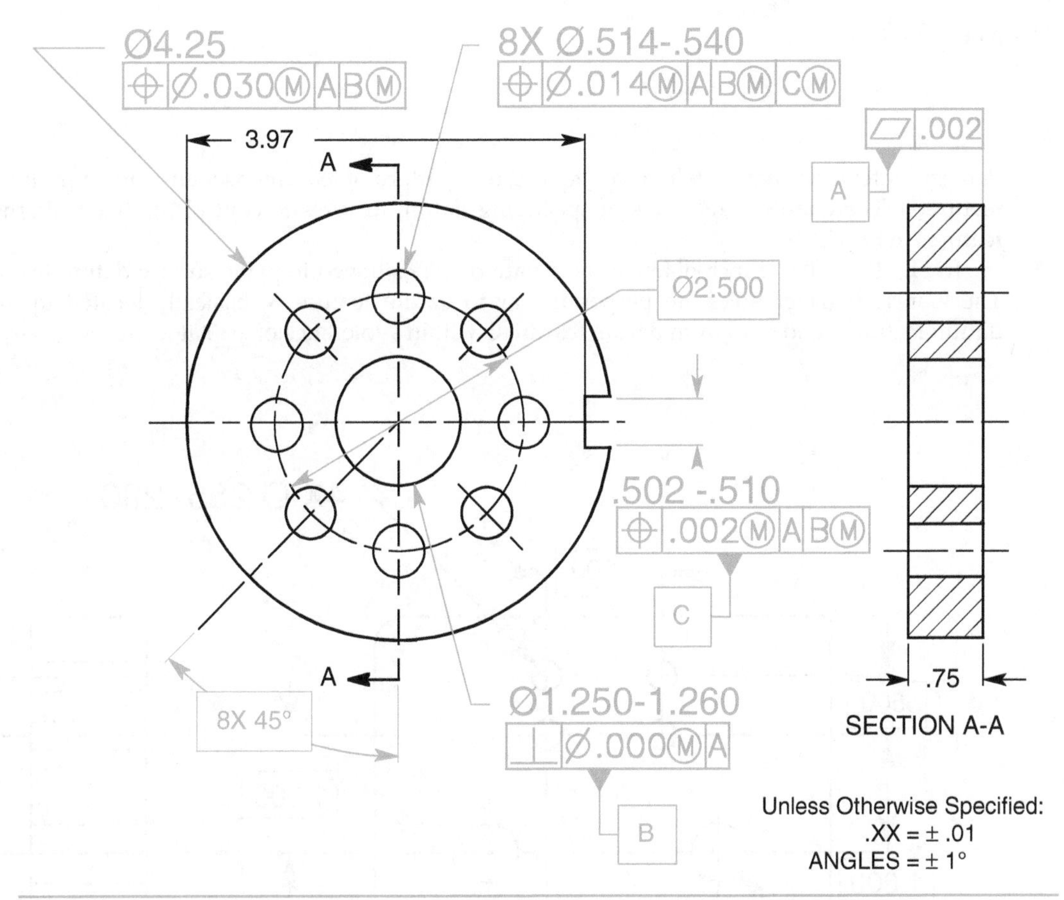

FIGURE 12-14 A pattern of holes located to a datum feature of size and clocked to a keyseat.

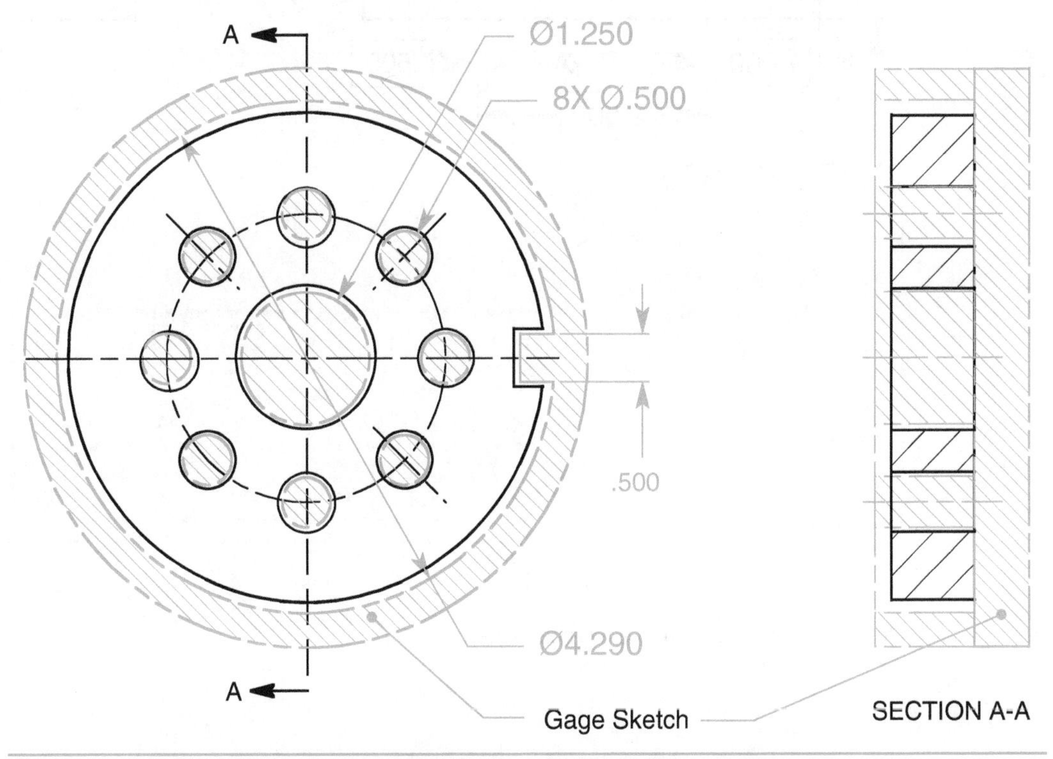

FIGURE 12-15 A gage sketched about the part in Fig. 12-14 illustrates shift tolerance.

Locating a Pattern of Features to a Second Pattern of Features

Patterns of features may be toleranced to a second pattern of features as well as to an individual feature(s). There are several ways of specifying datum features to control the two patterns of features in Fig. 12-16.

In Fig. 12-16, the ½-inch-diameter hole pattern is positioned to plane surface datum features. The ½-inch-diameter holes are perpendicular to datum feature A, basically located up from datum feature B and over from datum feature C within a tolerance of .030 at MMC.

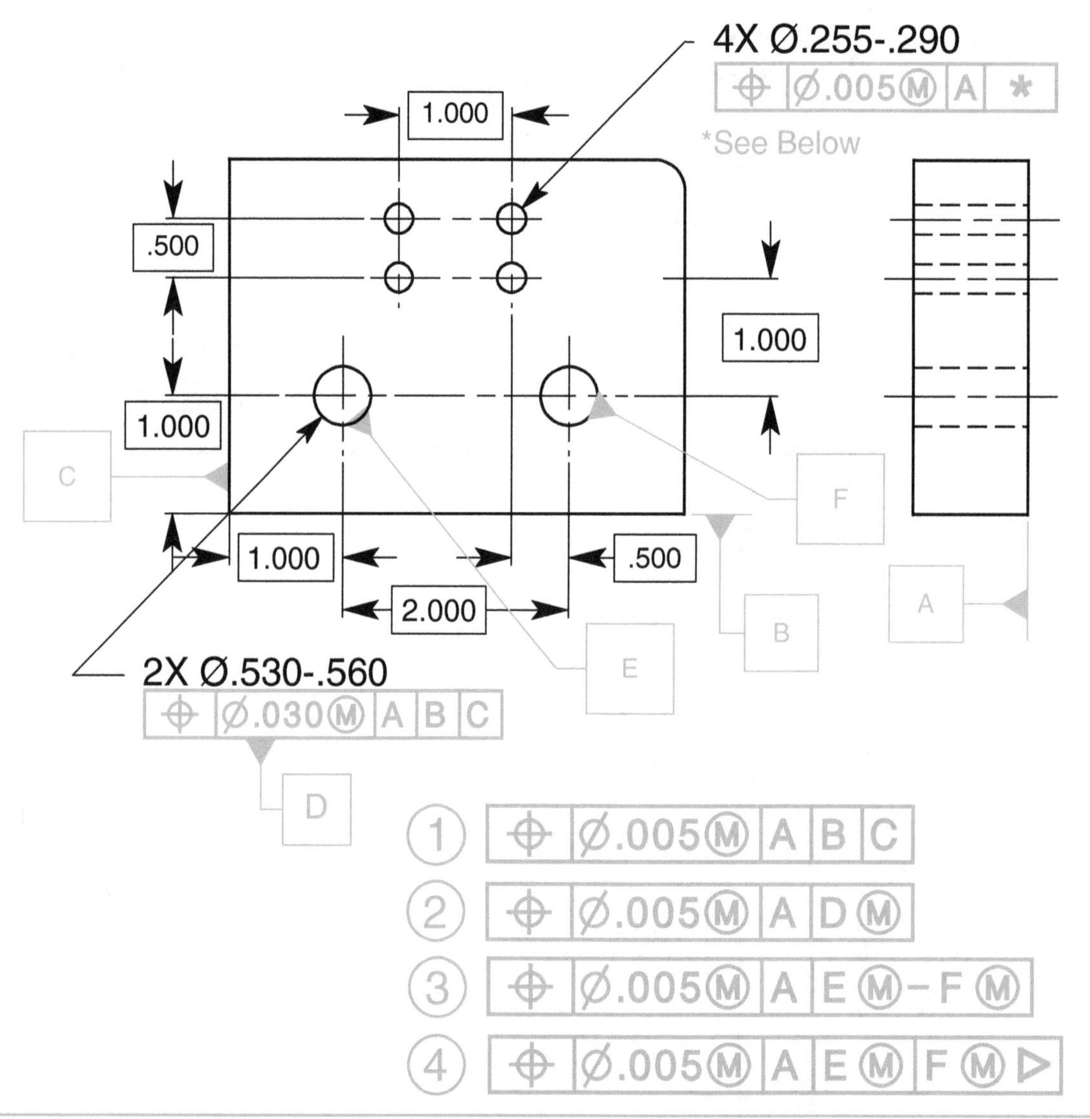

FIGURE 12-16 A four-hole pattern located to a two-hole pattern.

Now that the ½-inch-diameter hole pattern is positioned, the simplest and most straightforward way of tolerancing the ¼-inch-diameter hole pattern is to control it to datum features A, B, and C, the established datum reference frame shown in feature control frame #1 in Fig. 12-16. Where possible, it is best to use only one datum reference frame. In Fig. 12-17, a part with two patterns of holes controlled to each other through a common datum reference frame is shown in a gage designed to inspect the location of both patterns relative to datum features A, B, and C. If both patterns are toleranced to the same datum features, in the same order of precedence, and at the same material conditions, the patterns are to be considered one composite pattern of features. Since one pattern has cylindrical tolerance zones .030 in diameter at MMC and the other has cylindrical tolerance zones of .005 in diameter at MMC, the two patterns will be located to each other within a cylindrical tolerance of .035 at MMC. With bonuses included, the combined tolerances could be more than .085. If the tolerance between patterns must be smaller, the tolerance can be reduced.

If a large location tolerance between the two-hole pattern and datum features A, B, and C and a small location tolerance between the four-hole pattern and the two-hole pattern are required, one of the patterns may be the locating datum feature. In feature control frames #2 and #3 in Fig. 12-16, the two-hole pattern is identified as the locating datum feature.

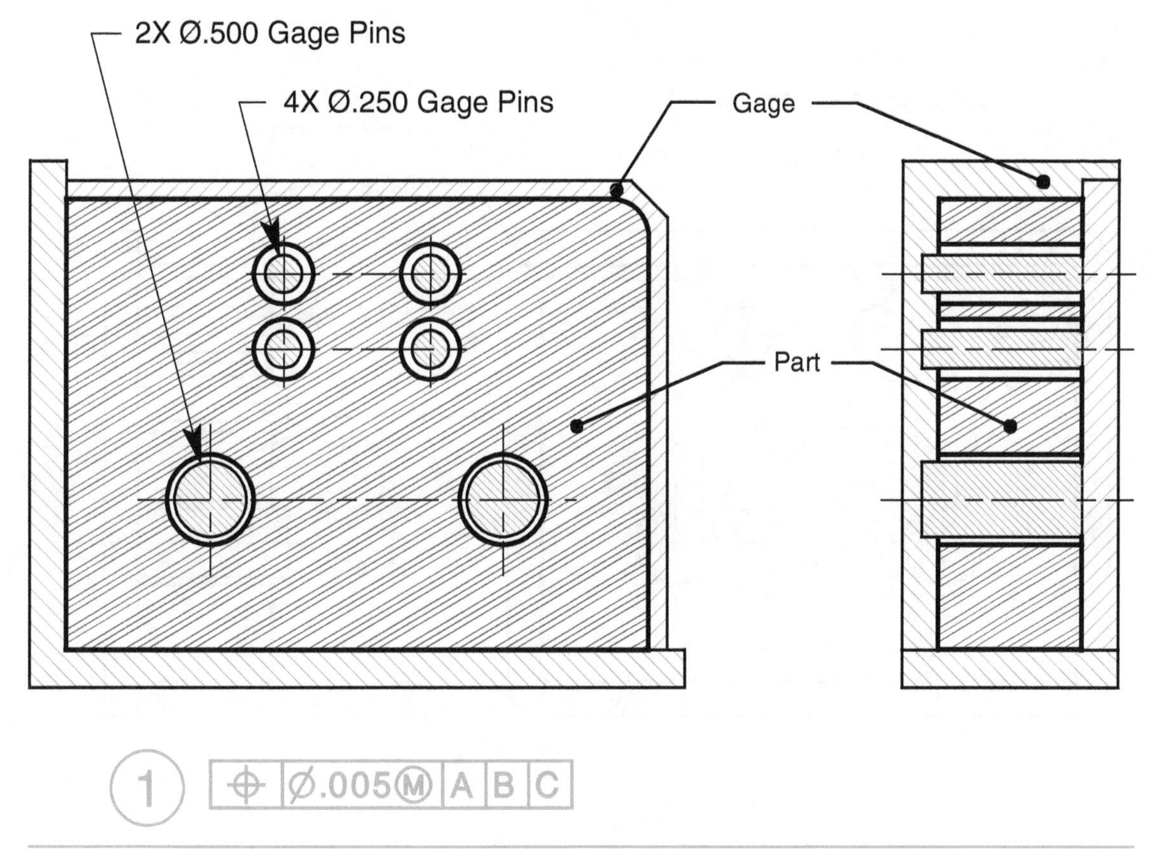

FIGURE 12-17 Gaging two patterns of features located to surface datum features A, B, and C.

If the four-hole pattern is controlled with feature control frame #2, the tolerance zone of each hole in the four-hole pattern is to be perpendicular to datum feature A, located to each other, and located to datum feature D at MMB within a tolerance of .005 at MMC. That is, both holes in the two-hole pattern act as one datum feature controlling the location and clocking of the four-hole pattern. This part is shown in Fig. 12-18 in a gage designed to inspect the four-hole pattern perpendicular to datum feature A, located to each other, and to the two-hole pattern, datum feature D at MMB.

Feature control frame #3 in Fig. 12-16 is equivalent to feature control frame #2. If the four-hole pattern is controlled with feature control frame #3, the tolerance zone of each hole in the four-hole pattern must be perpendicular to datum feature A, located to each other, and located to datum feature E at MMB—F at MMB within a tolerance of .005 at MMC. Datum features E and F are of equal value.

Datum feature E at MMB—F at MMB in the feature control frame for the four-hole pattern can be inspected with the same gage as datum feature D at MMB. The gage in Fig. 12-18 can be used to inspect the four-hole pattern controlled with either feature control frame #2 or #3.

Feature control frame #4 in Fig. 12-16 is similar to feature control frame #3 except that datum feature E, the secondary datum feature, is more important than datum feature F, the tertiary datum feature, because datum feature E precedes datum feature F in separate compartments. As a result, datum feature E is the locating feature, and datum feature F is the clocking feature.

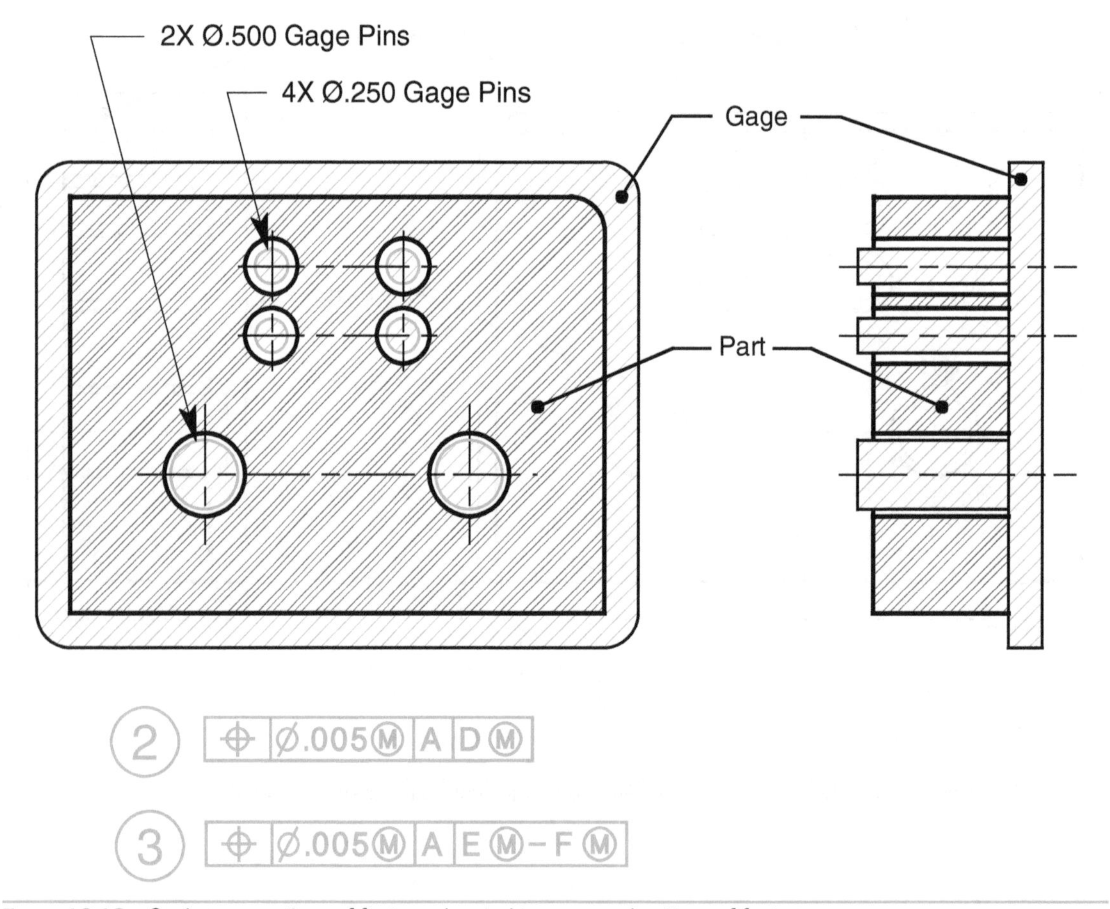

FIGURE 12-18 Gaging one pattern of features located to a second pattern of features.

That is, the function of datum feature F is only to prevent the part from rotating about datum feature E. The translation modifier symbol was introduced in the 2009 *Dimensioning and Tolerancing* standard and has been included in this feature control frame following datum feature F. This symbol unlocks the basic 2.000 dimension between the two datum feature simulators on the gage and allows datum feature simulator F to translate within the specified geometric tolerance to fully engage datum feature simulator E illustrated on the gage in Fig. 12-19. If the translation modifier symbol had not been included in the fourth feature control frame, the basic 2.000 dimension would lock datum feature simulator F in place requiring a gage, such as the one in Fig. 12-18, to inspect the part. Consequently, without the translation modifier in feature control frame #4, the four-hole pattern would be located to both datum features E and F, equivalent to feature control frames #2 and #3. (Gages are used here and throughout this book for illustration purposes only; parts toleranced with geometric dimensioning and tolerancing may be inspected with any appropriate inspection technique.)

Of the tolerancing techniques discussed above, the most straightforward is the plane surface datum reference frame, feature control frame #1 in Fig. 12-16. If the two holes are the locating datum feature, use the datum feature scheme shown in feature control frame #2. If only a single hole is the locating datum feature, specify that hole as the secondary datum feature and specify a surface such as datum feature B as a tertiary datum feature for clocking. If the clocking datum feature must be a hole or a pin, specify the translation modifier shown in feature control frame #4.

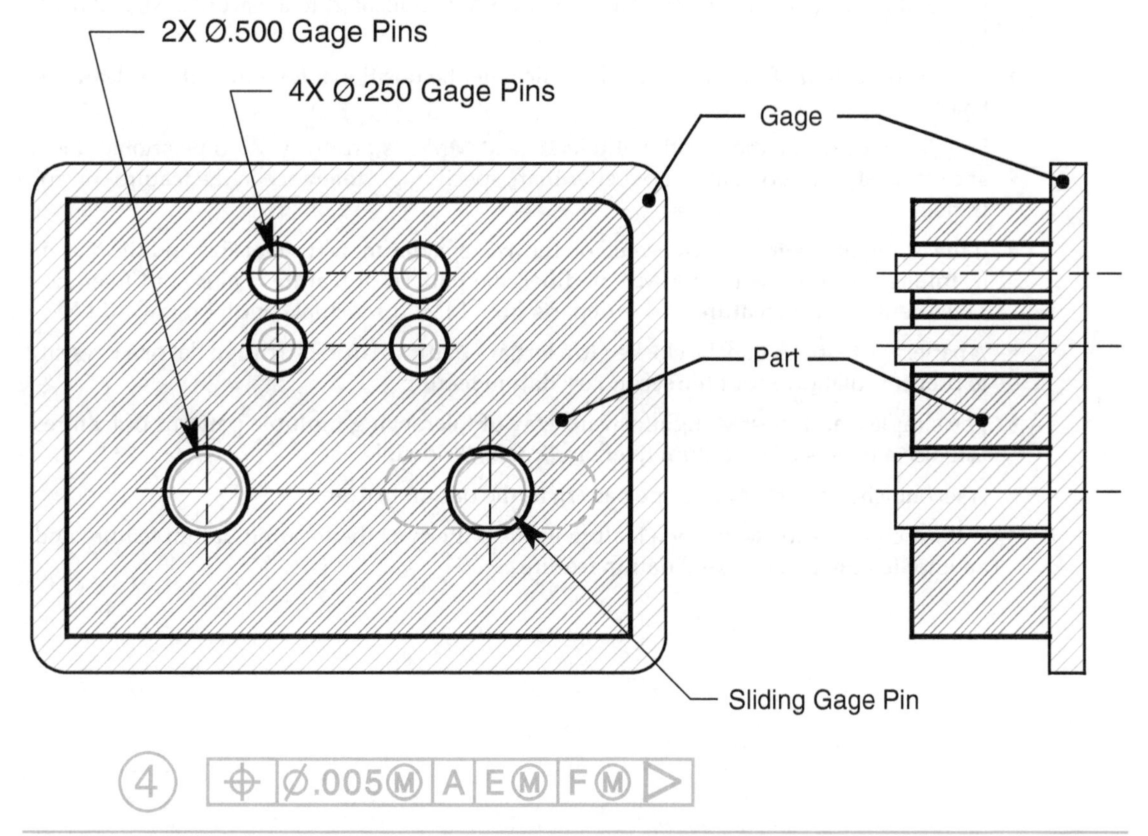

FIGURE 12-19 Gaging a pattern of features located to a feature of size and clocked to a second feature of size specified with the translation symbol.

Summary

- The designer must determine the attributes of each feature and the relationship between features.
- The first step in tolerancing a drawing is to specify the datum features.
- Identify the datum features and determine the order in which they appear in the feature control frame.
- The primary datum feature is the most important datum feature and is not controlled to any other feature. If Rule #1 doesn't sufficiently control the form of the primary datum feature, a form tolerance must be specified.
- Orientation controls of the secondary and tertiary datum features must be specified if the title block angularity tolerance is not adequate.
- Determine whether the size tolerance adequately controls the toleranced feature's form, Rule #1, or a form tolerance is required.
- The same tolerancing techniques specified for a single feature also apply to a pattern of features.
- Composite geometric tolerancing is employed when the tolerance between the datum features and the pattern is not as critical as the tolerance between features within the pattern.
- Another common geometry with industrial applications is a pattern of holes located to a feature of size, such as an inside or an outside diameter. Typically, the pattern is perpendicular to a flat surface, datum feature A, and located to a feature of size, datum feature B.
- One of the fastener formulas is used to calculate the positional tolerance of the clearance holes.
- For clearance holes, the positional tolerance at MMC is arbitrary. Zero positional tolerance at MMC is as good as, if not better than, specifying a tolerance in the feature control frame.
- If the center of a bolt circle is located on the axis of a datum feature of size and the datum feature is specified with an MMB modifier, the pattern of features gains shift tolerance as the center datum feature of size departs from MMC or virtual condition toward LMC.
- A pattern of features may be clocked to a tertiary datum feature, such as a flat or a keyseat, to prevent rotation about the secondary datum feature.
- The simplest and most straightforward way of tolerancing multiple patterns of features is to use a plain surface datum reference frame, if possible.
- A second choice is to specify one pattern as the datum feature.
- A third choice is to choose one feature in the pattern as the locating datum feature and another feature as a clocking datum feature.

Chapter Review

1. What type of geometric tolerances applies to the primary datum feature in a drawing like the drawing in Fig. 12-20? ____________

2. What geometric tolerance applies to the primary datum feature in the drawing in Fig. 12-20?

____

____

3. The primary datum feature controls ____________ of the feature being controlled.

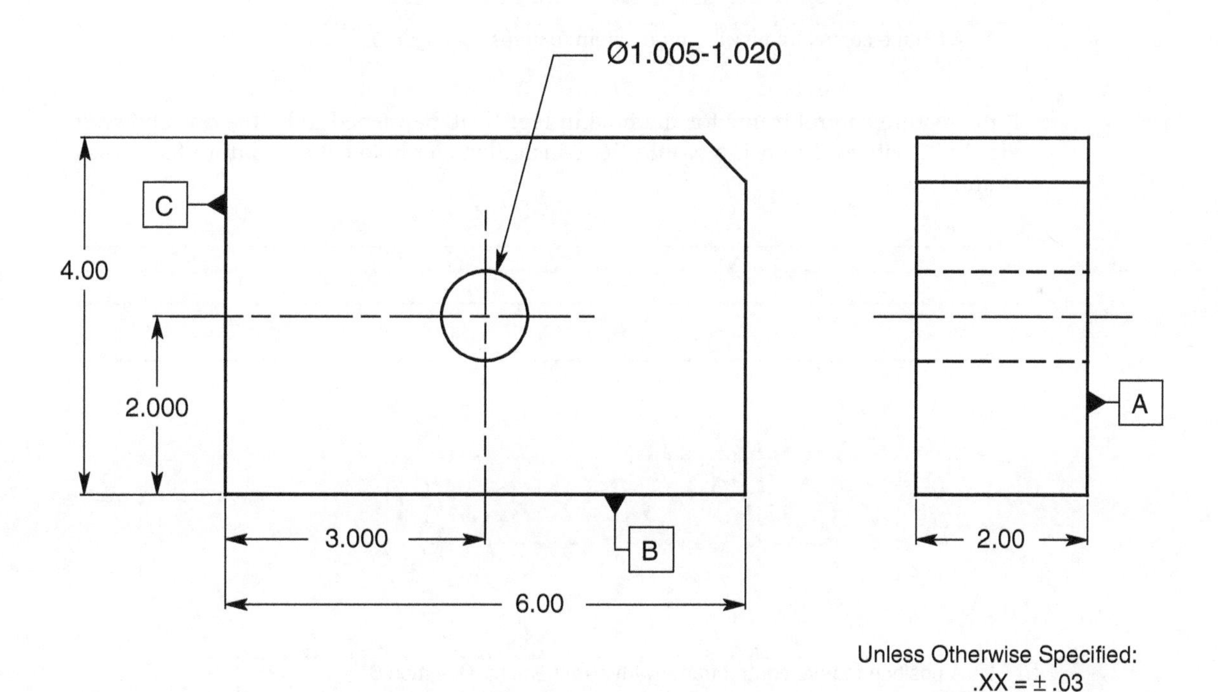

FIGURE 12-20 A hole located and oriented to datum features A, B, and C: Questions 1 through 6.

FIGURE 12-21 A feature control frame with two datum features: Question 4.

4. If the feature control frame for the hole in Fig. 12-20 happened to be the one shown in Fig. 12-21, what relationship would the 1-inch-diameter hole have to datum features B and C?

FIGURE 12-22 A feature control frame with three datum features: Question 5.

5. If the feature control frame for the hole in Fig. 12-20 happened to be the one shown in Fig. 12-22, what relationship would the 1-inch-diameter hole have to datum features A, B, and C?

FIGURE 12-23 A position feature control frame with a refinement: Question 6.

6. Complete the feature control frame in Fig. 12-23 so that it will refine the orientation of the hole in Fig. 12-20 within a cylindrical tolerance of .000 at MMC.

7. Draw a feature control frame to locate a pattern of holes within cylindrical tolerance zones .125 in diameter at MMC to their datum features, A, B, and C. Refine the feature-to-feature relationship perpendicular to datum feature A and located to each other within a .000 positional tolerance at MMC.

4X Ø.500-.530

8. What is the orientation tolerance for the pattern of holes specified in the answer for Question 7? ____________________

____

9. Keeping in mind that the primary datum feature controls orientation, explain how you would select a primary datum feature on a part. ____________________

____

____

____

____

10. How would you determine which datum feature should be secondary and which should be tertiary?

____

____

____

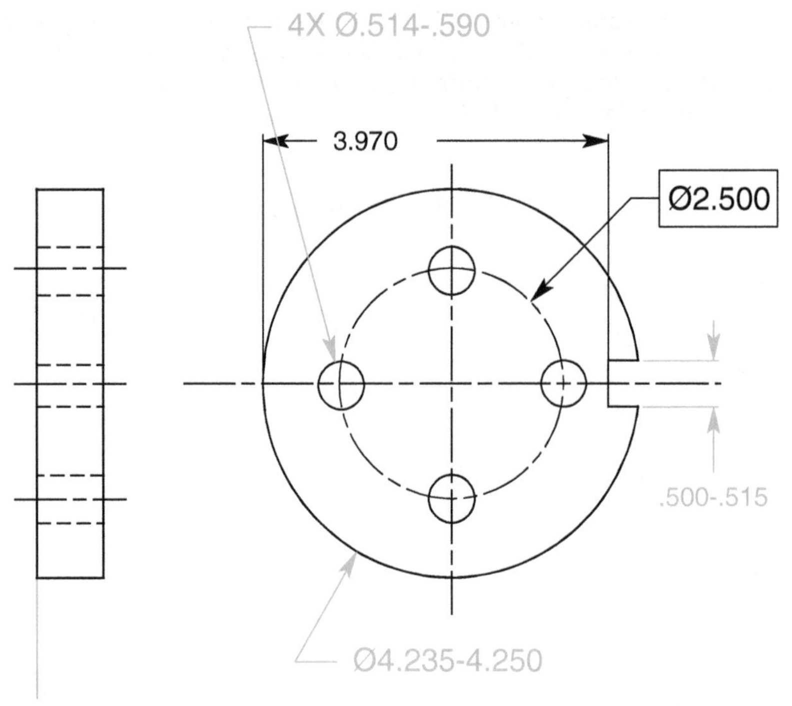

FIGURE 12-24 Four-hole pattern of features.

Refer to Fig. 12-24 to answer Questions 11 through 17.

11. Select a primary datum feature and specify a form control for it.

12. Select a secondary datum feature and specify an orientation control for it. The virtual condition of the mating part is a diameter of 4.255.

13. Tolerance the keyseat for a ½-inch key.

14. Tolerance the ½-inch clearance holes for ½-inch floating fasteners.

15. Are there other ways this part could be toleranced?

____

____

16. If the outside diameter is actually produced at 4.240, how much shift tolerance is available?

____

____

17. If the outside diameter is actually produced at 4.240 and the keyseat is actually produced at .505, how much can this part actually shift? Sketch a gage about the part. ______________

____

____

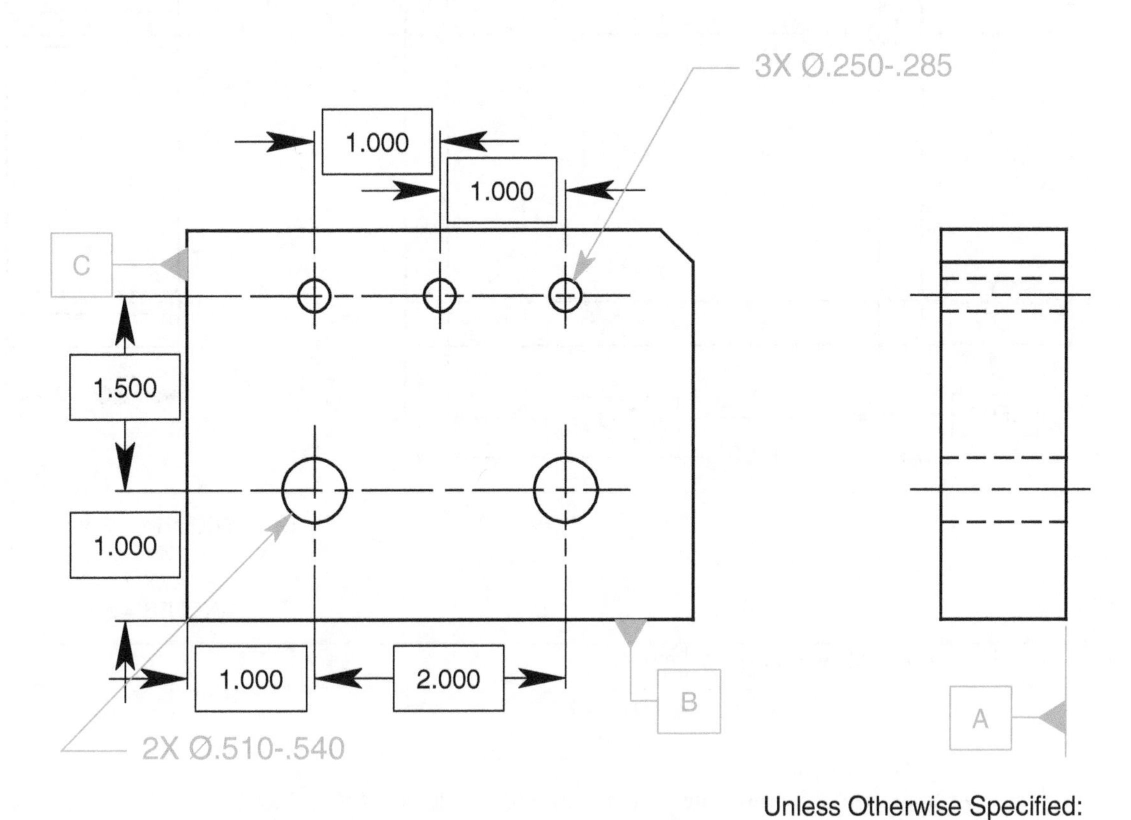

FIGURE 12-25 A pattern of features located to a second pattern of features.

Refer to Fig. 12-25 to answer Questions 18 through 20.

18. Locate the two-hole pattern to the surface datum features with a positional tolerance of .085 at MMC. Locate the same two holes to each other and orient them to datum feature A within a cylindrical tolerance of .010 at MMC.

19. Locate the three-hole pattern to the two-hole pattern within a .000 positional tolerance.

20. The two-hole pattern is specified as a datum feature at MMB. At what size does each of the two holes apply?

Problems

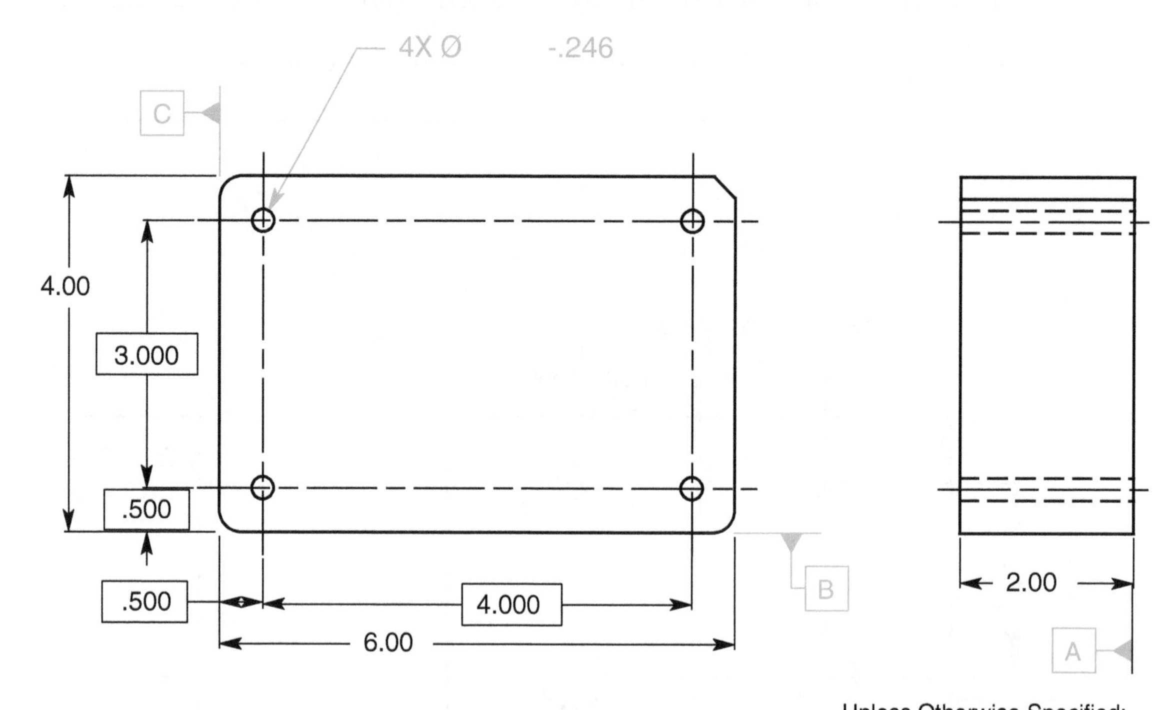

FIGURE 12-26 Tolerancing fixed fasteners: Prob. 1.

1. Tolerance the four-hole pattern in Fig. 12-26 for #10 (Ø.190) cap screws as fixed fasteners. Specify a tolerance of .010 at MMC for the clearance holes and 60% of the total tolerance for the threaded holes in the mating part.

   - How flat is datum feature A? ____
   - How perpendicular are datum features B and C to datum feature A and to each other?

     ____

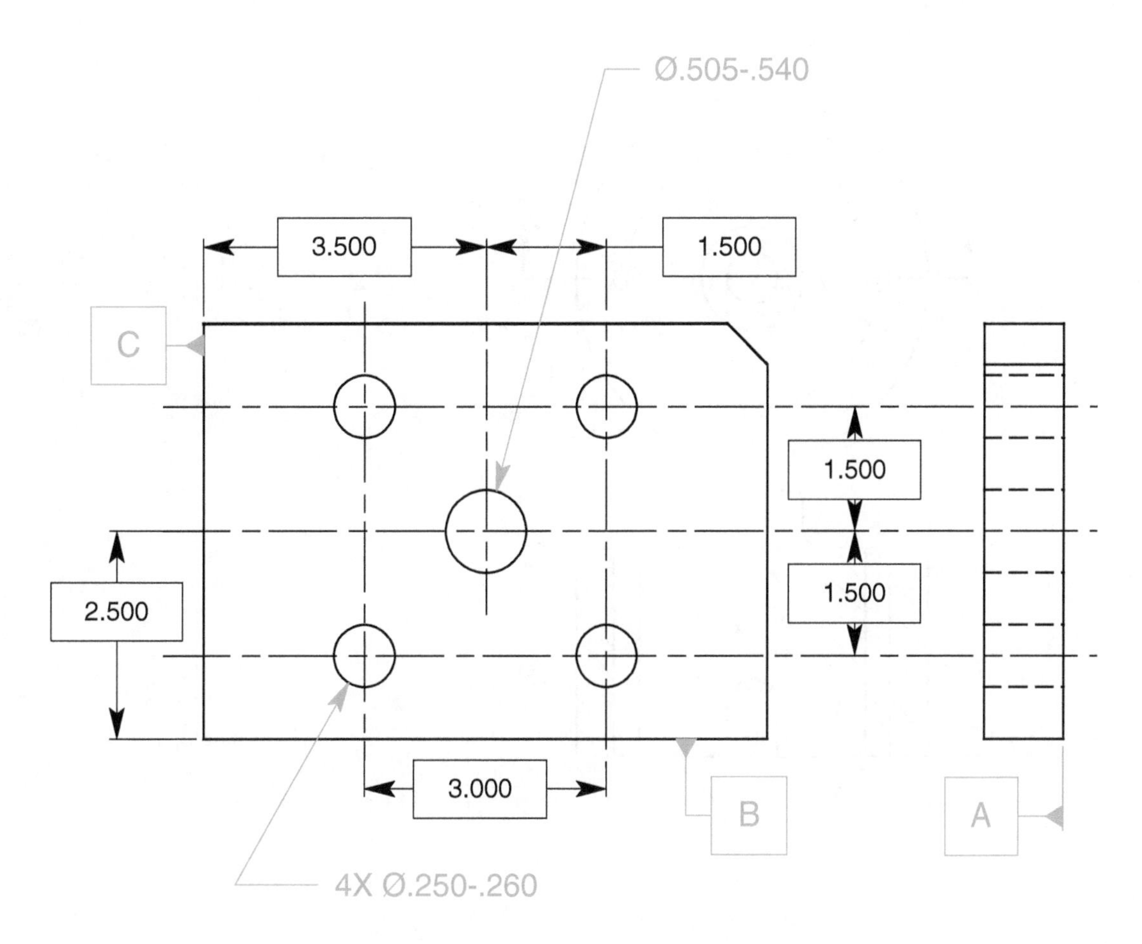

FIGURE 12-27 Tolerancing a pattern of features to a datum feature of size: Prob. 2.

2. In Fig. 12-27, tolerance the center hole to the outside edges with a tolerance of .060. Refine the orientation of the ½-inch hole to the back of the part within .005. Locate the four-hole pattern to the center hole. Clock the pattern to a surface. The four-hole pattern mates with a part having four pins with a virtual condition of .250 in diameter. Give each feature all of the tolerance possible.

   - At what size does the center hole apply for the purposes of positioning the four-hole pattern? ________________
   - If the center hole is produced at a diameter of .535, how much shift of the four-hole pattern is possible? ________________

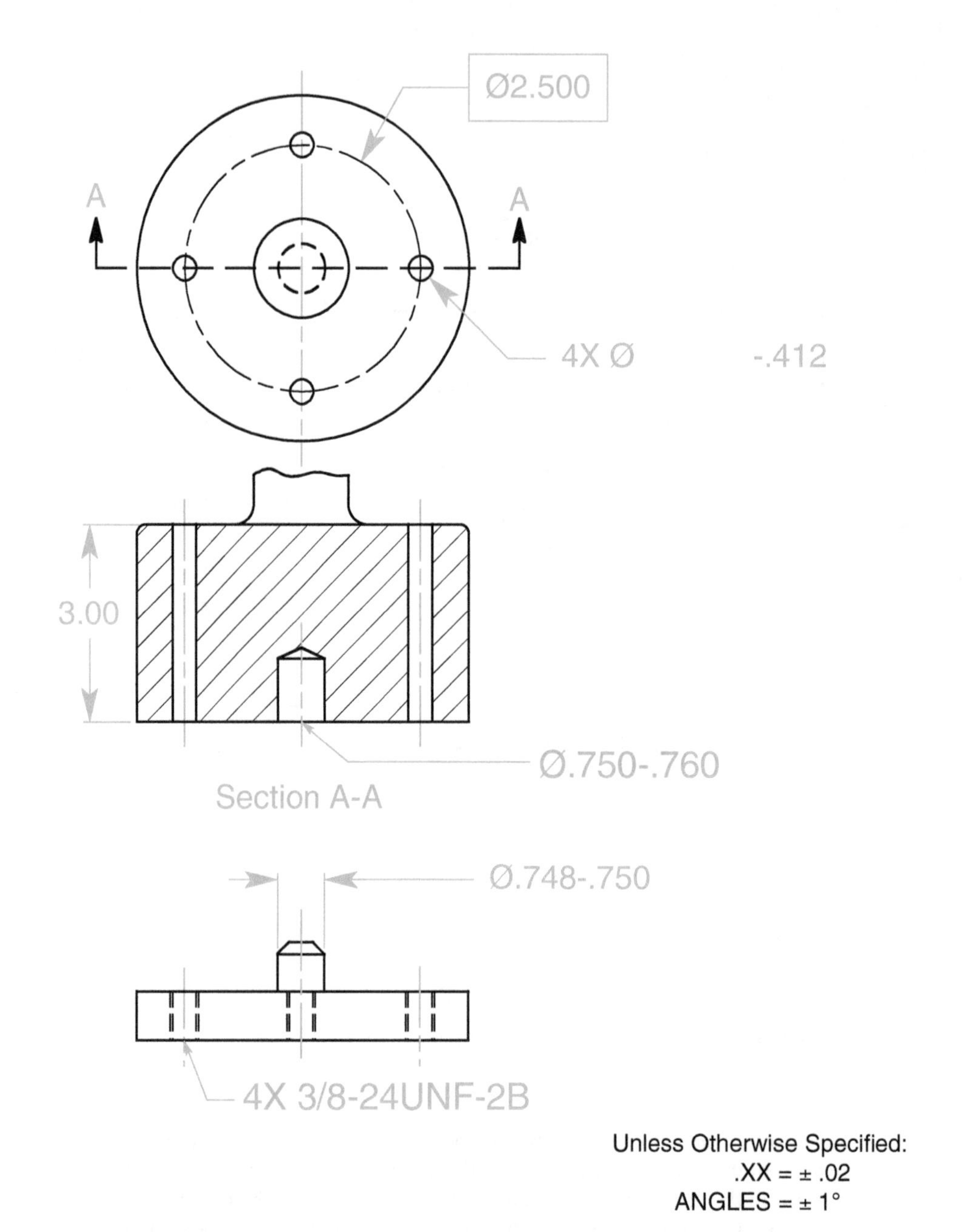

FIGURE 12-28 Tolerance two parts for assembly: Prob. 3.

3. Tolerance the two parts on the drawing in Fig. 12-28. Specify a flatness control of .002 on each of the primary datum features. Specify the appropriate orientation control to control the relationships between the primary and secondary datum features. Finally, complete the location tolerances for the hole patterns using a positional tolerance of .010 for the clearance holes. Use MMC and MMB wherever possible.

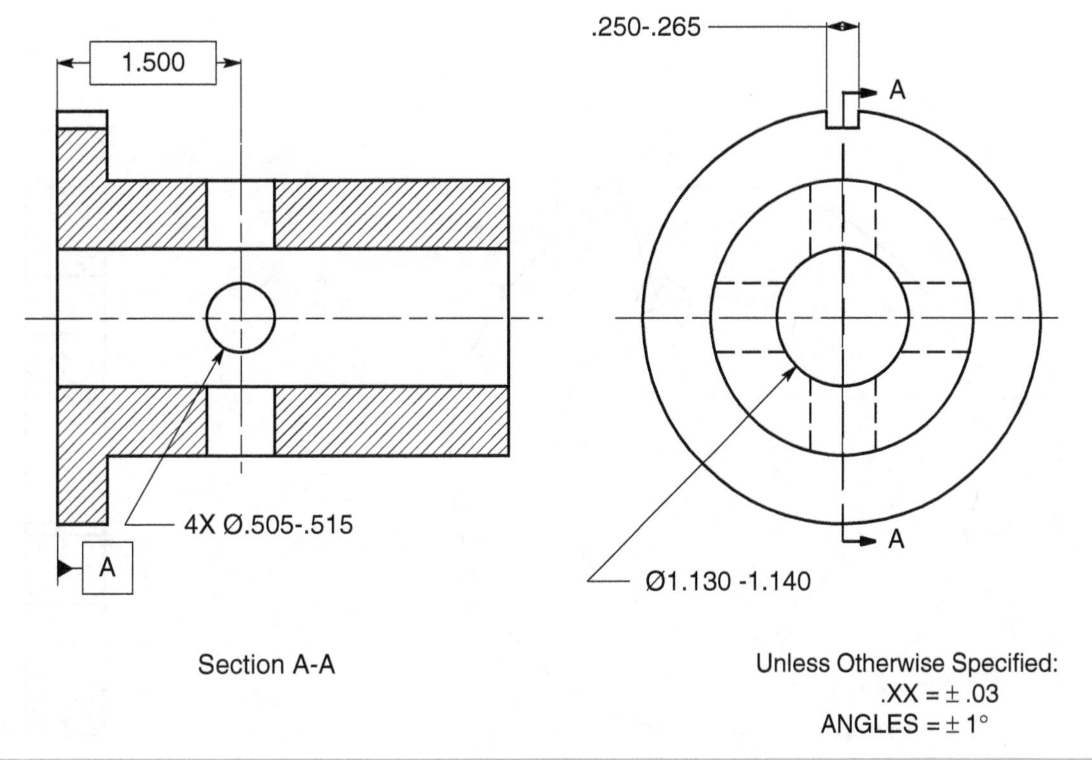

FIGURE 12-29 Tolerancing a four-hole pattern with a composite positional tolerance: Prob. 4.

4. In Fig. 12-29, orient the center bore to datum feature A with a geometric tolerance. The mating shaft is 1.125 in diameter. Orient the ¼-inch keyseat to datum feature A and locate it to the center bore. Position the four-hole pattern within a tolerance of .020 parallel to datum feature A, centered on the 1.130 bore, and clocked to the keyseat. Refine the location of the four holes in the four-hole pattern to each other and refine the orientation of the four-hole pattern parallel to datum feature A, perpendicular to the center bore, and parallel and perpendicular to the center plane of the keyseat within .005. Use MMC and MMB wherever possible.

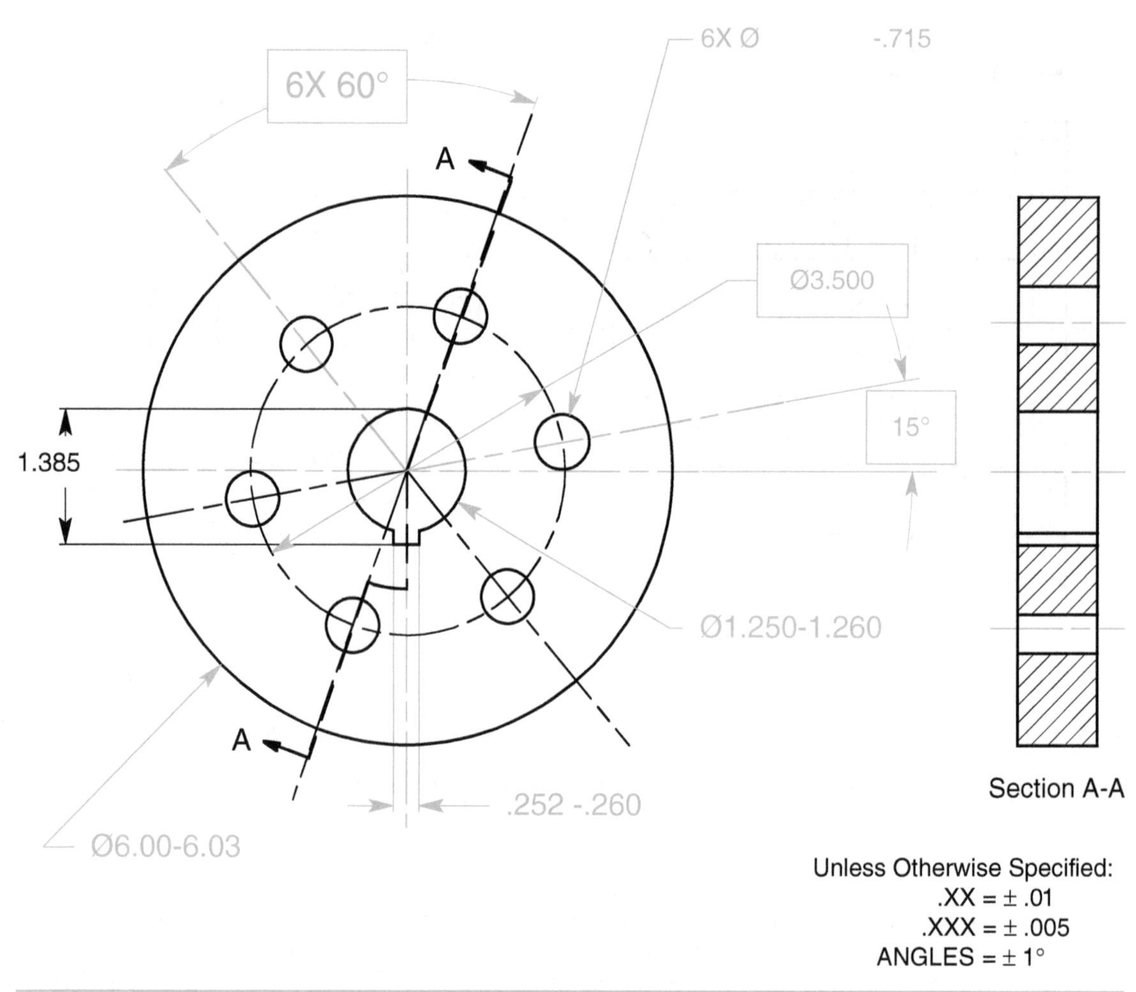

FIGURE 12-30 Tolerancing a pattern of features to a datum feature of size and a keyway: Prob. 5.

5. In Fig. 12-30, the inside diameter and the back are mating features. Select the primary datum feature. (Consider a form control.) The virtual condition of the mating shaft is 1.250 in diameter. Control the relationship between the primary and secondary datum features. Locate the keyway for a ¼-inch key. Locate the six-hole pattern for 5/8-inch (Ø.625) cap screws with a positional tolerance of .014 as fixed fasteners with the mating part. Use MMC and MMB wherever possible.

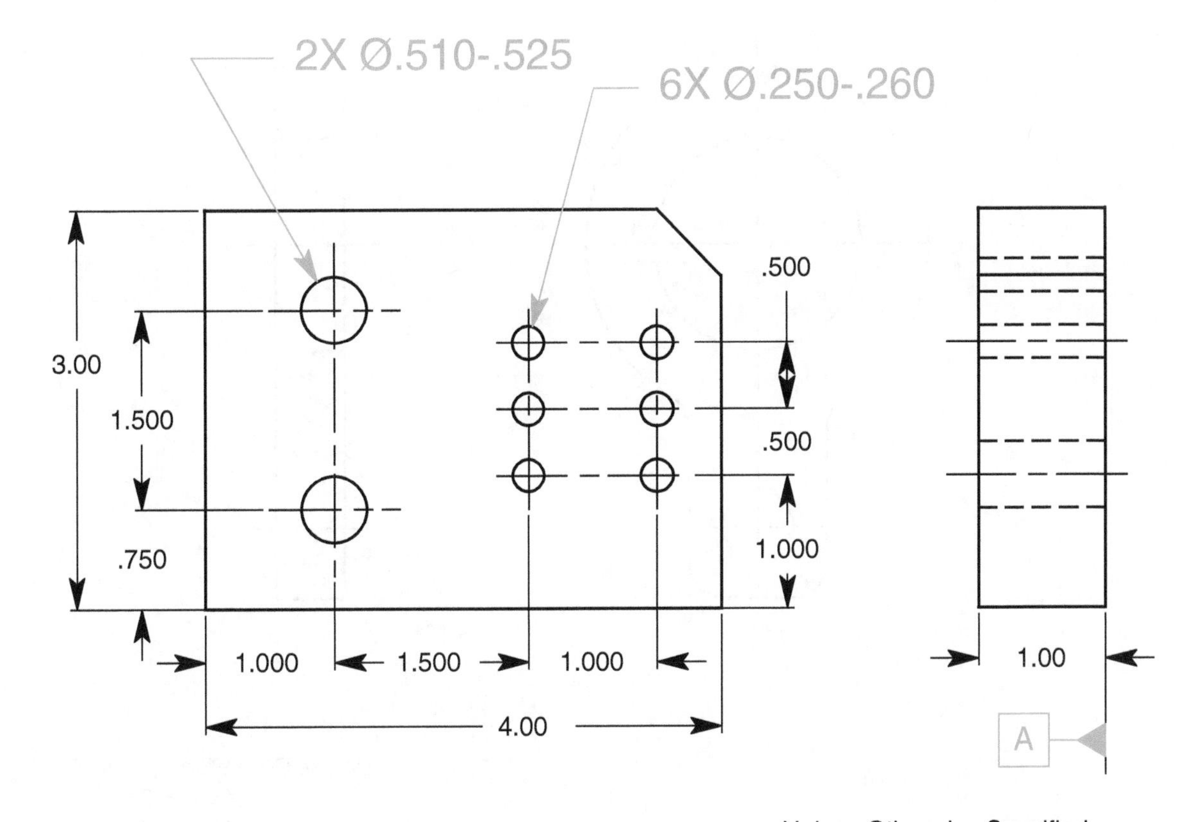

FIGURE 12-31 Tolerance a pattern of features to a second pattern of features: Prob. 6.

6. In Fig. 12-31, the location of the hole patterns to the outside edges is not critical; a tolerance of .060 at MMC is adequate. The location between the two ½-inch holes and their orientation to datum feature A must be within .010 at MMC. Control the six-hole pattern to the two-hole pattern within .000 at MMC. The mating part has virtual condition pins of .500 and .250 in diameter. Complete the tolerance.

   - At what size does the two-hole pattern apply for the purposes of positioning the six-hole pattern? ______________________
   - If the two large holes are produced at a diameter of .520, how much shift of the six-hole pattern is possible? ______________________

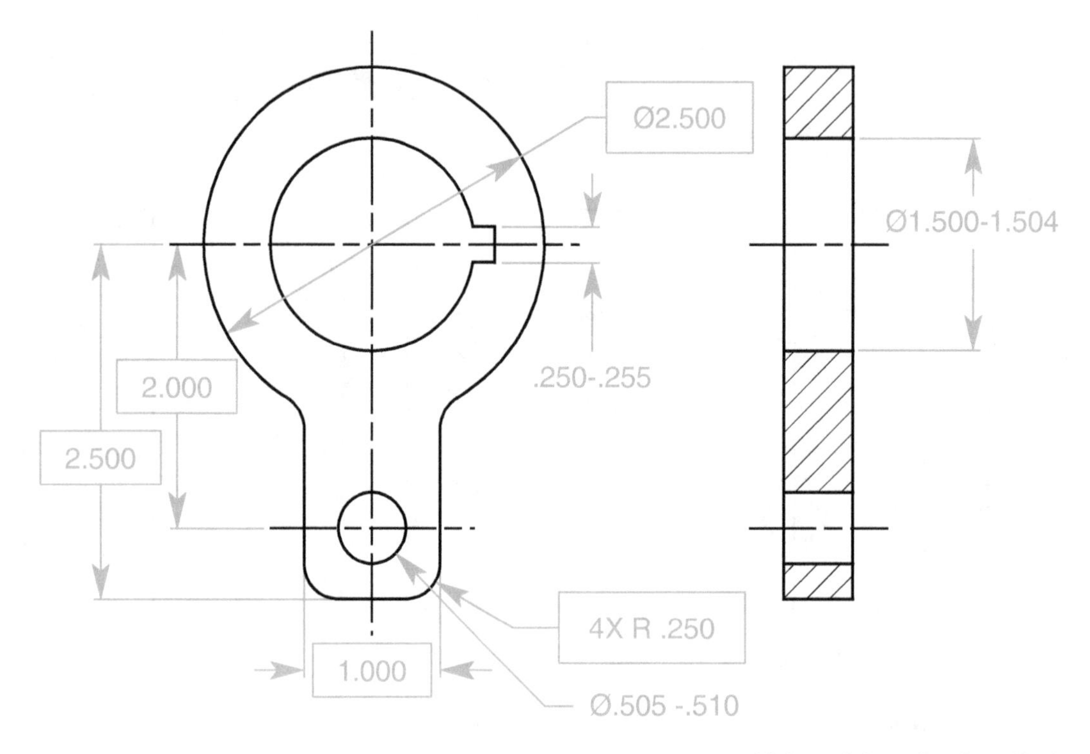

FIGURE 12-32 Tolerance a feature of size to a second feature of size: Prob. 7.

7. In Fig. 12-32, the inside diameter and the back are mating features. Select the primary datum feature. (Consider a form control.) The virtual condition of the mating shaft is 1.500 in diameter. Locate the keyway for a ¼-inch key. Locate the ½-inch-diameter hole for a mating pin .498 to .500 in diameter. Control the outside edge of the part with a tolerance of .030.

CHAPTER 13

Graphic Analysis

Graphic analysis, sometimes referred to as paper gaging, is a technique that effectively translates coordinate measurements into graphic geometry that can be easily analyzed. It provides the benefit of functional gaging without the time and expense required to design and manufacture a traditional functional gage.

Chapter Objectives

After completing this chapter, the learner will be able to:

- *Identify* the advantages of graphic analysis
- *Explain* the accuracy of graphic analysis
- *Perform* an inspection analysis of a composite geometric tolerance
- *Perform* an inspection analysis of a pattern of features controlled to a datum feature of size specified at maximum material boundary (MMB)

Advantages of Graphic Analysis

The graphic analysis approach to gaging has many advantages compared to gaging with traditional functional gages. A partial list of the advantages of using graphic analysis would include the following:

1. **Provides functional acceptance:** Most hardware is designed to provide interchangeability of parts. As machined features depart from their maximum material condition (MMC) size, the location tolerance of the features can be increased while maintaining functional interchangeability. The graphic analysis technique provides an evaluation of these added functional tolerances in the acceptance process. It also shows how an unacceptable part can be reworked.
2. **Reduces cost and time:** The high cost and long lead time required for the design and manufacture of a functional gage can be eliminated in favor of graphic analysis. Inspectors can conduct an immediate, inexpensive, functional inspection at their workstations.
3. **Eliminates gage tolerance and wear allowance:** Functional gage design allows 10% of the tolerance assigned to the part to be used for gage tolerance. Often, an additional wear allowance of up to 5% will be designed into a functional gage. This could allow up to 15% of the parts tolerance to be assigned to the functional gage. The graphic analysis technique dos not require any portion of the product tolerance to be assigned to the verification process. Graphic analysis does not require a wear allowance since there is no wear.

4. **Allows function verification of MMC/maximum material boundary (MMB), regardless of feature size (RFS)/regardless of material boundary (RMB), and least material condition (LMC)/least material boundary (LMB):** Functional gages are designed primarily to verify parts toleranced with the MMC and MMB modifier. In most cases, it is not practical to design functional gages to verify parts specified at RFS and RMB. Functional gages cannot be used to inspect parts toleranced with the LMC and LMB modifier. With the graphic analysis technique, features specified with any one of the material condition modifiers can be verified with equal ease.
5. **Allows verification of any shape tolerance zone:** Virtually any shape tolerance zone, (round, square, rectangular, etc.) can easily be constructed with graphic analysis methods. On the other hand, hardened steel functional gaging elements of nonconventional configurations are very difficult and expensive to produce.
6. **Provides a visual record for the material review board:** Material review board meetings are postmortems that examine rejected parts. Decisions on the disposition of nonconforming parts are usually influenced by what the most senior engineer thinks or the notions of the most vocal member present rather than the engineering information available. On the other hand, graphic analysis can provide a visual record of the part data and the extent that it is out of compliance.
7. **Minimizes storage required:** Inventory and storage of functional gages can be a problem. Functional gages can corrode if they are not properly stored. Graphic analysis graphs and overlays can be easily stored in drawing files or drawers.

The Accuracy of Graphic Analysis

The overall accuracy of graphic analysis is affected by such factors as the accuracy of the graph and overlay gage, the accuracy of the inspection data, the completeness of the inspection process, and the ability of the drawing to provide common drawing interpretations.

An error equal to the difference in the coefficient of thermal expansion of the materials used to generate the data graph and the tolerance zone overlay gage may be encountered if the same materials are not used for both sheets. Paper will also expand with the increase of humidity, and its use should be governed by that understanding. Mylar is a relatively stable material. When mylar is used for both the data graph and the overlay gage, any expansion or contraction error will be nullified.

Layout of the data graph and overlay gage will allow a small percent of error in the positioning of lines. This error is minimized by the magnification scale selected for the data graph.

Analysis of a Composite Geometric Tolerance

A pattern of features controlled with composite tolerancing can be inspected with a set of functional gages. Each segment of the feature control frame is represented by a gage. The gage used to inspect the location of the pattern of holes in Fig. 13-1, the pattern-locating control, the upper segment of the feature control frame, consists of three mutually perpendicular intersecting datum planes and four virtual condition pins .242 in diameter. The gage used to inspect the feature-to-feature location, the feature-relating control, the lower segment of the feature control frame, consists of only datum plane A and four virtual condition pins .250 in diameter. These two gages are required to inspect this pattern. If gages are not available, graphic analysis can be used. The procedure for inspecting composite tolerancing with graphic analysis is presented below.

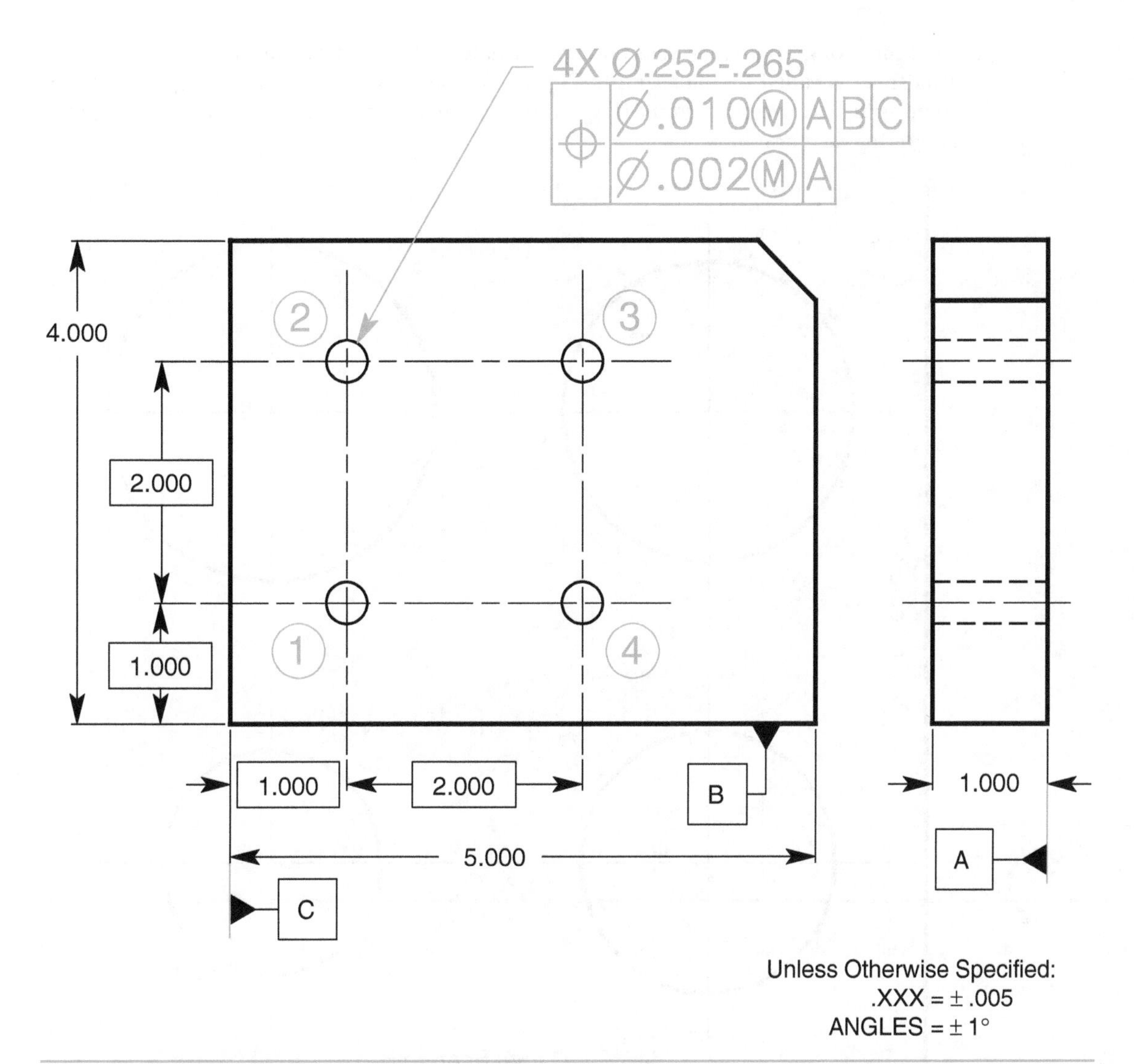

FIGURE 13-1 A pattern of features controlled with a composite tolerance.

The following is a sequence of steps required for generating the data graph for the graphic analysis of a composite tolerance:

1. Collect the inspection data shown in Table 13-1.

Feature Number	Feature Location from Datum Feature C, X-Axis	Feature Location from Datum Feature B, Y-Axis	Feature Size	Departure from MMC (Bonus)	Datum-to-Pattern Tolerance Zone Size	Feature-to-Feature Tolerance Zone Size
1	.997	1.003	Ø.256	.004	Ø.014	Ø.006
2	1.004	3.004	Ø.258	.006	Ø.016	Ø.008
3	3.006	2.998	Ø.260	.008	Ø.018	Ø.010
4	3.002	.998	Ø.254	.002	Ø.012	Ø.004

TABLE 13-1 Inspection Data Derived from a Part Made from Specifications on the Drawing in Fig. 13-1

2. On a piece of graph paper, select an appropriate scale and draw datum planes B and C. This sheet is called the data graph (Fig. 13-2). The drawing, the upper segment of the composite feature control frame (Fig. 13-3), and the inspection data dictate the configuration of the data graph.

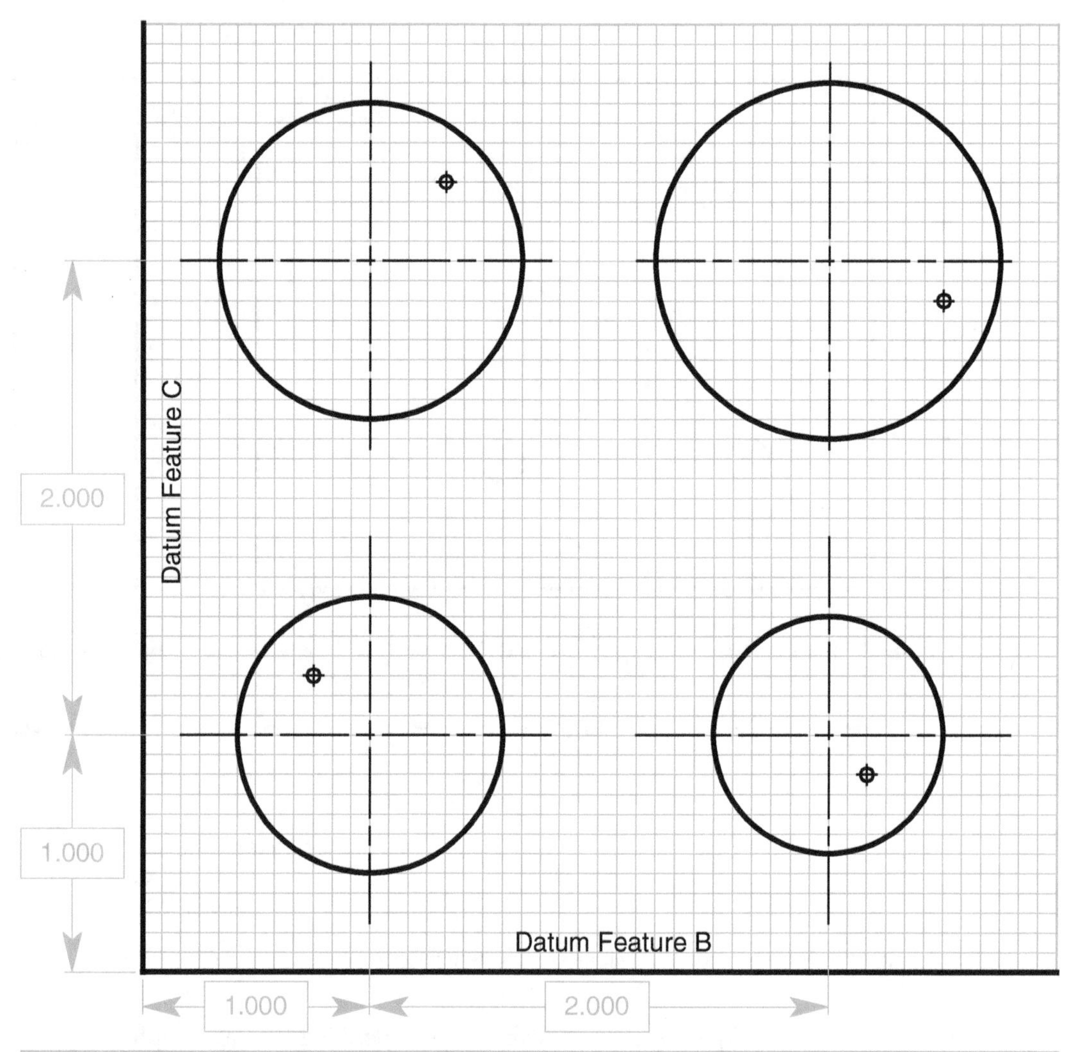

FIGURE 13-2 The data graph with tolerance zones and feature axes from the data in Table 13-1.

FIGURE 13-3 The upper segment of the composite feature control frame in Fig. 13-1.

3. From the drawing, determine the true position of each feature and draw the centerlines on the data graph.
4. Since tolerances are in the magnitude of thousandths of an inch, a second scale, called the deviation scale, is established. Typically, one square on the graph paper equals .001 of an inch on the deviation scale.

5. Draw the appropriate diameter tolerance zone around each true position using the deviation scale. For the drawing in Fig. 13-1, each tolerance zone is a circle with a diameter of .010 plus its bonus tolerance. The datum-to-pattern tolerance zone diameters are listed in Table 13-1.
6. Draw the actual location of each feature axis on the data graph. If the location of any one of the feature axes falls outside its respective cylindrical tolerance zone, the datum-to-pattern relationship is out of tolerance, and the part is rejected. If all of the axes fall inside their respective tolerance zones, the datum-to-pattern relationship is in tolerance, but the pattern must be further evaluated to satisfy the feature-to-feature relationships.

The following is a sequence of steps required for generating the overlay gage for the graphic analysis evaluation of a composite tolerance:

1. Place a piece of tracing paper over the data graph. This sheet is called the overlay gage. Trace the true position axes on the tracing paper. The drawing, the lower segment of the feature control frame (Fig. 13-4), and the inspection data dictate the configuration of the overlay gage shown in Fig. 13-5.

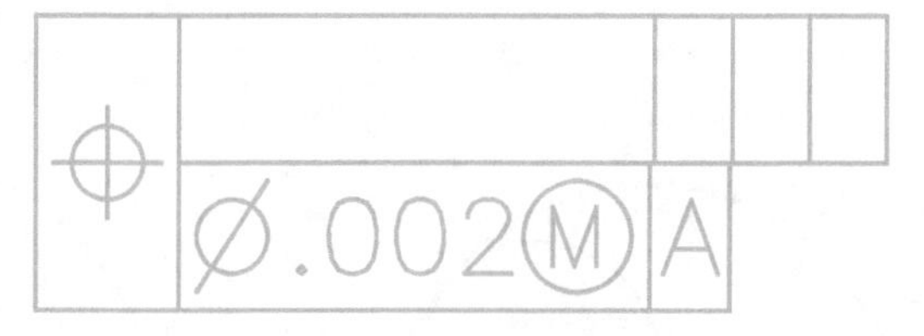

FIGURE 13-4 The lower segment of the composite feature control frame in Fig. 13-1.

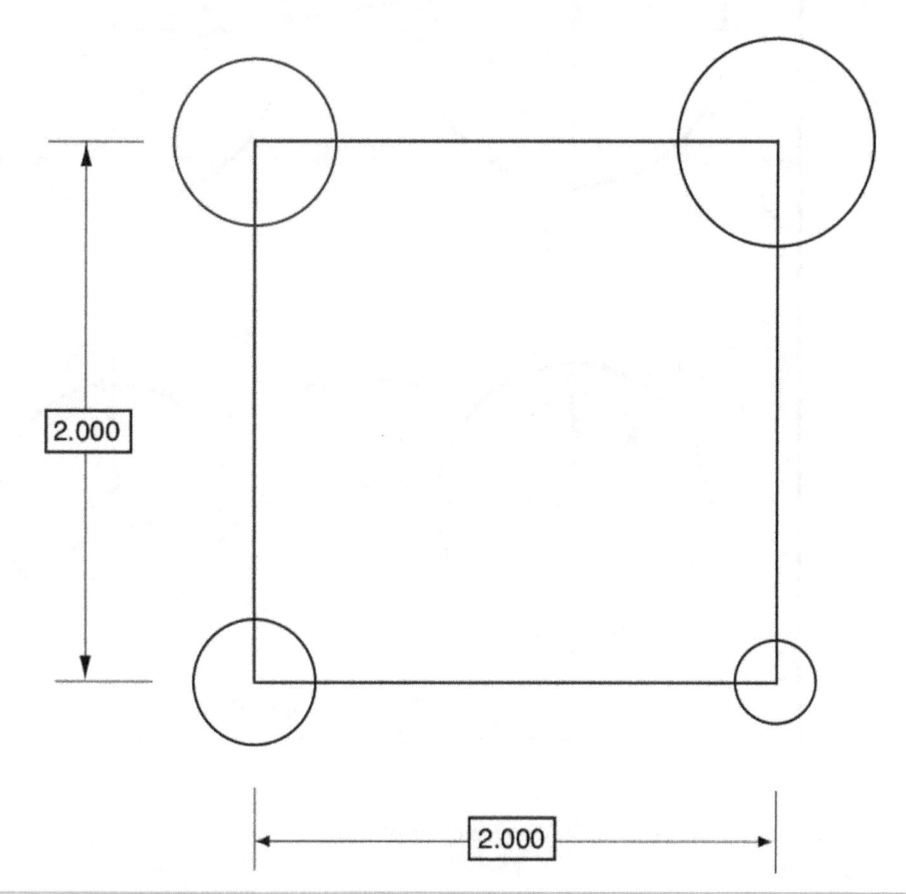

FIGURE 13-5 The overlay gage.

2. Draw the appropriate feature-to-feature positional tolerance zones around each true position axis on the tracing paper. Each tolerance zone is a circle with a diameter of .002 plus its bonus tolerance. The feature-to-feature tolerance zone diameters are listed in Table 13-1.
3. If the tracing paper can be adjusted to include all actual feature axes within the tolerance zones on the tracing paper, the feature-to-feature relationships are in tolerance. The pattern is acceptable if each feature axis simultaneously falls inside both of its respective tolerance zones.

When the overlay gage is placed over the data graph in Fig. 13-6, the axes of holes 1 through 3 can be placed inside their respective tolerance zones. The axis of the fourth hole, however, will not fit inside the fourth tolerance zone. Therefore, the pattern is not acceptable. However, it is easy to see on the data graph that this hole can be reworked. Simply enlarging the fourth hole by about .004 will make the pattern of holes acceptable.

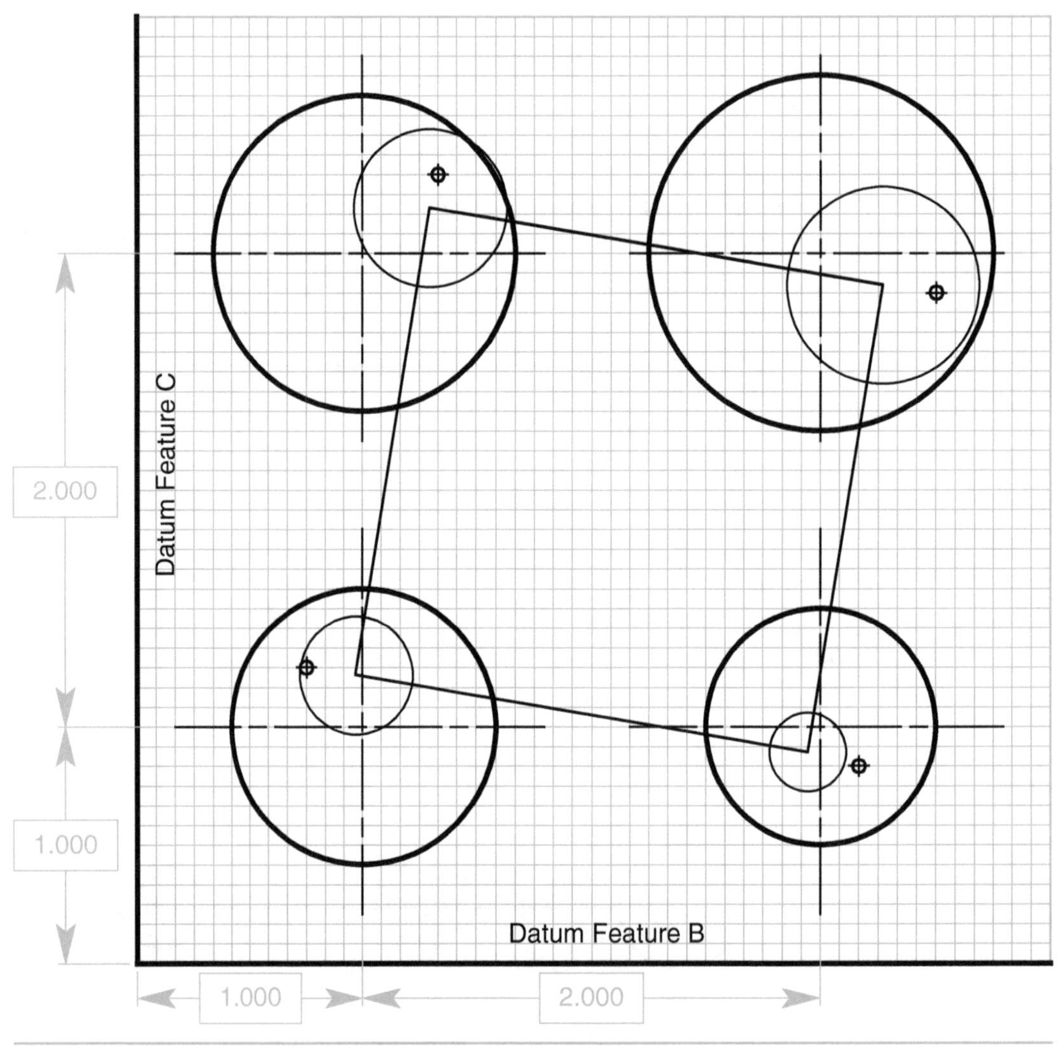

FIGURE 13-6 The overlay gage is placed on top of the data graph.

Analysis of a Pattern of Features Controlled to a Datum Feature of Size

A pattern of features controlled to a datum feature of size specified at MMC is a very complicated geometry that can be easily inspected with graphic analysis.

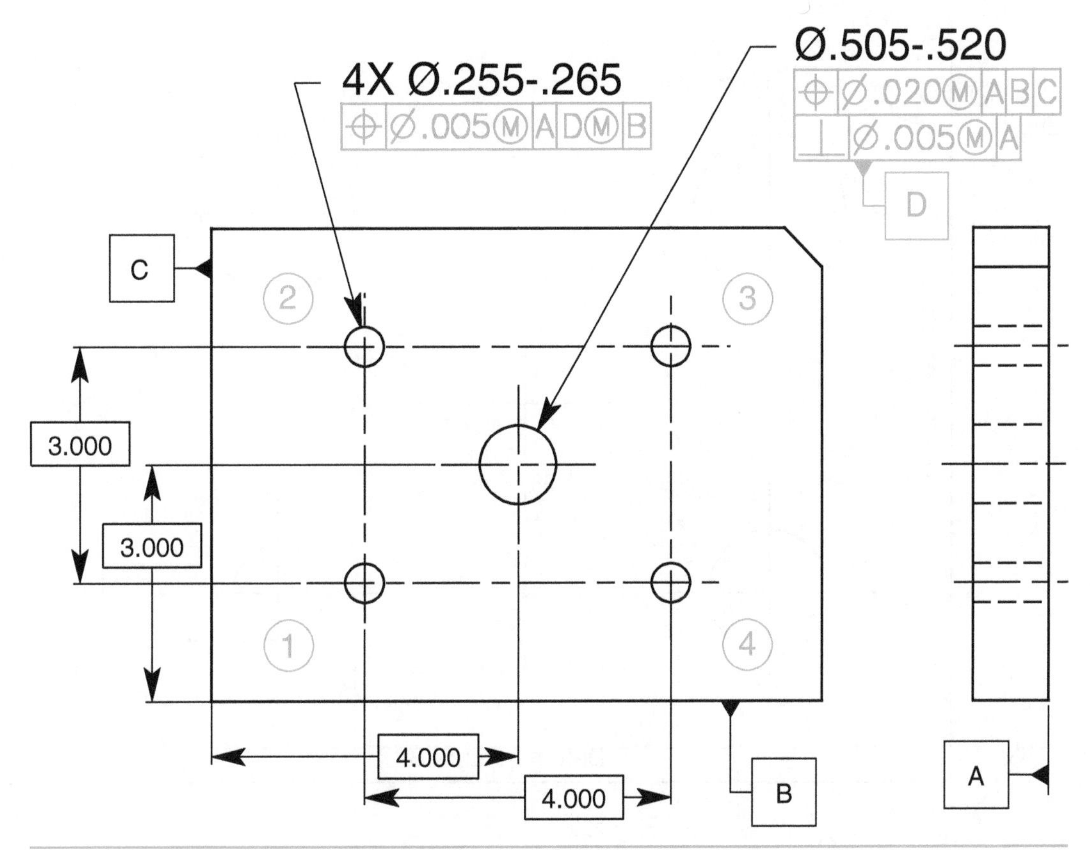

FIGURE 13-7 The drawing of a pattern of features controlled to a datum feature of size.

The following is a sequence of steps required for generating the data graph for the graphic analysis evaluation of a pattern of features controlled to a datum feature of size:

1. Collect the inspection data shown in Table 13-2.

Feature Number	Feature Location from Datum Feature D, X-Axis	Feature Location from Datum Feature D, Y-Axis	Actual Feature Size	Departure from MMC (Bonus)	Total Geometric Tolerance
1	−1.997	−1.498	Ø.258	.003	Ø.008
2	−1.998	1.503	Ø.260	.005	Ø.010
3	2.005	1.504	Ø.260	.005	Ø.010
4	2.006	−1.503	Ø.256	.001	Ø.006
Datum Feature			Ø.510	Shift Tolerance = .510 − .500 = .010	

TABLE 13-2 Inspection Data Derived from a Part Made from Specifications on the Drawing in Fig. 13-7

2. On a piece of graph paper, select an appropriate scale and draw datum planes B and C. This sheet is called the data graph (Fig. 13-8). The drawing, the feature control frame controlling the hole pattern (Fig. 13-9), and the inspection data dictate the configuration of the data graph.

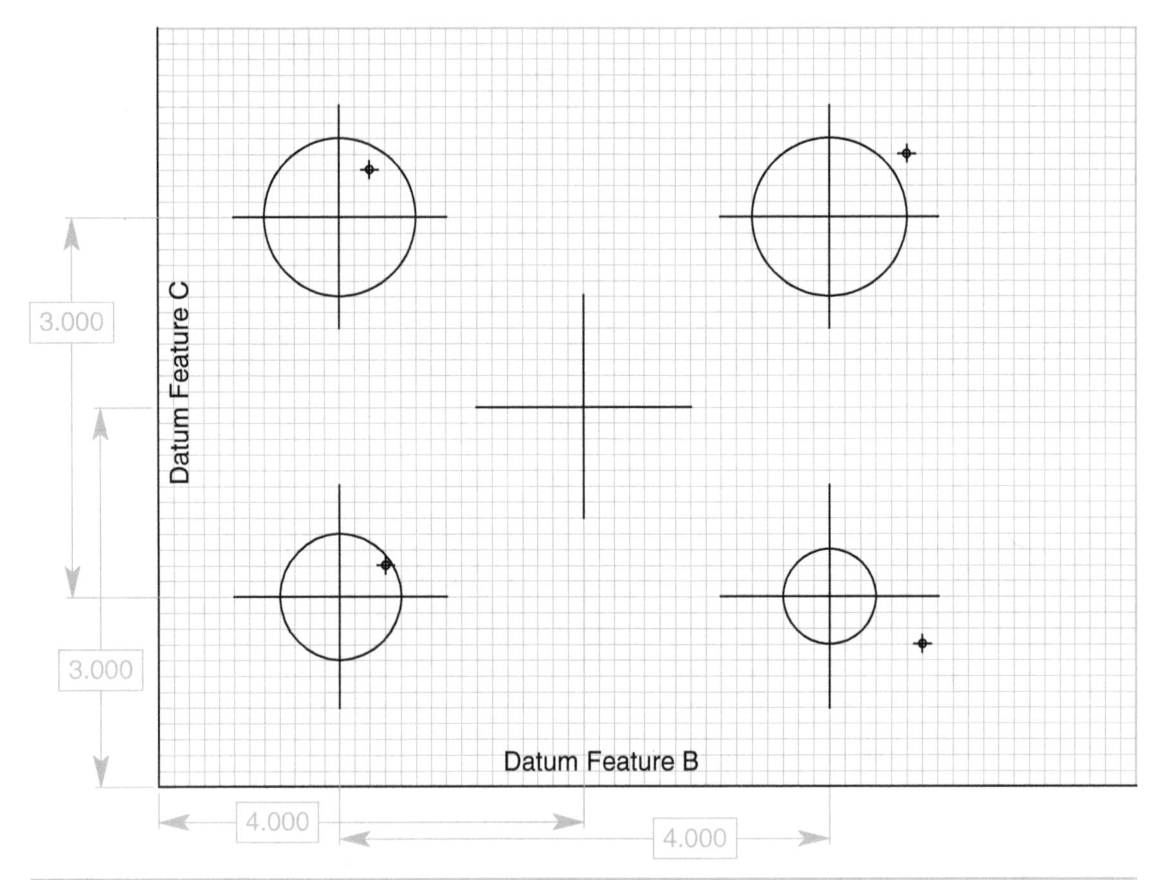

FIGURE 13-8 The data graph with feature axes and cylindrical tolerance zones representing the data in Table 13-2.

FIGURE 13-9 The feature control frame controlling the four-hole pattern in Fig. 13-7.

3. From the drawing, determine the true position of the datum feature of size, datum feature D, and the true position of each hole in the pattern. Draw their centerlines on the data graph.

4. Establish a deviation scale. The deviation scale is a second scale used to determine the size of the tolerance zones and the actual location of the feature axes. Typically, one square on the graph paper equals .001 of an inch on the deviation scale.

5. Draw the appropriate diameter tolerance zone around the true position of each hole using the deviation scale. For the drawing in Fig. 13-7, each tolerance zone is a circle with a diameter of .005 plus its bonus tolerance. The diameters of the tolerance zones are listed in Table 13-2.
6. If each feature axis falls inside its respective tolerance zone, the part is in tolerance. If one or more feature axes fall outside its respective tolerance zone, the part still may be acceptable if there is enough shift tolerance to shift all axes into their respective tolerance zones.

If further analysis is required to evaluate the part, the following sequence of steps is required to generate the overlay gage:

1. Place a piece of tracing paper over the data graph. This sheet is called the overlay gage shown in Fig. 13-10.
2. Trace the actual location of each feature axis onto the overlay gage.
3. Trace datum feature B onto the overlay gage.
4. Trace the true position axis of datum feature D onto the overlay gage.

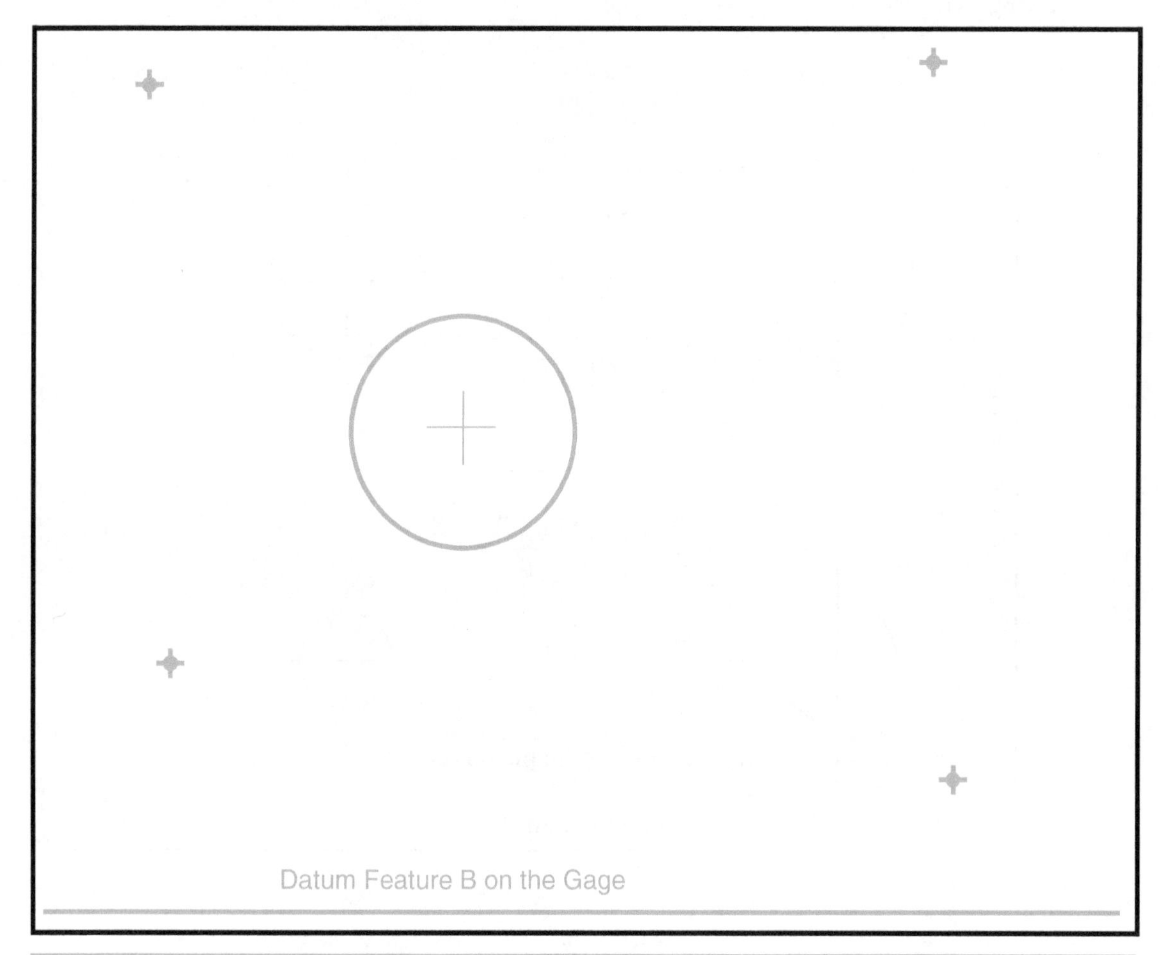

FIGURE 13-10 The overlay gage includes the actual axis of each feature in the pattern, the cylindrical shift tolerance zone, and the clocking datum feature, datum plane B.

5. Calculate the shift tolerance allowed; draw the appropriate cylindrical tolerance zone around datum axis D. The shift tolerance equals the difference between the actual datum feature size and the size at which the datum feature applies. Datum feature D applies at its virtual condition with respect to datum feature A shown in Fig. 13-7. Consequently, datum feature D at MMC minus the perpendicularity tolerance with respect to datum feature A equals the virtual condition (.505 – .005 = .500). According to the inspection data, the hole, datum feature D, is produced at a diameter of .510. The shift tolerance equals .510 minus .500, or a diameter of .010.
6. If the tracing paper can be adjusted to include all feature axes on the overlay gage within their respective tolerance zones on the data graph and datum axis D contained within its shift tolerance zone while orienting datum feature B on the overlay gage parallel to datum feature B on the data graph, the pattern of features is in tolerance. The graphic analysis in Fig. 13-11 indicates that the four-hole pattern of features is acceptable.

Graphic analysis is a powerful graphic tool for analyzing part configuration. This graphic tool is easy to use, accurate, and repeatable. It should be in every inspector's toolbox. Graphic analysis is also a powerful analytical tool that engineers can use to better understand how tolerances on drawings will behave.

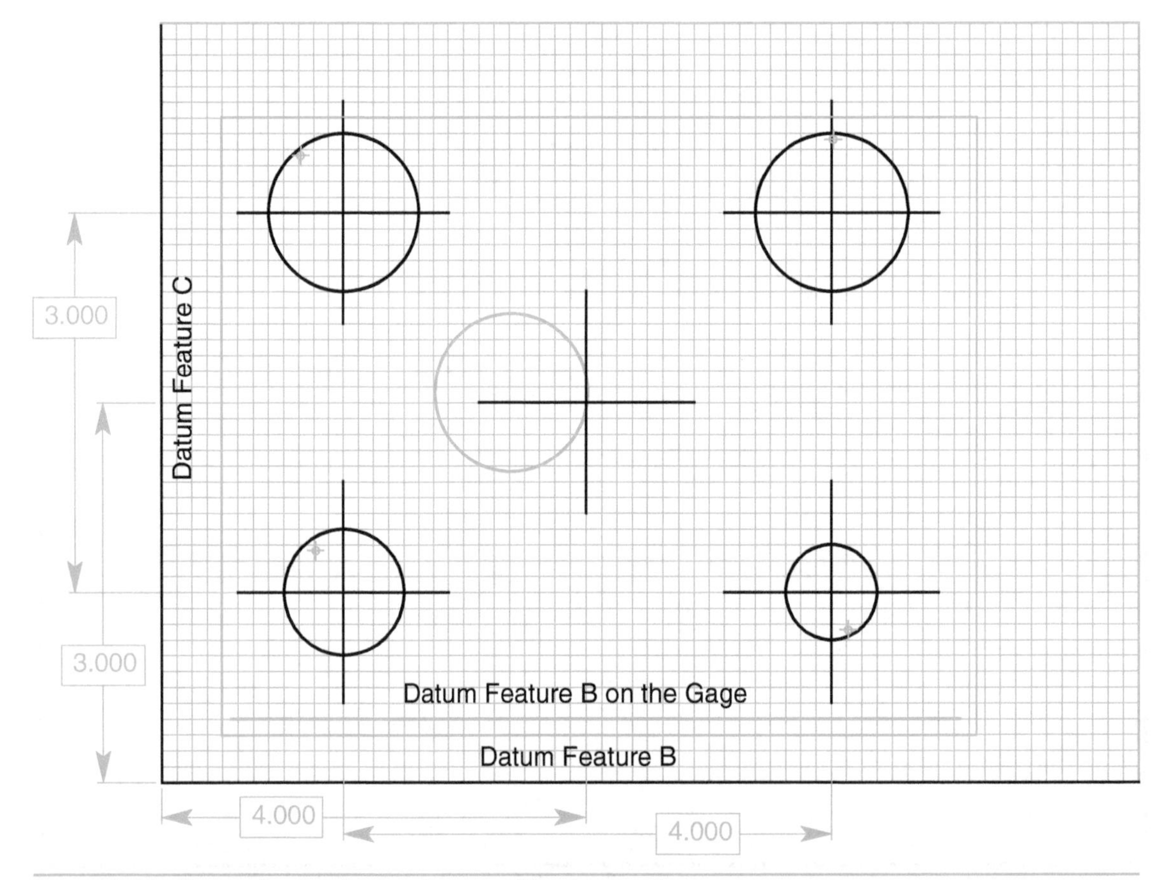

FIGURE 13-11 The overlay gage is placed on top of the data graph.

Summary

- The advantages of graphic analysis:
  1. Provides functional acceptance
  2. Reduces time and cost
  3. Eliminates gage tolerance and wear allowance
  4. Allows functional verification of RFS and RMB, LMC and LMB, as well as MMC and MMB
  5. Allows verification of a tolerance zone of any shape
  6. Provides visual record for the material review board
  7. Minimizes storage required for gages
- The accuracy of graphic analysis:
  Accuracy is affected by such factors as the accuracy of the data graph and overlay gage, the accuracy of the inspection data, the completeness of the inspection process, and the ability of the drawing to provide common drawing interpretations.
- Sequence of steps for the analysis of a composite geometric tolerance:
  1. Draw the datum features, the true positions, the datum to pattern tolerance zones, and the actual feature locations on the data graph.
  2. On a piece of tracing paper placed over the data graph, trace the true positions and construct the feature-to-feature tolerance zones. This sheet is called the overlay gage.
  3. Adjust the overlay gage to fit over the actual feature locations. If each actual feature location falls inside both of its respective tolerance zones, the pattern of features is in tolerance.
- Sequence of steps for the analysis of a pattern of features controlled to a datum feature of size:
  1. Draw the datum features, the true positions, the tolerance zones, and the actual feature locations on the data graph. If the actual feature locations fall inside the tolerance zones, the part is good, and no further analysis is required. If not, continue to step 2 to utilize the available shift tolerance.
  2. On a piece of tracing paper placed over the data graph, trace the actual feature locations, the clocking datum feature, and the true position of the datum feature of size. Then draw the shift tolerance zone about the true position of the datum feature of size. This sheet is called the overlay gage.
  3. Adjust the overlay gage so that the actual feature locations fit inside their respective tolerance zones on the data graph while at the same time keeping the shift tolerance zone over the axis on the data graph and the clocking datum features parallel. If each actual feature location falls inside its respective tolerance zone, the pattern of features is in tolerance.

Chapter Review

1. List the advantages of graphic analysis.

____

____

____

____

____

____

____

2. List the factors that affect the accuracy of graphic analysis.

____

____

____

____

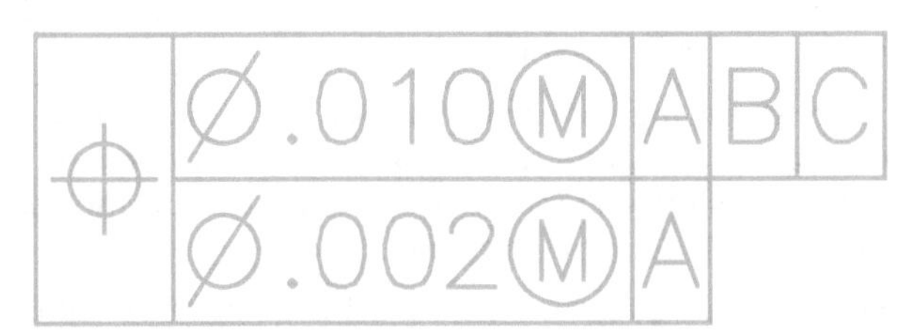

FIGURE 13-12 A composite feature control frame.

Refer to Fig. 13-12 to answer Questions 3 through 7.

3. A piece of graph paper with datum features, true positions, tolerance zones, and actual feature locations drawn on it is called a _______________________.

4. The upper segment of the composite feature control frame, the drawing, and the inspection data dictate the configuration of the _______________________.

5. A piece of tracing paper with datum features, true positions, tolerance zones, and actual feature locations traced or drawn is called a ______________________________.

6. The lower segment of the feature control frame, the drawing, and the inspection data dictate the configuration of the ______________________________.

7. If the tracing paper can be adjusted to include all feature axes within the ______________ ______________ on the tracing paper, the feature-to-feature relationships are in tolerance.

FIGURE 13-13 A datum feature of size specified at MMB.

Refer to Fig. 13-13 to answer Questions 8 through 11.

8. To inspect a pattern of features controlled to a datum feature of size, the feature control frame, the drawing, and the inspection data dictate the configuration of the ______________ ______________________________.

9. Draw the actual location of each feature on the data graph. If each feature axis falls inside its respective tolerance zone, the part is ______________________________.

10. If any of the feature axes fall outside its respective tolerance zone, ______________ ______________________________.

11. If the tracing paper can be adjusted to include all feature axes on the overlay gage within their respective tolerance zones on the data graph and datum axis D contained within its shift tolerance zone while orienting datum feature B on the overlay gage parallel to datum feature B on the data graph, the pattern of features is ______________ ______________________________.

Problems

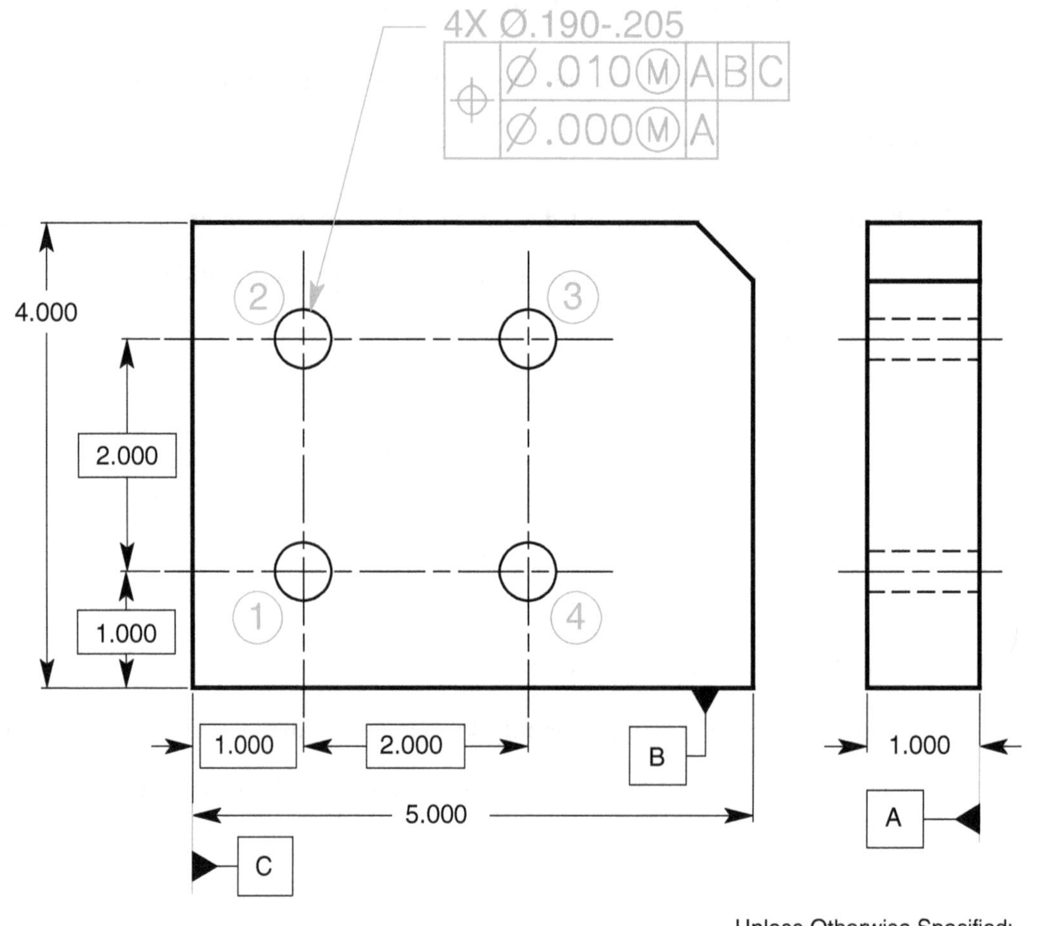

FIGURE 13-14 A pattern of features controlled with a composite tolerance: Prob. 1.

Feature Number	Location from Datum Feature D, X-Axis	Location from Datum Feature D, Y-Axis	Feature Size	Departure from MMC (Bonus)	Datum-to-Pattern Tolerance Zone Size	Feature-to-Feature Tolerance Zone Size
1	1.002	1.003	Ø.200			
2	1.005	3.006	Ø.198			
3	3.005	3.002	Ø.198			
4	3.003	.998	Ø.196			

TABLE 13-3 Inspection Data for the Graphic Analysis of Prob. 1

1. A part was made from the drawing in Fig. 13-14, and the inspection data have been tabulated in Table 13-3. Perform a graphic analysis of the part. Is the pattern within tolerance?

   ____

   If it is not in tolerance, can it be reworked? If so, how? ______________________________

   ____

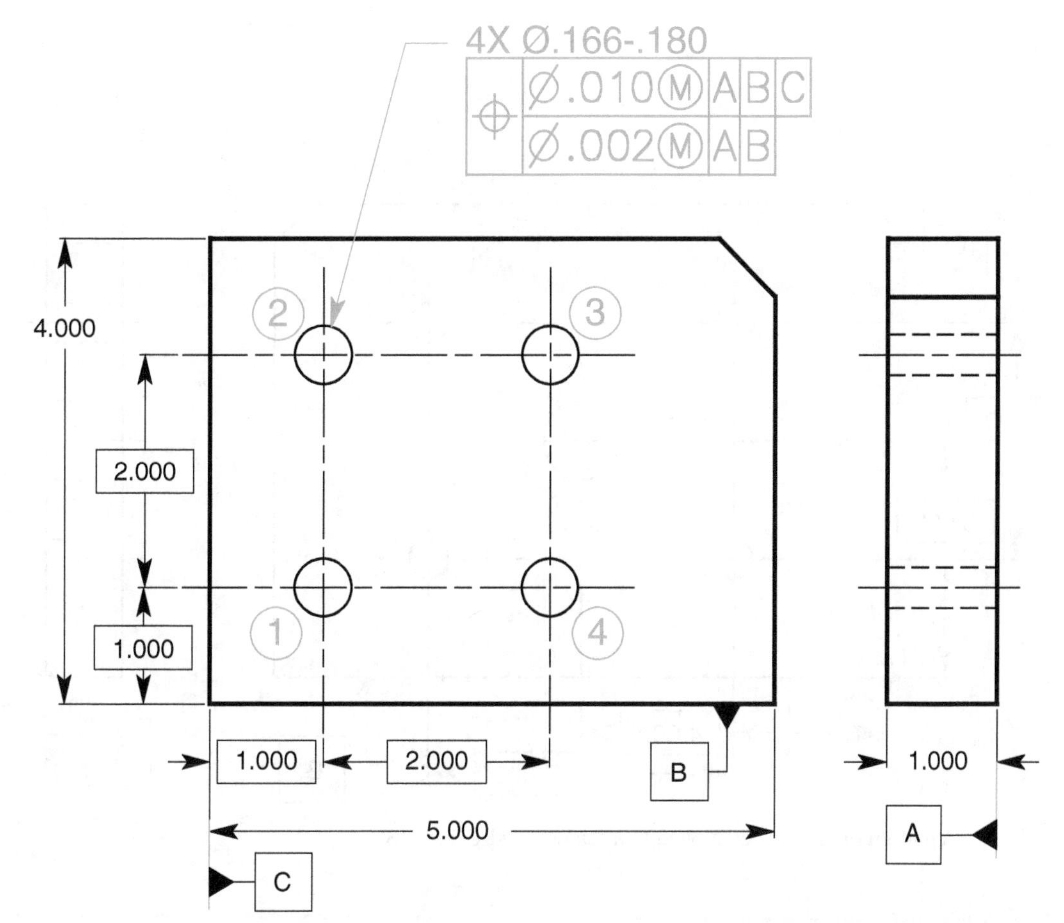

FIGURE 13-15 A pattern of features controlled with a composite tolerance: Prob. 2.

Feature Number	Location from Datum Feature D, X-Axis	Location from Datum Feature D, Y-Axis	Feature Size	Departure from MMC (Bonus)	Datum-to-Pattern Tolerance Zone Size	Feature-to-Feature Tolerance Zone Size
1	1.004	.998	Ø.174			
2	.995	3.004	Ø.174			
3	3.000	3.006	Ø.172			
4	3.006	1.002	Ø.176			

TABLE 13-4 Inspection Data for the Graphic Analysis of Prob. 2

2. A part was made from the drawing in Fig. 13-15, and the inspection data were tabulated in Table 13-4. Perform a graphic analysis of the part. Is the pattern within tolerance?

____

If it is not in tolerance, can it be reworked? If so, how? ______________________________

____

____

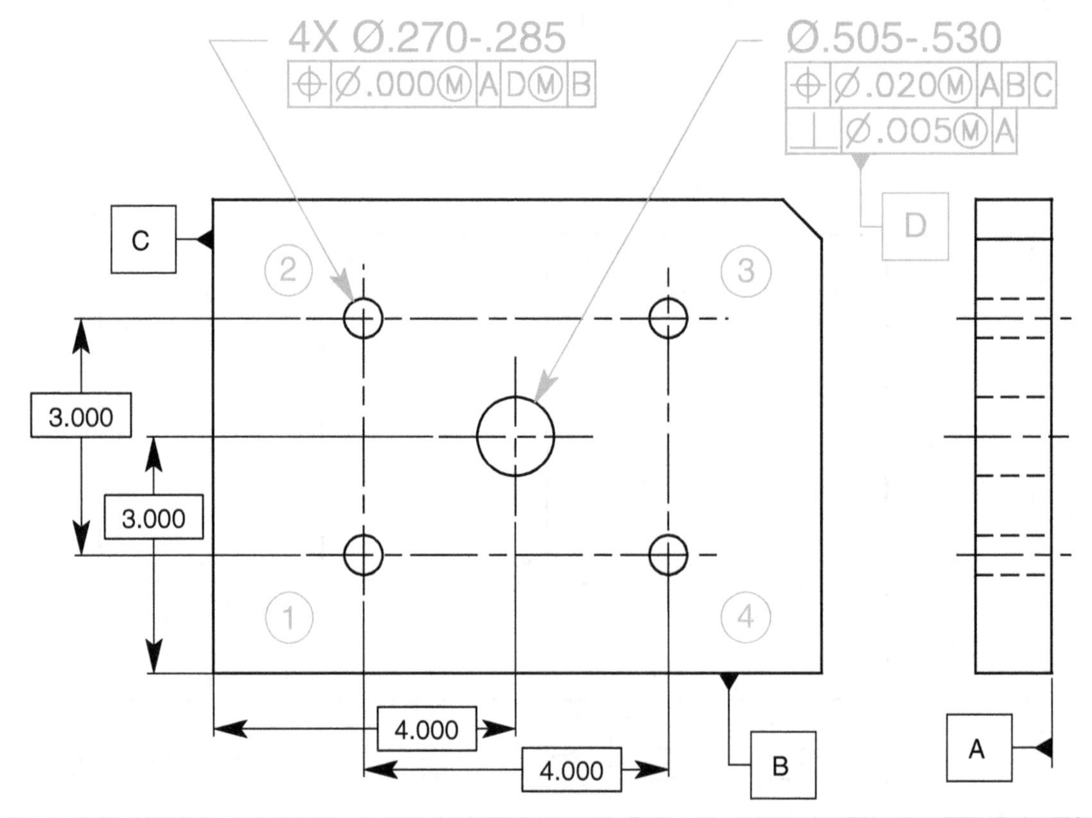

FIGURE 13-16 A pattern of features controlled to a feature of size: Prob. 3.

Feature Number	Location from Datum Feature D, X-Axis	Location from Datum Feature D, Y-Axis	Actual Feature Size	Departure from MMC (Bonus)	Total Geometric Tolerance
1	−1.992	−1.493	Ø.278		
2	−1.993	1.509	Ø.280		
3	2.010	1.504	Ø.280		
4	2.010	−1.490	Ø.282		
Datum			Ø.520	Shift Tolerance =	

TABLE 13-5 Inspection Data for the Graphic Analysis of Prob. 3

3. A part was made from the drawing in Fig. 13-16, and the inspection data were tabulated in Table 13-5. Perform a graphic analysis of the part. Is the pattern within tolerance?

____

If it is not in tolerance, can it be reworked? If so, how? ______________________________

____

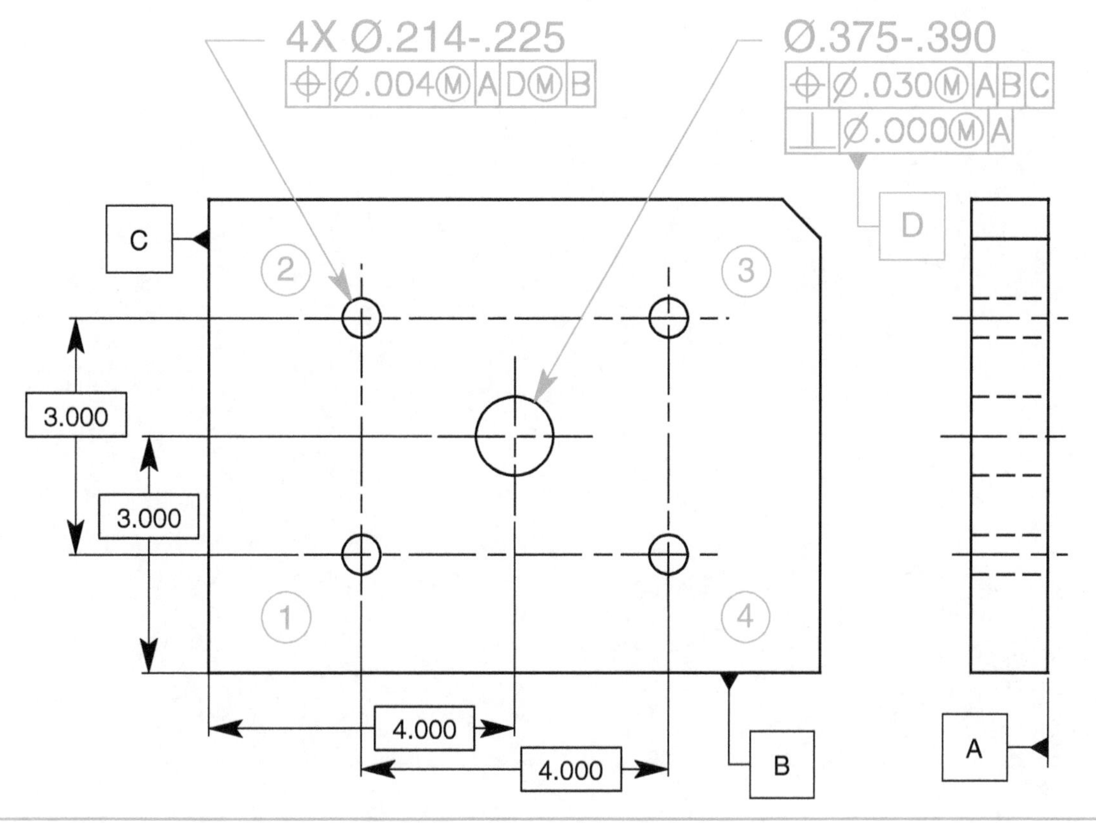

FIGURE 13-17 A pattern of features controlled to a feature of size: Prob. 4.

Feature Number	Location from Datum Feature D, X-Axis	Location from Datum Feature D, Y-Axis	Actual Feature Size	Departure from MMC (Bonus)	Total Geometric Tolerance
1	−1.995	−1.495	Ø.224		
2	−1.996	1.503	Ø.218		
3	2.005	1.497	Ø.220		
4	1.997	−1.506	Ø.222		
Datum			Ø.380	Shift Tolerance =	

TABLE 13-6 Inspection Data for the Graphic Analysis of Prob. 4

4. A part was made from the drawing in Fig. 13-17, and the inspection data were tabulated in Table 13-6. Perform a graphic analysis of the part. Is the pattern within tolerance?

____

If it is not in tolerance, can it be reworked? If so, how? ______________

____

APPENDIX A

Concentricity and Symmetry

Note *The concentricity and symmetry controls have been deleted from the ASME Y 14.5-2018 standard. However, many readers of this text continue to use or are required to read drawings toleranced to an earlier version of the standard. Consequently, the chapter in the previous (second) edition on concentricity and symmetry (Chap. 10) appears here in its entirety as App. A.*

Both concentricity and symmetry controls are reserved for a few unique tolerancing applications. Both employ the same tolerancing concept, but they are just applied to different geometries. The concentricity tolerance controls features constructed about a datum axis, and the symmetry tolerance controls features constructed about a datum center plane. Both concentricity and symmetry locate features by controlling their median points within a specified tolerance zone. They are typically used where it is important to accurately balance the mass of a part about its axis or center plane.

Chapter Objectives

After completing this chapter, the learner will be able to:

- *Define* concentricity and symmetry
- *Specify* concentricity and symmetry on drawings
- *Describe* the inspection process of concentricity and symmetry
- *Explain* applications of concentricity and symmetry

Concentricity

Definition

Concentricity is that condition where the median points of all diametrically opposed elements of a surface of revolution are congruent with the axis or center point of a datum feature. Concentricity also applies to correspondingly located points of two or more radially disposed features such as the flats on a regular hexagon or opposing lobes on features such as an ellipse.

Specifying Concentricity

Concentricity is a location control. It has a cylindrical (or spherical) shaped tolerance zone that is coaxial with the datum axis. Concentricity tolerance only applies on a regardless of feature size (RFS) basis; it must have at least one datum feature that applies only at regardless of material boundary (RMB). The feature control frame is usually placed beneath the size dimension or attached to an extension of the dimension line, as shown in Fig. A-1. The concentricity tolerance has no relationship to the size of the feature being controlled and may be either larger or smaller than the size tolerance. If a concentricity tolerance is specified to control the location of a sphere, the tolerance zone is spherical, and its center point is basically located from the datum feature(s).

Interpretation

The concentricity tolerance controls all median points of all diametrically opposed points on the surface of the toleranced feature. The aggregate of all median points, sometimes described as a "cloud of median points," must lie within a cylindrical (or spherical) shaped tolerance zone whose axis (or center point) is coincident with the datum axis. The concentricity tolerance is independent of both size and form. Differential measurement excludes size, shape, and form while controlling the median points of the feature. The feature control frame in Fig. A-2 specifies a cylindrical tolerance zone .005 in diameter and coaxial with the datum axis. Differential measurements are taken along and around the toleranced feature to determine the location of its median points. If all median points fall inside the tolerance zone, the feature is in tolerance.

Inspection

Concentricity can be inspected, for acceptance only, by placing a dial indicator on a toleranced surface of revolution and rotating the part about its datum axis. If the full indicator movement (FIM) on the dial indicator does not exceed the specified tolerance, the feature is acceptable. This technique is a simple first check that will accept parts but will not reject them, and it can be used only on surfaces of revolution. If the measurement does exceed the FIM, the part is not necessarily out of tolerance.

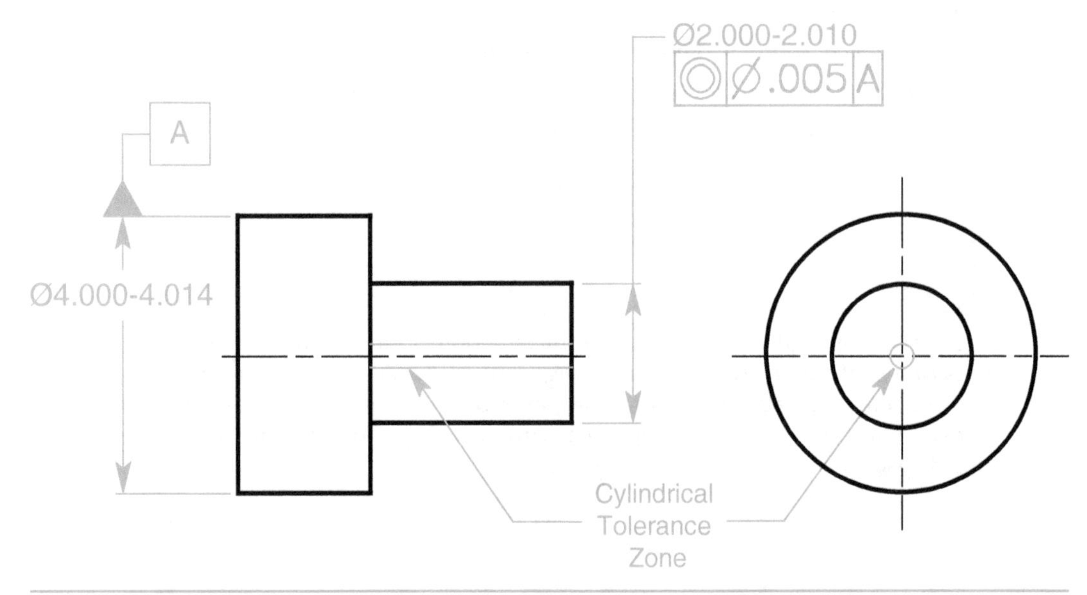

FIGURE A-1 Concentricity is a location control that employs a cylindrical tolerance zone.

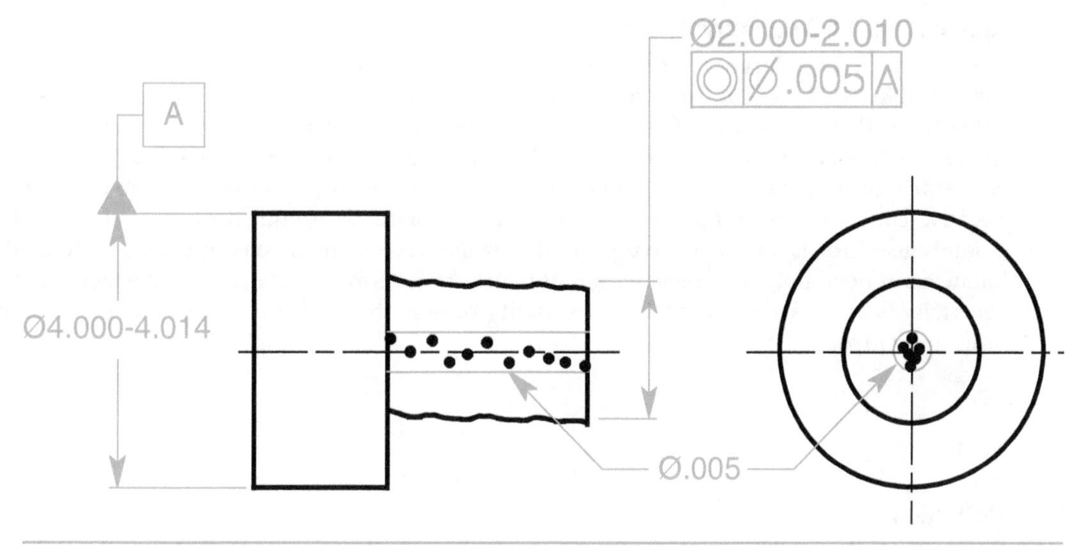

FIGURE A-2 A concentricity tolerance locating a coaxial feature.

To reject parts and to inspect features such as regular polygons and ellipses, the traditional method of differential measurements is employed, as shown in Fig. A-3. That is, the datum feature is placed in a chucking device that will rotate the part about its datum axis. A point on the surface of the toleranced feature is measured with a dial indicator. The part is then rotated 180° so that the diametrically opposed point can be measured. The difference between the measurements of the two points determines the location of the median point. This process is repeated a predetermined number of times. If all median points fall within the tolerance zone, the feature is in tolerance. The size and form, Rule # 1, are measured separately.

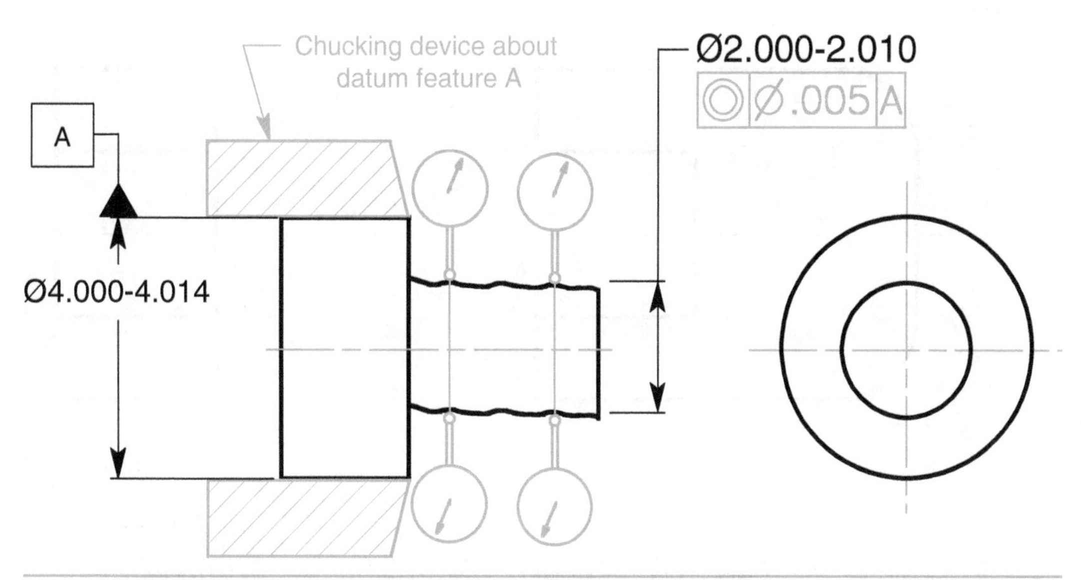

FIGURE A-3 Inspecting a part using the differential measurements technique.

Applications of Concentricity

The concentricity tolerance is often used to accurately control balance for high-speed rotating parts. Runout also controls balance, but it controls form and surface imperfections at the same time. Runout is relatively easy and inexpensive to inspect, but manufacturing parts toleranced with a runout control is more difficult and more expensive. Parts toleranced with concentricity are time consuming and expensive to inspect but less expensive to manufacture because concentricity is not as rigorous a requirement as runout. Concentricity is appropriately used for large, expensive parts that must have a small coaxial tolerance to achieve balance but need not have the same small tolerance for form and surface imperfections. Concentricity is also used to control the coaxiality of noncircular features, such as regular polygons and ellipses.

Symmetry

Definition

Symmetry is that condition where the median points of all opposed or correspondingly located elements of two or more feature surfaces are congruent with the axis or center plane of a datum feature.

Specifying Symmetry

Symmetry is a location control. It has a tolerance zone that consists of two parallel planes evenly disposed about the center plane or axis of its datum feature. Symmetry tolerance applies only at RFS; it must have at least one datum feature that may apply only at RMB. The feature control frame is usually placed beneath the size dimension or attached to an extension of the dimension line, as shown in Fig. A-4. The symmetry tolerance has no relationship to the size of the feature being controlled and may be either larger or smaller than the size tolerance.

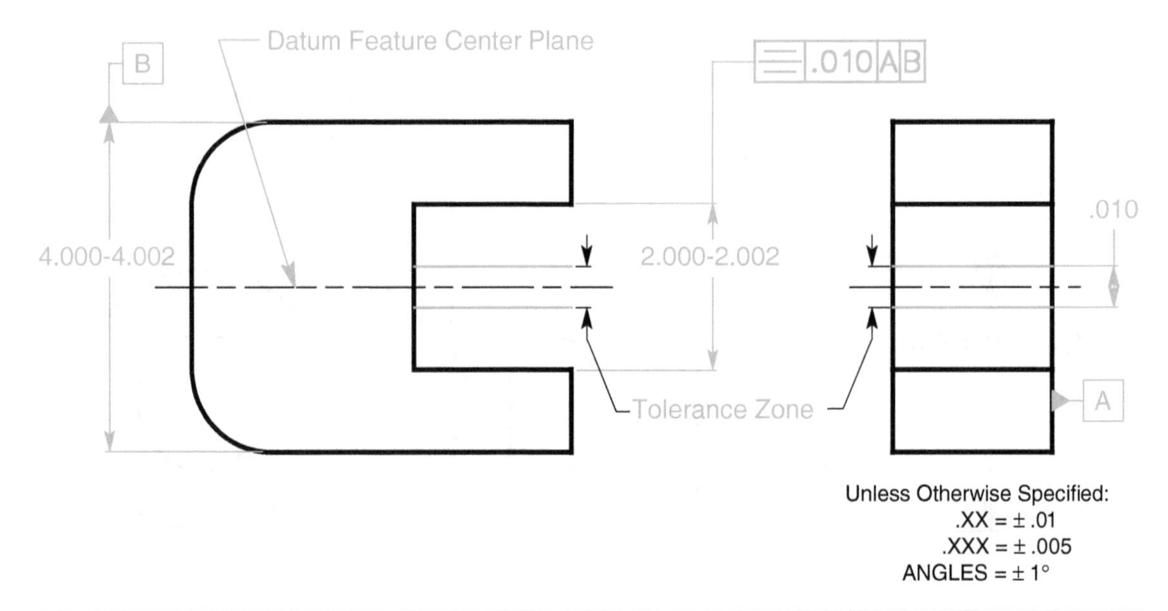

FIGURE A-4 Symmetry controls location with a tolerance zone that consists of two parallel planes.

Interpretation

The symmetry tolerance controls the median points of all opposed or correspondingly located points of two or more surfaces. The aggregate of all median points, sometimes described as a "cloud of median points," must lie within a tolerance zone defined by two parallel planes equally disposed about the center plane or axis of the datum feature. That is, half of the tolerance is on one side of the center plane, and half is on the other side. The symmetry tolerance is independent of both size and form. Differential measurement excludes size, shape, and form while controlling the median points of the feature. The feature control frame in Fig. A-5 specifies a tolerance zone consisting of two parallel planes .010 apart, perpendicular to datum feature A, and equally disposed about the center plane of datum feature B. Differential measurements are taken between the two surfaces to determine the location of the median points. If all median points fall inside the tolerance zone, the feature is in tolerance.

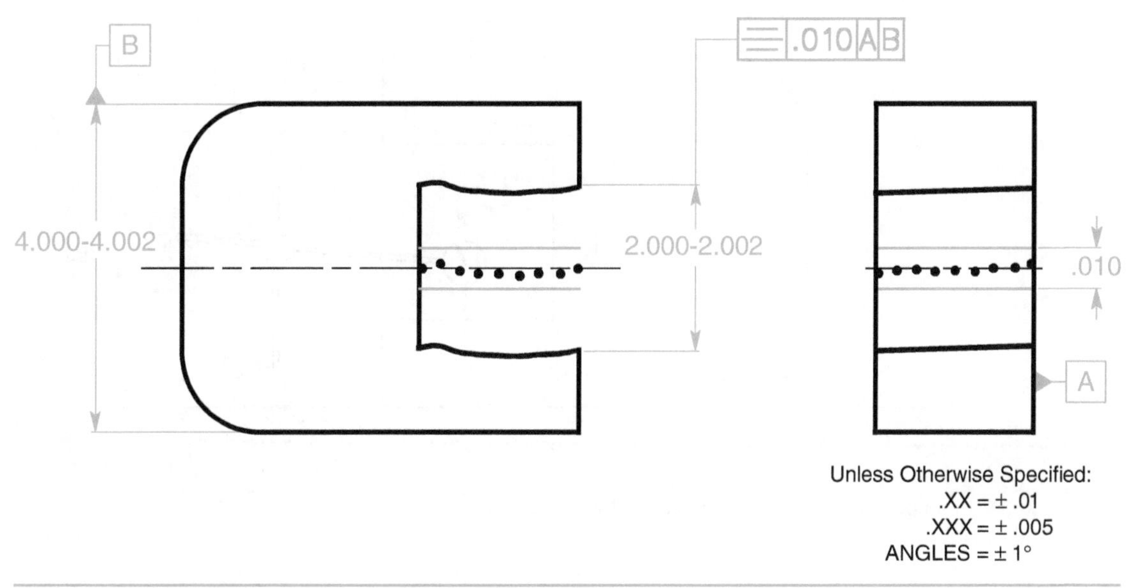

FIGURE A-5 A symmetry tolerance locating a symmetrical feature.

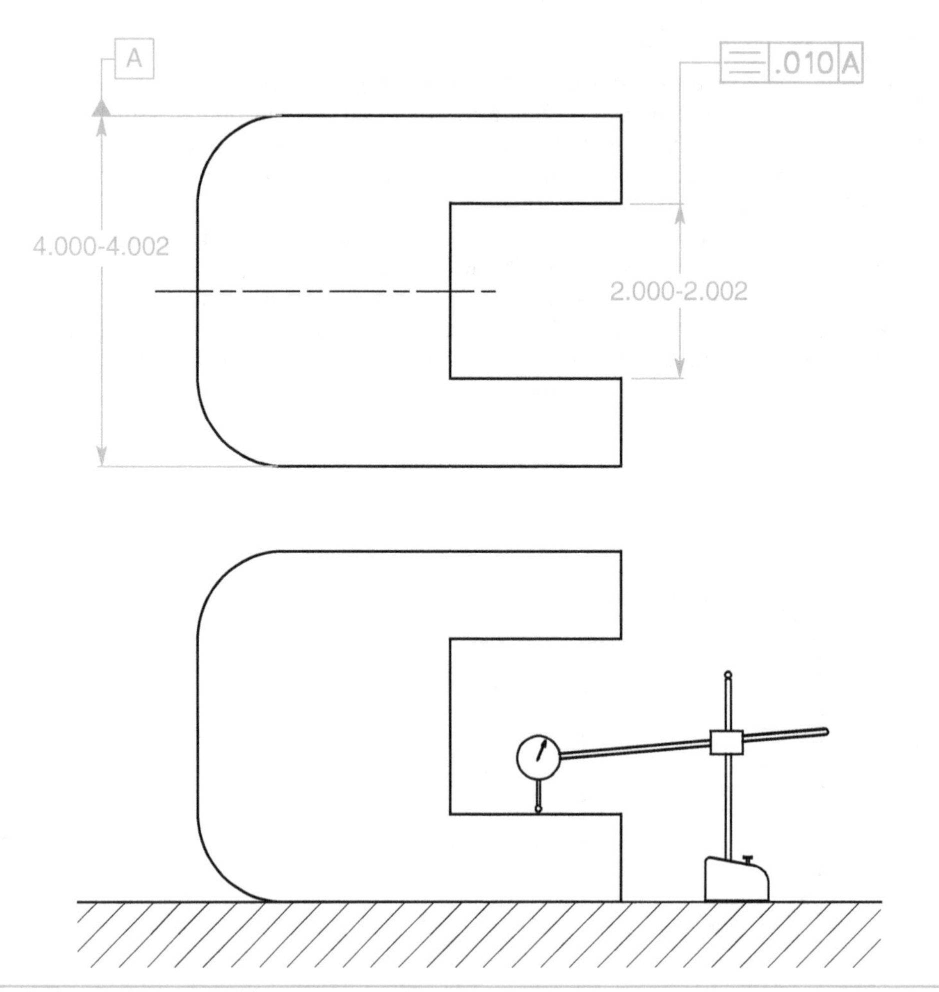

FIGURE A-6 Inspecting a part with a symmetry tolerance.

Inspection

A simple method of measuring symmetry is shown in Fig. A-6. This method can be used only if the datum surfaces are very closely parallel compared to the symmetry tolerance. In this example, one of the datum surfaces is placed on the surface plate. A dial indicator is used to measure a number of points on the surface of the slot. These measurements are recorded. The part is turned over, and the process is repeated. The measurements are compared to determine the location of the median points and whether the feature is in tolerance or not. Size and form, Rule # 1, are measured separately.

Applications of Symmetry

The symmetry tolerance is often used to accurately control balance for rotating parts or to ensure equal wall thickness. Specify symmetry only when it is necessary because it is time consuming and expensive to manufacture and inspect. The symmetry control is appropriately used for large, expensive parts that require a small symmetry tolerance to balance mass. If the restrictive symmetry control is not required, profile of a surface or the more versatile position tolerance may be used to control a symmetrical relationship. See Chap. 8 for a discussion of the application of the position control to tolerance symmetrical features.

Summary

- Concentricity is that condition where the median points of all diametrically opposed points of a surface of revolution are congruent with the axis of a datum feature.
- Concentricity is a location control that has a cylindrical tolerance zone that is coaxial with the datum axis.
- The concentricity tolerance applies only at RFS, and the datum reference applies only at RMB.
- The aggregate of all median points must lie within a cylindrical tolerance zone whose axis is coincident with the axis of the datum feature.
- The concentricity tolerance is independent of both size and form.
- Differential measurement excludes size, shape, and form while controlling the median points of the feature.
- The concentricity tolerance is often used to accurately control balance for high-speed rotating parts.
- Symmetry is that condition where the median points of all opposed or correspondingly located points of two or more feature surfaces are congruent with the axis or center plane of a datum feature.
- Symmetry is a location control that has a tolerance zone that consists of two parallel planes evenly disposed about the center plane or axis of the datum feature.
- The symmetry tolerance applies only at RFS.
- The symmetry control must have at least one datum feature that may apply only at RMB.
- The aggregate of all median points must lie within a tolerance zone defined by two parallel planes equally disposed about the center plane of the datum feature.
- The symmetry tolerance is independent of both size and form.
- The symmetry tolerance is often used to accurately control balance for rotating parts or to ensure equal wall thickness.
- Specify symmetry only when it is necessary because it is time consuming and expensive to manufacture and inspect.

Chapter Review

1. Both concentricity and symmetry controls are reserved for a few ______________________

    ____.

2. Both concentricity and symmetry employ the same tolerancing ______________________;

    they just apply to different ____.

3. Concentricity is that condition where the median points of all diametrically opposed elements of a surface of revolution are congruent with ______________________

    ____.

4. Concentricity is a ______________ control. It has a ______________ shaped tolerance zone that is coaxial with ______________.

5. Concentricity tolerance applies only on ______________ basis. It must have at least ______________ that applies only at ______________.

6. For concentricity, the aggregate of all ______________ must lie within a ______________ tolerance zone whose axis is coincident with the axis of the ______________.

7. Concentricity can be inspected, for acceptance only, by placing a ______________ on the toleranced surface of revolution and rotating the part about its ______________.

8. To reject parts and to inspect features such as regular polygons and ellipses, the traditional method of ______________ is employed.

9. The concentricity tolerance is often used to accurately control ______________ for high-speed rotating parts.

10. Concentricity is time consuming and expensive to ______________ but less expensive to ______________ than the runout tolerance.

11. Symmetry is that condition where the ______________ of all opposed or correspondingly located elements of two or more feature surfaces are ______________ with the ______________ of a datum feature.

12. Symmetry is a ______________ control.

13. Symmetry has a tolerance zone that consists of ______________ evenly disposed about the ______________ of the datum feature.

14. Symmetry tolerance applies only at ______________.

15. Symmetry must have at least one ______________ that only applies at ______________.

16. The aggregate of all ______________________________

    must lie within a tolerance zone defined by ______________________________

    equally disposed about the center plane or axis of the ______________________________.

17. The symmetry tolerance is independent of both ______________________________.

18. Differential measurement excludes ______________________________

    while controlling the ______________________________ of the feature.

19. The symmetry tolerance is often used to accurately control ______________________________

    for rotating parts or to ensure equal ______________________________.

20. Specify symmetry only when it is necessary because it is ______________________________

    ______________________________ to manufacture and inspect.

Problems

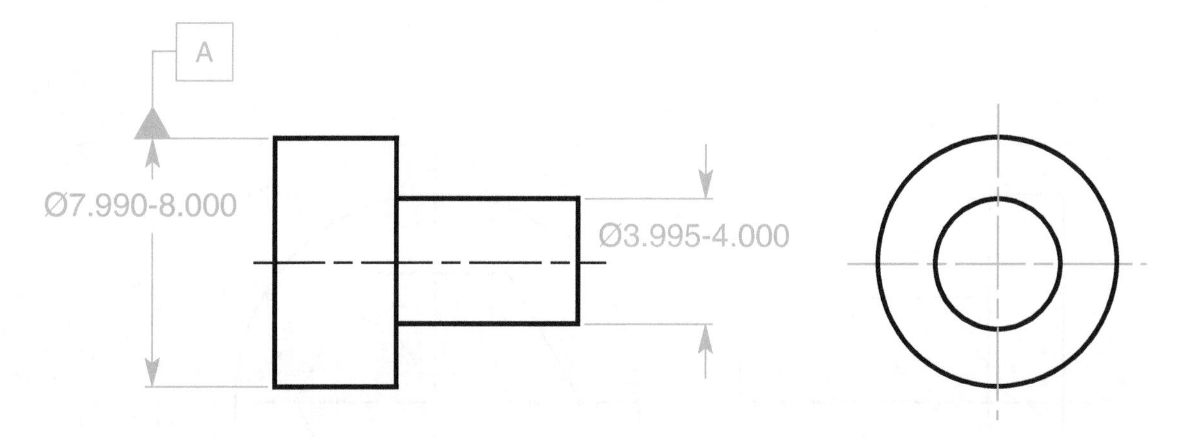

Unless Otherwise Specified:
.XX = ± .01
.XXX = ± .005
ANGLES = ± 1°

FIGURE A-7 Controlling the coaxiality of a cylinder: Prob. 1.

1. The mass of this high-speed rotating part in Fig. A-7 must be accurately balanced. The form of the surface is sufficiently controlled by the size tolerance. Specify a coaxiality control for the axis of the 4-inch diameter within a cylindrical tolerance of .001 at RFS to datum feature A at RMB.

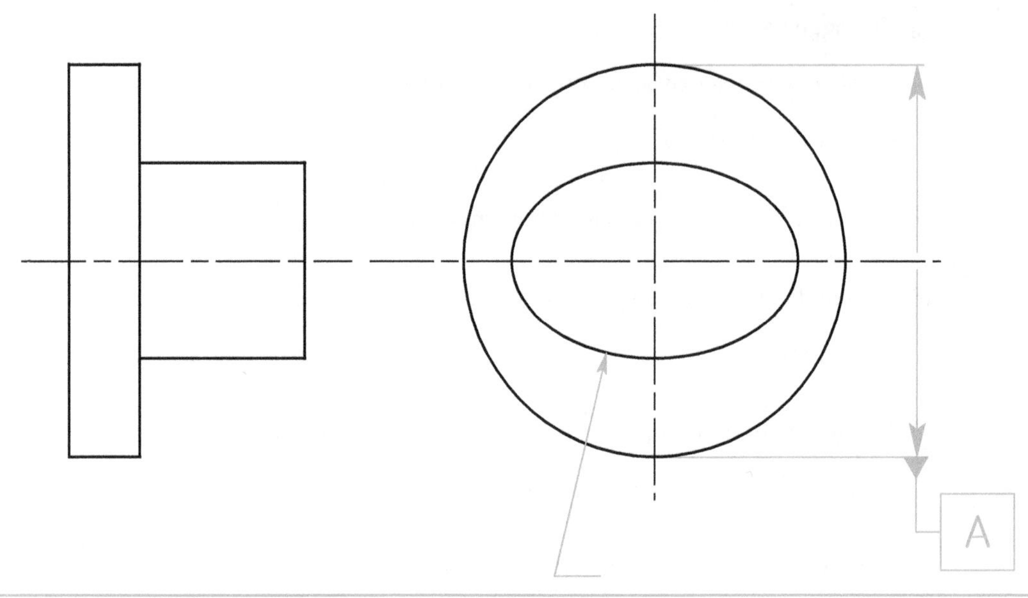

FIGURE A-8 Controlling the coaxiality of an ellipse: Prob. 2.

2. The mass of the ellipse in Fig. A-8 must be accurately balanced. Specify a coaxiality control that will locate the median points of the ellipse within a cylindrical tolerance of .004 at RFS to datum feature A at RMB.

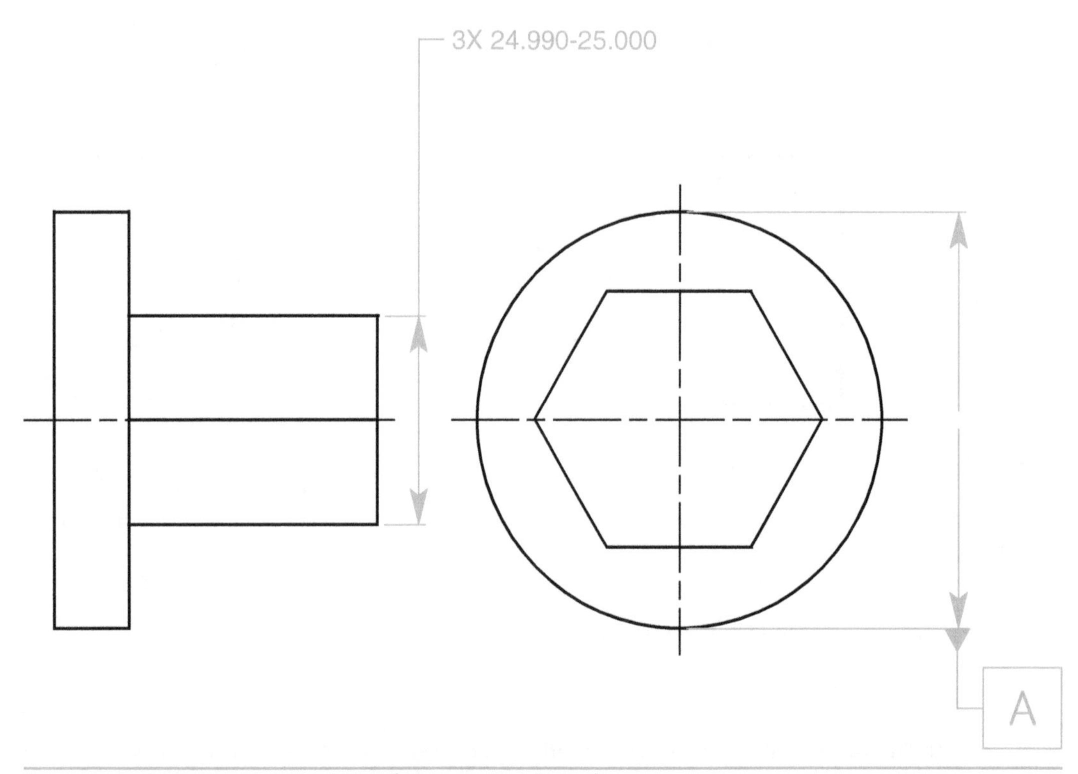

FIGURE A-9 Controlling the coaxiality of the hexagon: Prob. 3.

3. The mass of the hexagon in Fig. A-9 must be accurately balanced. Specify a coaxiality control for the median points of the hexagon within a cylindrical tolerance of .005 at RFS to datum feature A at RMB.

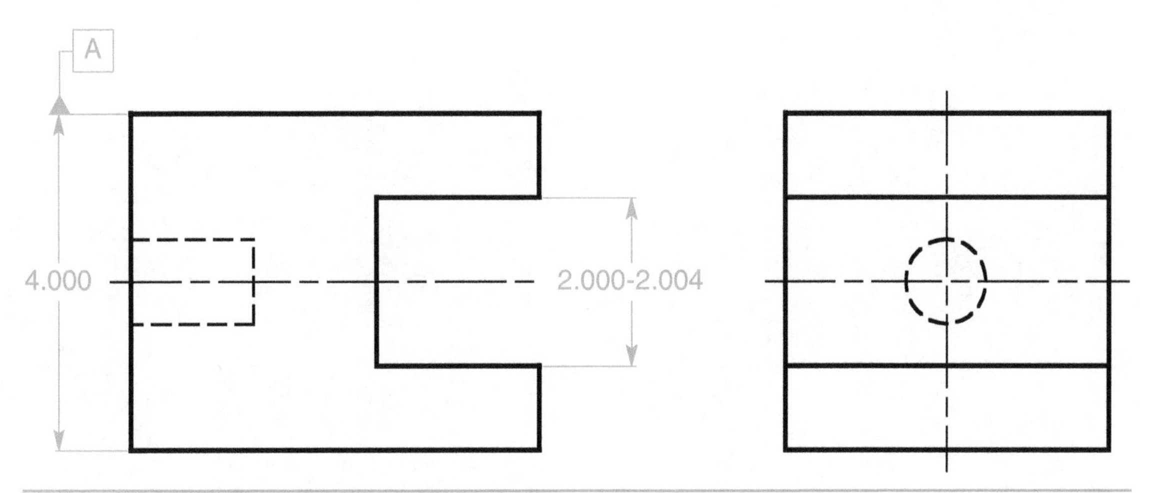

FIGURE A-10 Controlling the symmetry of a slot: Prob. 4.

4. The mass of the part in Fig. A-10 must be accurately balanced. Specify a geometric tolerance that will centrally locate the slot in this part within a tolerance of .005 at RFS to datum feature A at RMB.

APPENDIX B
Reference Tables

Drill Diameter, Inch	Amount Oversize, Inch		
	Average Max.	Mean	Average Min.
1/16	.002	.0015	.001
1/8	.0045	.003	.001
1/4	.0065	.004	.0025
1/2	.008	.005	.003
3/4	.008	.005	.003
1	.009	.007	.004

Source: Machinery's Handbook, Revised 28th Edition, Industrial Press Inc., New York, 2008

TABLE B-1 Oversize Diameters in Drilling

Fastener	Nominal Size	Decimal Diameter	A	Clearance Hole LMC*
Hex Head Machine Screws	#8	.1640	.244	.204
	#10	**.1900**	**.305**	**.247**
	1/4	**.2500**	**.425**	**.337**
	5/16	.3125	.484	.398
	3/8	.3750	.544	.460
	7/16	.4375	.603	.520
	1/2	**.5000**	**.725**	**.612**
	5/8	.6250	.906	.765
	3/4	.7500	1.088	.919
Socket Head Cap Screws	#4	.1120	.176	.144
	#6	.1380	.218	.178
	#8	.1640	.262	.213
	#10	**.1900**	**.303**	**.246**
	1/4	**.2500**	**.365**	**.307**
	5/16	.3125	.457	.384
	3/8	.3750	.550	.462
	7/16	.4375	.642	.540
	1/2	**.5000**	**.735**	**.617**
	5/8	.6250	.921	.773
	3/4	.7500	1.107	.928
Round Head Machine Screws	#4	.1120	.193	.152
	#6	.1380	.240	.189
	#8	.1640	.287	.226
	#10	**.1900**	**.334**	**.262**
	1/4	**.2500**	**.443**	**.346**
	5/16	.3125	.557	.434
	3/8	.3750	.670	.522
	7/16	.4375	.707	.572
	1/2	**.5000**	**.766**	**.633**
	5/8	.6250	.944	.784
	3/4	**.7500**	**1.185**	**.967**

*Clearance hole diameter at LMC is calculated with the formula (fastener + fastener head)/2.

TABLE B-2 Machine and Cap Screw Sizes

Drill No.	Fract.	Deci.	Drill No.	Fract.	Deci.	Drill Ltr.	Fract.	Deci.	Fract.	Deci.
60	–	.0400	29	–	.1360	B	–	.238	**7/16**	**.438**
59	–	.0410		9/64	.140	C	–	.242	29/64	.453
58	–	.0420	28	–	.141	D	–	.246	15/32	.469
57	–	.0430	27	–	.144		**1/4**	**.250**	31/64	.484
56	–	.0465	26	–	.147	E	–	.250	**1/2**	**.500**
	3/64	.0469	25	–	.150	F	–	.257	33/64	.516
55	–	.0520	24	–	.152	G	–	.261	17/32	.531
54	–	.0550	23	–	.154		17/64	.266	35/64	.547
53	–	.0595		5/32	.156	H	–	.266	**9/16**	**.562**
	1/16	**.0625**	22	–	.157	I	–	.272	37/64	.578
52	–	.0635	21	–	.159	J	–	.277	19/32	.594
51	–	.0670	20	–	.161		9/32	.281	39/64	.609
50	–	.0700	19	–	.166	K	–	.281	**5/8**	**.625**
49	–	.0730	18	–	.170	L	–	.290	41/64	.641
48	–	.0760		11/64	.172	M	–	.295	21/32	.656
	5/64	.0781	17	–	.173		19/64	.297	43/64	.672
47	–	.0785	16	–	.177	N	–	.302	**11/16**	**.688**
46	–	.0810	15	–	.180		**5/16**	**.313**	45/64	.703
45	–	. 0820	14	–	.182	O	–	.316	23/32	.719
44	–	.0860	13	–	.185	P	–	.323	47/64	.734
43	–	.0890		**3/16**	**.188**		21/64	.328	**3/4**	**.750**
42	–	.0935	12	–	.189	Q	–	.332	49/64	.766
	3/32	.0938	11	–	.191	R	–	.339	25/32	.781
41	–	.0960	10	–	.194		11/32	.344	51/64	.797
40	–	.0980	9	–	.196	S	–	.348	**13/16**	**.813**
39	–	.0995	8	–	.199	T	–	.358	53/64	.828
38	–	.1015	7	–	.201		23/64	.359	27/32	.844
37	–	.1040		13/64	.203	U	–	.368	55/64	.859
36	–	.1065	6	–	.204		**3/8**	**.375**	**7/8**	**.875**
	7/64	.1094	5	–	.206	V	–	.377	57/64	.891
35	–	.1100	4	–	.209	W	–	.386	29/32	.906
34	–	.1110	3	–	.213		25/64	.391	59/64	.922
33	–	.1130		7/32	.219	X	–	.397	**15/16**	**.938**
32	–	.1160	2	–	.221	Y	–	.404	61/64	.953
31	–	.1200	1	–	.228		13/32	.406	31/32	.969
	1/8	**.1250**	A	–	.234	Z	–	.413	63/64	.984
30	–	.1290		15/64	.234		27/64	.422	**1**	**1.000**

TABLE B-3 Number, Letter, and Fractional Drill Size Chart

Inches			Inches		
Fractions	Decimals	Millimeters	Fractions	Decimals	Millimeters
	.00394	0.1	23/64	.359375	9.1281
	.00787	0.2	**3/8**	.375	9.525
	.01181	0.3	25/64	.390625	9.9219
1/64	.015625	0.3969		.3937	10.00
	.0157	0.4	13/32	.40625	10.3188
	.01969	0.5	27/64	.421875	10.7156
	.02362	0.6		.43307	11.00
	.02756	0.7	7/16	.4375	11.1125
1/32	.03125	0.7938	29/64	.453125	11.5094
	.0315	0.8	15/32	.46875	11.9063
	.03543	0.9		.47244	12.00
	.03937	1.00	31/64	.484375	12.3031
3/64	.046875	1.1906	**1/2**	.5000	12.70
1/16	.0625	1.5875		.51181	13.00
5/64	.078125	1.9884	33/64	.515625	13.0969
	.07874	2.00	17/32	.53125	13.4938
3/32	.09375	2.3813	35/64	.546875	13.8907
7/64	.109375	2.7781		.55118	14.00
	.11811	3.00	9/16	.5625	14.2875
1/8	.125	3.175	37/64	.578125	14.6844
9/64	.140625	3.5719		.59055	15.00
5/32	.15625	3.9688	19/32	.59375	15.0813
	.15748	4.00	39/64	.609375	15.4782
11/64	.171875	4.3656	**5/8**	.625	15.875
3/16	.1875	4.7625		.62992	16.00
	.19685	5.00	41/64	.640625	16.2719
13/64	.203125	5.1594	21/32	.65625	16.6688
7/32	.21875	5.5563		.66929	17.00
15/64	.234375	5.9531	43/64	.671875	17.0657
	.23633	6.00	11/16	.6875	17.4625
1/4	.2500	6.35	45/64	.703125	17.8594
17/64	.265625	6.7469		.70866	18.00
	.27559	7.00	23/32	.71875	18.2563
9/32	.28125	7.1438	47/64	.734375	18.6532
19/64	.296875	7.5406		.74803	19.00
5/16	.3125	7.9375	**3/4**	.7500	19.05
	.31496	8.00	49/64	.765625	19.4469
21/64	.328125	8.3344	25/32	.78125	19.8438
11/32	.34375	8.7313		.7874	20.00
	.35433	9.00	51/64	.796875	20.2407

TABLE B-4 Inch to Millimeter Conversion Chart

Inches		Millimeters	Inches		Millimeters
Fractions	Decimals		Fractions	Decimals	
13/16	.8125	20.6375	29/32	.90625	23.0188
	.82677	21.00	59/64	.921875	23.4157
53/64	.828125	21.0344	15/16	.9375	23.8125
27/32	.84375	21.4313		.94488	24.00
55/64	.859375	21.8282	61/64	.953125	24.2094
	.86614	22.00	31/32	.96875	24.063
7/8	.875	22.225		.98425	25.00
57/64	.890625	22.6219	63/64	.984375	25.0032
	.90551	23.00	**1**	1.0000	25.4001

TABLE B-4 Inch to Millimeter Conversion Chart (*Continued*)

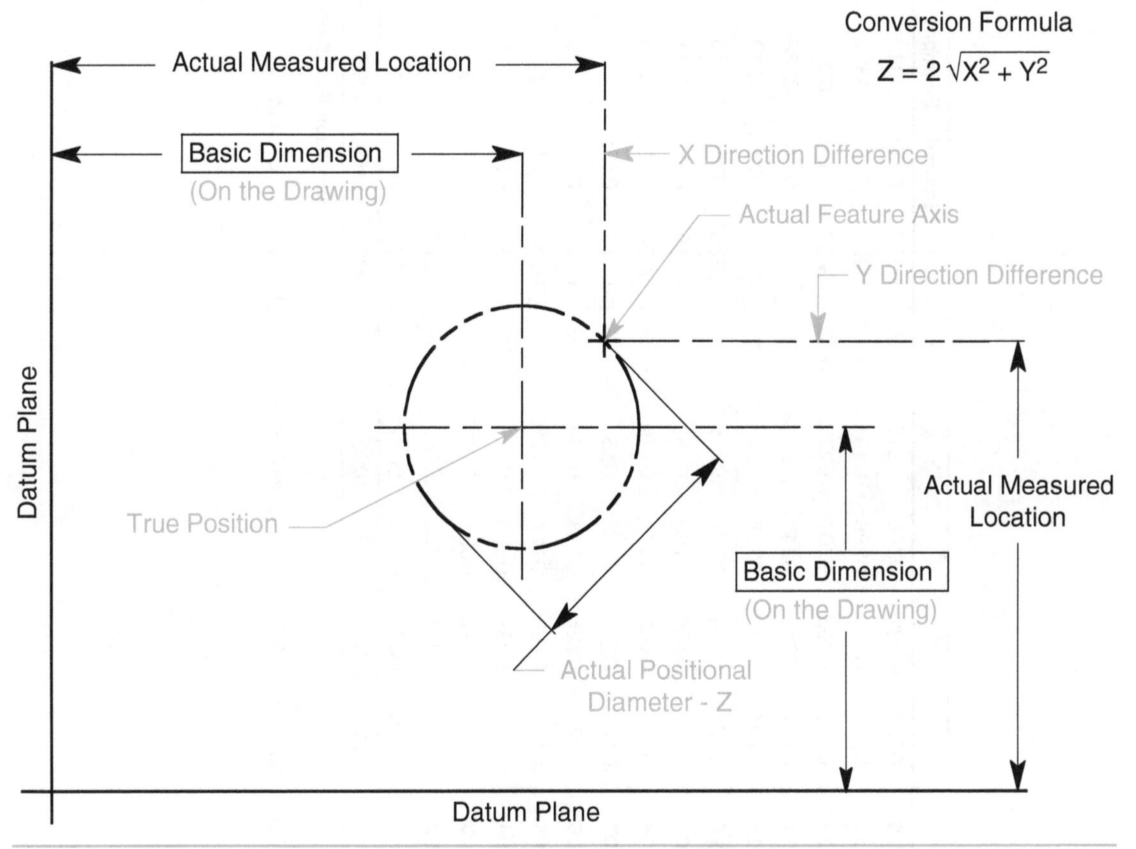

FIGURE B-1 Conversion diagram from coordinate measurements to cylindrical tolerance.

Y Direction Deviation	Z Diameter Positional Tolerance														
.015	.0301	.0303	.0306	.0310	.0316	.0323	.0331	.0340	.0350	.0360	.0372	.0384	.0397	.0410	.0424
.014	.0281	.0283	.0286	.0291	.0297	.0305	.0313	.0322	.0333	.0344	.0356	.0369	.0382	.0396	.0410
.013	.0261	.0263	.0267	.0272	.0278	.0286	.0295	.0305	.0316	.0328	.0340	.0354	.0368	.0382	.0397
.012	.0241	.0243	.0247	.0253	.0260	.0268	.0278	.0288	.0300	.0312	.0325	.0339	.0354	.0369	.0384
.011	.0221	.0224	.0228	.0234	.0242	.0250	.0261	.0272	.0284	.0297	.0311	.0325	.0340	.0356	.0372
.010	.0201	.0204	.0209	.0215	.0224	.0233	.0244	.0256	.0269	.0283	.0297	.0312	.0328	.0344	.0360
.009	.0181	.0184	.0190	.0197	.0206	.0216	.0228	.0241	.0254	.0269	.0284	.0300	.0316	.0333	.0350
.008	.0161	.0165	.0171	.0179	.0189	.0200	.0213	.0226	.0241	.0256	.0272	.0288	.0305	.0322	.0340
.007	.0141	.0146	.0152	.0161	.0172	.0184	.0198	.0213	.0228	.0244	.0261	.0278	.0295	.0313	.0331
.006	.0122	.0126	.0134	.0144	.0156	.0170	.0184	.0200	.0216	.0233	.0250	.0268	.0286	.0305	.0323
.005	.0102	.0108	.0117	.0128	.0141	.0156	.0172	.0189	.0206	.0224	.0242	.0260	.0278	.0297	.0316
.004	.0082	.0089	.0100	.0113	.0128	.0144	.0161	.0179	.0197	.0215	.0234	.0253	.0272	.0291	.0310
.003	.0063	.0072	.0085	.0100	.0117	.0134	.0152	.0171	.0190	.0209	.0228	.0247	.0267	.0286	.0306
.002	.0045	.0056	.0072	.0089	.0108	.0126	.0146	.0165	.0184	.0204	.0224	.0243	.0263	.0283	.0303
.001	.0028	.0045	.0063	.0082	.0102	.0122	.0141	.0161	.0181	.0201	.0221	.0241	.0261	.0281	.0301
	.001	**.002**	**.003**	**.004**	**.005**	**.006**	**.007**	**.008**	**.009**	**.010**	**.011**	**.012**	**.013**	**.014**	**.015**
	X Direction Deviation														

TABLE B-5 Conversion of Coordinate Measurement (X and Y) to Cylindrical Tolerance (Z)

Index

Printed in the USA
CPSIA information can be obtained
at www.ICGtesting.com
LVHW061630301124
797961LV00007B/893

* 9 7 8 1 2 6 5 8 2 1 4 5 6 *